PROCESS INTEGRATION

PROCESS SYSTEMS ENGINEERING
A Series edited by George Stephanopoulos and Efstratios Pistikopoulos

PROCESS INTEGRATION

Mahmoud M. El-Halwagi
The Artie McFerrin Department of Chemical Engineering
Texas A & M University
Texas, USA

AMSTERDAM • BOSTON • HEIDELBERG • LONDON • NEW YORK • OXFORD
PARIS • SAN DIEGO • SAN FRANCISCO • SINGAPORE • SYDNEY • TOKYO

Academic Press is an imprint of Elsevier

ELSEVIER

ACADEMIC
PRESS

ELSEVIER B.V.
Radarweg 29
P.O. Box 211, 1000 AE
Amsterdam, The Netherlands

ELSEVIER Inc.
525 B Street, Suite 1900
San Diego, CA 92101-4495
USA

ELSEVIER Ltd
The Boulevard, Langford Lane
Kidlington, Oxford OX5 1GB
UK

ELSEVIER Ltd
84 Theobalds Road
London WC1X 8RR
UK

First edition 2006

Library of Congress Cataloging in Publication Data
A catalog record is available from the Library of Congress.

British Library Cataloguing in Publication Data
A catalogue record is available from the British Library.

ISBN-13: 978 0 12 370532 7
ISBN-10: 0 12 370532 0

♾ The paper used in this publication meets the requirements of ANSI/NISO Z39.48-1992 (Permanence of Paper).

Working together to grow
libraries in developing countries
www.elsevier.com | www.bookaid.org | www.sabre.org

ELSEVIER   BOOK AID International   Sabre Foundation

Transferred to Digital Printing in 2011

To my parents, my wife, and my sons with love and gratitude

■ PREFACE

Processing facilities are among the most important contributors to the World's economy and sustainable development. Examples of these processes include chemical, petroleum, gas, petrochemical, pharmaceutical, food, microelectronics, metal, textile, and forestry-product industries. These industries are characterized by the processing of enormous quantities of material and energy resources. As such, there is a significant need for continuous process improvement by targeting key goals such as conserving natural resources, enhancing productivity, and mitigating the discharge of ecologically harmful materials. The question is how can engineers achieve these goals methodically and efficiently? For a given process with numerous units and streams, how can engineers maneuver through the complexities of the process, identify opportunities, and pursue them?

Chemical engineering education and training have provided an arsenal of generally applicable design and analysis tools for individual processing units (e.g., reactors, heat exchangers, distillation columns, absorbers, strippers, pumps, compressors, etc.). A typical process engineer is expected to address a wide variety of applications for these individual units by using a consistent set of fundamental and generic design and analysis tools. Until recently, these powerful tools have been *limited to the level of individual units*. Regrettably, the much-encountered problem of synthesizing, designing, or optimizing a whole process (or parts of the process) with multiple units and streams has been mostly handled as a subjective exercise which is primarily guided by experience and conjecture. It is worth noting that the solution of individual problems must not be done in isolation of the rest of the process but rather in the context of how such problems interact with the rest of the process. Unfortunately, the knowledge and skill sets needed to address open-ended problems involving multiple units and streams are seldom covered in engineering education and professional training. This has been a key limitation that restricted the power of chemical engineering education and training in addressing real-life process-improvement objectives.

Since the process operates as an integrated system of units and streams, it must be understood as such and it must be treated as such. Recent research in the area of process systems engineering has led to a remarkable breakthrough referred to as process integration. This breakthrough has expanded the power of engineering design to methodically and insightfully address systems of multiple units and streams using a consistent basis of fundamental concepts of chemical and systems engineering. Process integration is defined as a holistic approach to design and operation that emphasizes the unity of the process. There are two distinct hallmarks of process integration. First, there are insights and performance targets for the process that can be identified only by addressing the process as a whole system. These insights and targets are unseen by looking at the individual components as separate "silos". Second, the whole effect (of the integrated system) is greater than the sum of the individual effects (of individual units and streams). Process integration provides an excellent framework for benchmarking process performance, characterizing root causes of problems limiting the performance, identifying opportunities, methodically generating efficient strategies that achieve the desired performance targets, and inventing innovative solutions.

This textbook presents state-of-the-art fundamentals, tools, and applications of process integration. It describes how to effectively integrate the process units and streams and conserve natural resources in a systematic and generally applicable way. The book places a great deal of emphasis on targeting techniques that determine process potential, attractive opportunities, root causes of problems, and performance benchmarks ahead of detailed design and without commitment to the specific selection of units or technologies. The book also presents methodical tools and techniques that guide the engineers in synthesizing and detailing integrated solutions while enhancing creativity and incorporating relevant data and expertise. Graphical, algebraic, and mathematical approaches are presented to furnish a host of tools appropriate for various groups of readers and different problems with increasing level of complexity and sophistication. The techniques described throughout the book have proven their worth through numerous industrial applications with impressive track record and results.

In keeping with the tutorial theme of the book, the treatment of theoretical foundations is limited to that needed to provide the basis for the reasoning behind the presented materials. Additional references and suggested reading materials are listed throughout the book to provide more details on the theoretical foundations and mathematical proofs. Numerous examples are given to demonstrate the power and applicability of the tools and techniques.

This book is appropriate for senior-level design classes and introductory graduate classes. The book is also tailored to serve as a self-study sourcebook for process engineers and working professionals interested in optimizing process performance and managing natural resources. Finally, the book can be used as a resource for researchers in the area of process synthesis and integration.

Many extraordinary people have contributed to my career and to this book. Heartfelt thanks are due to all of my friends who helped me in so many ways throughout my life. I am very thankful to the many professional associates and leaders of the process systems engineering community whose contributions have impacted my technical interests and development. I am particularly grateful to Dr. Dennis Spriggs (President of Matrix Process Integration) who has significantly impacted my career through enjoyable collaborations and industrial projects that introduced me to many process-integration approaches and practices.

I am indebted to my colleagues at Texas A&M University. Many thanks are due to Dr. John Baldwin. I have learned a great deal from him while co-teaching the notorious design classes. I am also thankful to Drs. Juergen Hahn, Ken Hall, and Sam Mannan for the enriching collaboration, insightful discussions, and continuous support.

I am grateful to the numerous undergraduate students at Texas A&M University and Auburn University as well as attendees of my industrial workshops, short courses, and seminars whose invaluable input was instrumental in developing and refining the book.

A special tribute is due to my former and current graduate students. I have learned a lot from their proclivity to pose thoughtful questions and to devise innovative solution approaches. I have been truly fortunate to closely interact with this enthusiastic, well-accomplished, and impressive group which includes: Nasser Al-Azri, Hassan Alfadala (Qatar University), Abdul-Aziz Almutlaq (King Saud University), Meteab Al-Otaibi (SABIC), Saad Al-Sobhi (Qatar University), Musaed Al-Thubaiti (Aramco), Srinivas "B.K." Bagepalli (General Electric Industrial Systems), Abdullah Bin Mahfouz, Benjamin Cormier, Eric Crabtree (Parsons), Alec Dobson (Solutia), Russell Dunn (Polymer and Chemical Technologies), Brent Ellison (Light Ridge Resources), Fred Gabriel (Weyerhauser), Walker Garrison (Cinergy), Ian Glasgow (École Polytechnique de Montréal), Murali Gopalakrishnan (Lyondell Equistar), Daniel Grooms, Ahmad Hamad (Emerson Process), Dustin Harell (Intel),

Ronnie Hassanen, Ana Carolina Hortua, Vasiliki Kazantzi, Eva Lovelady, Rubayat Mahmud (Intel), Tanya Mohan, Lay Myint, Bahy Noureldin (Aramco), Grace Nworie, Madhav Nyapathi (Atkins), Gautham Parthasarathy (Solutia), Viet Pham, Xiaoyun Qin, Arwa Rabie, Jagdish Rao, Andrea Richburg (3M), Brandon Shaw, Mark Shelley (University of Houston), Chris Soileau (Veritech), Carol Stanley, Ragavan Vaidyanathan (Celanese), Anthony Warren (General Electric Plastics), Key Warren (Southern Company), Lakeshie Williams (Georgia Pacific), Matt Wolf (Allied Signal), Jose Zavala, and Mingjie Zhu (AtoFina).

My sincere thanks are due to the visiting scholars who joined my research group and contributed significantly. I am particularly thankful to Drs. Mario Eden (Auburn University), Amro El-Baz (Zagazig University), Dominic C. Y. Foo (University of Nottingham, Malaysia Campus), and B. J. Kim (Soongsil University).

The financial support of my process-integration research by various Federal, State, industrial, and international sponsors is gratefully acknowledged. I am also indebted to Mr. Artie McFerrin and his family for their generous endowment and enthusiastic support which allowed me to pursue exciting and exploratory research and to transfer the findings to the classroom.

I appreciate the editing and production work of Elsevier and CEPHA Imaging Private Limited, specially the help and support of Deirdre Clark and Aparna Shankar.

I am grateful to my mother for providing unconditional love and boundless support. She has always been a beacon of light in my life. I am indebted to my father, Dr. Mokhtar El-Halwagi, for serving as a magnificent role model and an impressive source of inspiration. He has introduced me to many concepts in systems integration and has guided my steps in chemical engineering and in life. I am also grateful to my grandfather, the late Dr. Mohamed El-Halwagi, for sharing his passion for chemical engineering and for steering me in the direction of seeking knowledge and passing it on. I am thankful to my brother, Dr. Baher El-Halwagi, for his constant support and continuous encouragement. I am forever obliged to my wife, Amal, for her genuine love, gentle heart, unwavering support, and unlimited patience. She has contributed significantly to my personal life and professional career. I have always sought her guidance and have always counted on her superb engineering sense and enlightening perspectives. Finally, I am grateful to my sons, Omar and Ali, for the joy they bring to my life through their fantastic sense of humor, pure love, unique talents, and the special bonds we share.

Mahmoud M. El-Halwagi

◼ CONTENTS

3 Graphical Techniques for Direct-Recycle Strategies

4 Synthesis of Mass Exchange Networks: A Graphical Approach

5 Visualization Techniques for the Development of Detailed Mass-Integration Strategies

6 Algebraic Approach to Targeting Direct Recycle

7 An Algebraic Approach to the Targeting of Mass Exchange Networks

8 Recycle Strategies Using Property Integration

1

INTRODUCTION TO PROCESS INTEGRATION

The process industries are among the most important manufacturing facilities. They span a wide range of industries including chemical, petroleum, gas, petrochemical, pharmaceutical, food, microelectronics, metal, textile, and forestry products. The performance of these industries is strongly dependent on their engineering and engineers. So, what are the primary responsibilities of process engineers in the process industries? Many process engineers would indicate that their role in the process industries is to design and operate industrial processes and make them work faster, better, cheaper, safer, and greener. All of these tasks lead to more competitive processes with desirable profit margins and market share. Specifically, these responsibilities may be expressed through the following specific objectives:

- Process innovation
- Profitability enhancement
- Yield improvement
- Capital-productivity increase
- Quality control, assurance, and enhancement
- Resource conservation
- Pollution prevention
- Safety
- Debottlenecking

1

These objectives are also closely related to the seven themes identified by Keller and Bryan (2000) as the key drivers for process-engineering research, development, and changes in the primary chemical process industries. These themes are:

- Reduction in raw-material cost
- Reduction in capital investment
- Reduction in energy use
- Increase in process flexibility and reduction in inventory
- Ever greater emphasis on process safety
- Increased attention to quality
- Better environmental performance

The question is how? What are the challenges, required methodologies, and enabling tools needed by engineers to carry out their responsibilities. In order to shed some light on these issues, let us consider the following motivating example.

1.1 GENERATING ALTERNATIVES FOR DEBOTTLENECKING AND WATER REDUCTION IN ACRYLONITRILE PROCESS

Consider the process shown in Figure 1-1a for the production of acrylonitrile (AN, C_3H_3N). The main reaction in the process involves the vapor phase

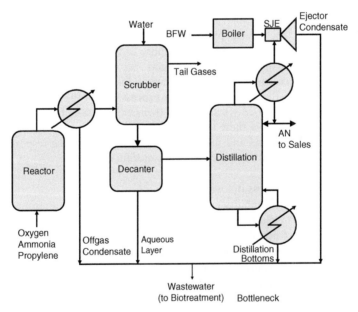

FIGURE 1-1a   PROCESS FOR AN MANUFACTURE (EL-HALWAGI 1997)

catalytic reaction of propylene, ammonia, and oxygen at 450°C and 2 atm. To produce AN and water, i.e.

$$C_3H_6 + NH_3 + 1.5O_2 \xrightarrow{\text{catalyst}} C_3H_3N + 3H_2O$$

The reaction products are quenched in an indirect-contact cooler/condenser which condenses a portion of the reactor off-gas. The remaining off-gas is scrubbed with water, then decanted into an aqueous layer and an organic layer. The organic layer is fractionated in a distillation column under slight vacuum which is induced by a steam-jet ejector. Wastewater is collected from four process streams: off-gas condensate, aqueous layer of decanter, distillation bottoms, and jet-ejector condensate. The wastewater stream is fed to the biotreatment facility. At present, the biotreatment facility is operating at full hydraulic capacity and, consequently, it constitutes a bottleneck for the plant. The plant has a sold-out profitable product and wishes to expand. Our task is debottleneck the process.

The intuitive response to debottlenecking the process is to construct an expansion to the biotreatment facility (or install another one). This solution focuses on the symptom of the problem: the biotreatment is filling up, therefore we must expand its capacity. A legitimate question is whether there are other solutions, probably superior ones, that will address the problem by making in-plant process modifications as opposed to "end-of-pipe" solution? Invariably, the answer in this case and most other process design problems is "yes". If so, how do we determine the root causes of the problem (not just the symptoms) and how can we generate superior solutions? Where do we start and how to address the problem?

For now, let us start with a conventional engineering approach involving a brainstorming session among a group of process engineers who will generate a number of ideas and evaluate them. Since the objective is to debottleneck the biotreatment facility, then an effective approach may be based on reducing the influent wastewater flowrate into biotreatment. One way of reducing wastewater flowrate is to adopt a wastewater recycle strategy in which it is desired to recycle some (or all) of the wastewater to the process. For instance, let us recycle some of the wastewater to the distillation column (Figure 1-1b). After analyzing this solution, it does not seem to be effective. The fresh water to the process is still the same, water generated by the main AN-producing reaction is the same, and therefore the wastewater leaving the plant will remain the same. So, let us employ a recycle strategy that replaces fresh water with wastewater. This way, the fresh water into the process is reduced and, consequently, the wastewater leaving the process is reduced as well. One option is to recycle the wastewater to the scrubber (Figure 1-1c) assuming that it is feasible to process the wastewater in the scrubber without negatively impacting the process performance. In such cases, both fresh water and wastewater will be reduced. Alternatively, it may be possible to recycle the wastewater to the boiler (Figure 1-1d). Along the same lines, the wastewater may be recycled to both the scrubber and the boiler (Figure 1-1e).

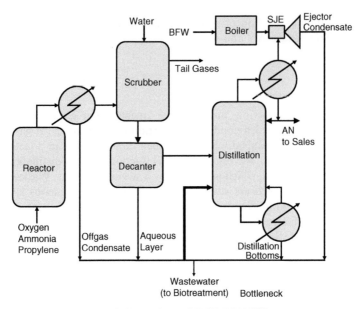

FIGURE 1-1b   RECYCLE TO THE DISTILLATION COLUMN

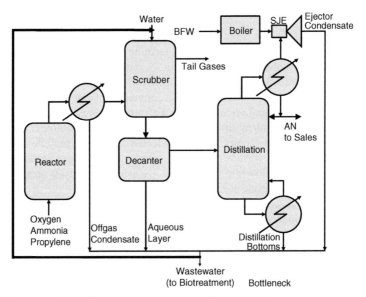

FIGURE 1-1c   RECYCLE TO REPLACE SCRUBBER WATER

How should the wastewater be distributed between the two units? One can foresee many possibilities for distribution (50-50, 51-49, 60-40, 99-1, etc.). Another alternative is to consider segregating (avoiding the mixing of) the wastewater streams. Segregation would prevent some wastewater streams

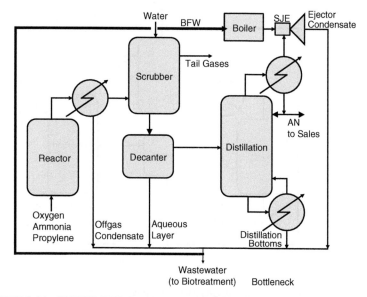

FIGURE 1-1d   RECYCLE TO SUBSTITUTE BOILER FEED WATER

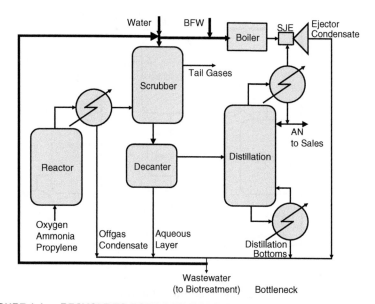

FIGURE 1-1e   RECYCLE TO BOTH SCRUBBER AND BOILER

from mixing with the more polluted streams, thereby enhancing their likelihood for recycle. For instance, the off-gas condensate and the decanter aqueous layer may be segregated from the two other wastewater streams and recycled to the scrubber and the boiler (Figure 1-1f). Clearly, there are many

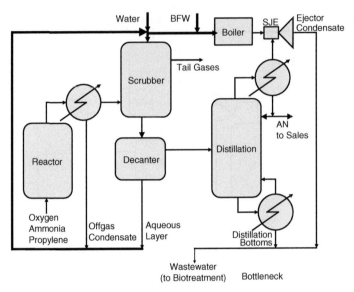

FIGURE 1-1f SEGREGATION OF WASTEWATER AND RECYCLE OF TWO SEGREGATED STREAMS

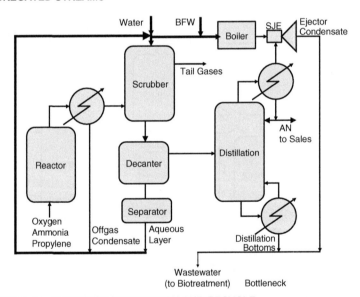

FIGURE 1-1g COMBINED SEPARATION AND RECYCLE

alternatives for segregation and recycle. In order to safeguard against the accumulation of impurities or the detrimental effects of replacing fresh water with waste streams, it may be necessary to consider the use of separation technologies to clean up the streams and render them in a condition acceptable for recycle. For example, a separator may be installed to treat the decanter wastewater (Figure 1-1g). But, what separation technologies

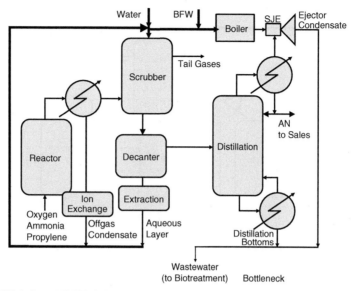

FIGURE 1-1h   DEFINING SEPARATION TECHNOLOGIES

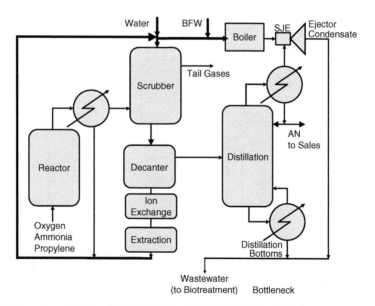

FIGURE 1-1i HYBRID SEPARATION TECHNOLOGIES FOR THE DECANTER WASTEWATER

should be used? To remove what? From which streams? Figures 1-1h–1-1j are just three possibilities (out of numerous alternatives) for the type and allocation of separation technologies. And so on! Clearly, there are *infinite number of alternatives* that can solve this problem. So many decisions have to be made on the rerouting of streams, the distribution of streams, the

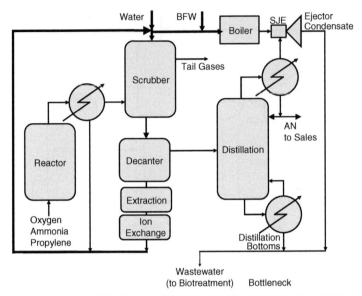

FIGURE 1-1j   SWITCHING THE ORDER OF SEPARATION TECHNOLOGIES

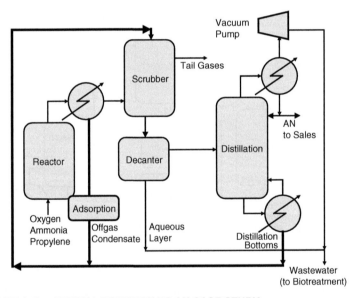

FIGURE 1-1k   OPTIMAL SOLUTION TO AN CASE STUDY

changes to be made in the process (including design and operating variables), the substitution of materials and reaction pathways, and the replacement or addition of units. It is worth describing the optimum solution (in terms of cost) to this problem as shown in Figure 1-1k. The development of this solution is shown in detail in Chapter Five.

The following observations may be inferred from the foregoing discussion:

- There are typically numerous alternatives that can solve a typical challenging process improvement problem
- The optimum solution may not be intuitively obvious
- One should not focus on the symptoms of the process problems. Instead, one should identify the root causes of the process deficiencies
- It is necessary to understand and treat the process as an integrated system
- There is a critical need to systematically extract the optimum solution from among the numerous alternatives without enumeration.

1.2 TRADITIONAL APPROACHES TO PROCESS DEVELOPMENT AND IMPROVEMENT

Until recently, there have been three primary conventional engineering approaches to address process development and improving problems:

- Brainstorming and Solution through Scenarios: A select few of the engineers and scientists most familiar with the process work together to suggest and synthesize several conceptual design scenarios (typically three to five). For instance, the foregoing exercise of generating alternatives for the AN case study falls under this category. Each generated scenario is then assessed (e.g., through simulation, techno-economic analysis, etc.) to examine its feasibility and to evaluate some performance metrics (e.g., cost, safety, reliability, flexibility, operability, environmental impact, etc.). These metrics are used to rank the generated scenarios and to select a recommended solution. This recommended solution may be inaccurately referred to as the "optimum solution" when in fact it is only optimum out of the few generated alternatives. Indeed, it may be far away for the true optimum solution.
- Adopting/Evolving Earlier Designs: In this approach, a related problem that has been solved earlier is identified. The problem may be at the same plant or another plant. Then, its solution is either copied, adopted, or evolved to suit the problem at hand and to aid in the generation of a similar solution.
- Heuristics: Over the years, process engineers have discovered that certain design problems may be categorized into groups or regions each having a recommended way of solution. Heuristics is the application of experience-derived knowledge and rules of thumb to a certain class of problems. It is derived from the Greek word "heuriskein" which means "to discover". Heuristics

have been used extensively in industrial applications (e.g., Harmsen 2004).

Over the years, these approaches have provided valuable solutions to industrial problems and are commonly used. Notwithstanding the usefulness of these approaches in providing solutions that typically work, they have several serious limitations (Sikdar and El-Halwagi 2001):

- Cannot enumerate the infinite alternatives: Since these approaches are based on brainstorming few alternatives or evolving an existing design, the generated alternatives are limited.
- Is not guaranteed to come close to optimum solutions: Without the ability to extract the optimum from the infinite alternatives, these approaches may not provide effective solutions (except for very simple cases, extreme luck, or near-exhaustive effort). Just because a solution works and is affordable does not mean that it is a good solution. Additionally, when a solution is selected from few alternatives, it should not be called an optimum solution. It is only optimum with respect to the few generated alternatives.
- Time and money intensive: Since each generated alternative should be assessed (at least from a techno-economic perspective), there are significant efforts and expenses involved in generating and analyzing the enumerated solutions.
- Limited range of applicability: Heuristics and rules of thumb are most effective when the problem at hand is closely related to the class of problems and design region for which the rules have been derived. However, they must be used with extreme care. Even subtle differences from one process to another may render the design rules invalid.
- Does not shed light on global insights and key characteristics of the process: In addition to solving the problem, it is beneficial to understand the underlying phenomena, root causes of the problem, and insightful criteria of the process. Trial and error as well as heuristic rules rarely provide these aspects.
- Severely limits groundbreaking and novel ideas: If the generated solutions are derived from the last design that was implemented or based exclusively on the experience of similar projects, what will drive the "out-of-the-box" thinking that leads to process innovation.

These limitations can be eliminated if these three conventional approaches are incorporated within a systematic and integrative framework. The good news is that recent advances in process design have led to the development of systematic, fundamental, and generally applicable techniques can be learned and applied to overcome the aforementioned limitations and methodically address process-improvement problems. This is possible through *process integration and its vital elements of process synthesis and analysis.*

1.3 WHAT IS PROCESS SYNTHESIS?

Synthesis involves putting together separate elements into a connected or a coherent whole. The term "process synthesis" dates back to the early 1970s and gained much attention with the seminal book of Rudd et al. (1973). Process synthesis may be defined as (Westerberg 1987): "the discrete decision-making activities of conjecturing (1) which of the many available component parts one should use, and (2) how they should be interconnected to structure the optimal solution to a given design problem". Process synthesis is concerned with the activities in which the various process elements are combined and the flowsheet of the system is generated so as to meet certain objectives. Therefore, the aim of process synthesis (Johns 2001) is: "to optimize the logical structure of a chemical process, specifically the sequence of steps (reaction, distillation, extraction, etc.), the choice of chemical employed (including extraction agents), and the source and destination of recycle streams". Hence, in process synthesis we know process inputs and outputs and are required to revise the structure and parameters of the flowsheet (for retrofitting design of an existing plant) or create a new flowsheet (for grassroot design of a new plant). This is shown in Figure 1-2.

Reviews on process synthesis techniques are available in literature (e.g., Westerberg 2004; Seider et al., 2003; Biegler et al., 1997; Smith 1995; Stephanopoulos and Townsend 1986).

The result of process synthesis is a flowsheet which represents the configuration of the various pieces of equipment and their interconnection. Next, it is necessary to analyze the performance of this flowsheet.

1.4 WHAT IS PROCESS ANALYSIS?

While synthesis is aimed at combining the process elements into a coherent whole, analysis involves the decomposition of the whole into its constituent elements for individual study of performance. Hence, process analysis can be contrasted (and complemented) with process synthesis. Once an alternative is generated or a process is synthesized, its detailed characteristics (e.g., flowrates, compositions, temperature, and pressure) are predicted using analysis techniques. These techniques include mathematical models, empirical correlations, and computer-aided process simulation tools. In addition, process analysis may involve predicting and validating performance using

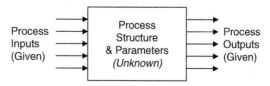

FIGURE 1-2   PROCESS SYNTHESIS PROBLEMS

experiments at the lab and pilot-plant scales, and even actual runs of existing facilities. Thus, in *process analysis* problems we know the process inputs along with the process structure and parameters while we seek to determine the process outputs (Figure 1-3).

1.5 WHY INTEGRATION?

Now, we turn our attention to a motivating example on coal pyrolysis process. A simplified flowsheet of the process is shown in Figure 1-4. The main products are different hydrocarbon cuts. Benzene is further processed in a dehydrogenation reactor to produce cyclohexane. A hydrogen-rich gas is produced out of the cyclohexane reactor and is currently flared. Medium and heavy distillates contain objectionable materials (primarily sulfur, but

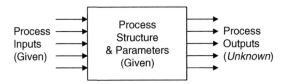

FIGURE 1-3   PROCESS ANALYSIS PROBLEM

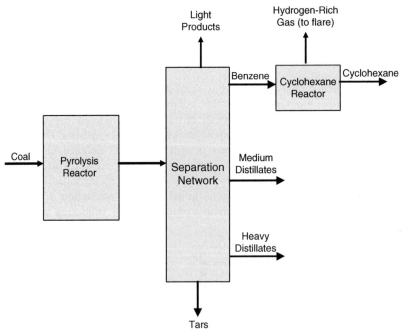

FIGURE 1-4   PYROLYSIS OF COAL

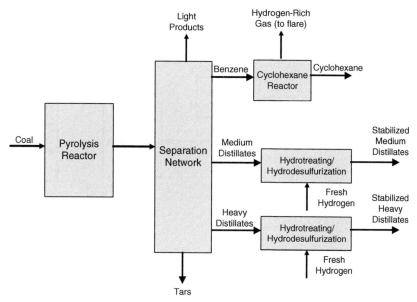

FIGURE 1-5 SYNTHESIZED FLOWSHEET FOR STABILIZATION OF MEDIUM AND HEAVY DISTILLATES

also nitrogen, oxygen, halides) that should be removed and unsaturated hydrocarbons (e.g. olefins and gum-forming unstable diolefins) that should be converted to paraffins. Our design objective is to synthesize a revised process to remove sulfur (and other objectionable materials) and stabilize olefins and diolefins. One way of addressing the problem is to synthesize a revised flowsheet that include hydrotreating and hydrodesulfurization units as shown in Figure 1-5. These units employ fresh hydrogen to remove the objectionable materials and stabilize the olefins and diolefins. This is a synthesized solution that will work, but what is wrong with this solution? There is *no integration* of mass (hydrogen). On one hand, fresh hydrogen is purchased and used in hydrotreating/hydrodesulfurization. On the other hand, hydrogen produced from benzene dehydrogenation is flared. Integration of mass is needed to conserve resources and reduce cost.

Reflecting back on the AN case study, the original flowsheet (Figure 1-1a) suffered from lack of water integration. While fresh water was used in the boiler and the scrubber, wastewater was discharged into biotreatment. The various alternatives to solve the AN example attempted to provide some level of water integration.

Next, let us consider the pharmaceutical processing facility (El-Halwagi 1997) illustrated in Figure 1-6. The feed mixture (C_1) is first heated to 550 K, then fed to an adiabatic reactor where an endothermic reaction takes place. The off-gases leaving the reactor (H_1) at 520 K are cooled to 330 K prior to being forwarded to the recovery unit. The mixture leaving the bottom of the reactor is separated into a vapor fraction and a slurry fraction.

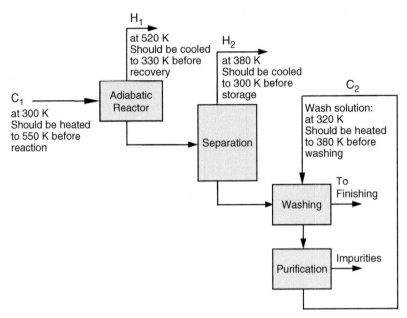

FIGURE 1-6   HEATING AND COOLING REQUIREMENTS FOR PHARMACEUTICAL PROCESS (EL-HALWAGI 1997)

The vapor fraction (H_2) exits the separation unit at 380 K and is to be cooled to 300 prior to storage. The slurry fraction is washed with a hot immiscible liquid at 380 K. The wash liquid is purified and recycled to the washing unit. During purification, the temperature drops to 320 K. Therefore, the recycled liquid (C_2) is heated to 380 K.

One alternative for synthesizing a solution that addresses the energy requirements for the pharmaceutical process is to add two heaters and two coolers that respectively employ a heating utility (e.g., steam, heating oil) and a cooling utility (e.g., cooling water, refrigerant). This solution (Figure 1-7) will work, but what is wrong with it? There is *no integration of heat*. There are two process hot streams to be cooled and two process cold streams to be heated. It seems advantageous to attempt to transfer heat from the process hot streams to the process cold streams before paying for external heating and cooling utilities. In fact, exchanging heat between process hot streams and process cold streams will result in a simultaneous reduction in the usage of external heating and cooling utilities. In addition to cost savings, the process will also conserve natural resources by virtue of decreasing the consumption of fuel and other energy sources needed for the generation of heating and cooling utilities.

The foregoing discussion highlights the need for an integration framework to guide and assist process synthesis activities and conserve process resources. This is the scope of process integration.

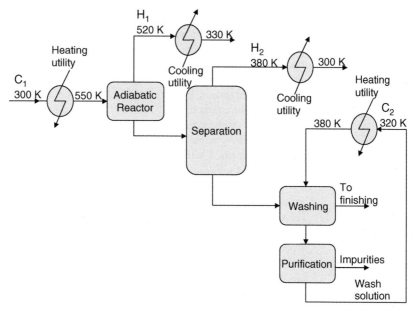

FIGURE 1-7   HEATING AND COOLING REQUIREMENTS FOR PHARMACEUTICAL PROCESS (EL-HALWAGI 1997)

1.6 WHAT IS PROCESS INTEGRATION?

A chemical process is an integrated system of interconnected units and streams. Proper understanding and solution of process problems should not be limited to symptoms of the problems but should identify the root causes of these problems by treating the process as a whole. Furthermore, effective improvement and synthesis of the process must account for this integrated nature. Therefore, integration of process resources is a critical element in designing and operating cost-effective and sustainable processes. *Process integration is a holistic approach to process design, retrofitting, and operation which emphasizes the unity of the process* (El-Halwagi 1997). In light of the strong interaction among process units, resources, streams, and objectives, process integration offers a unique framework for fundamentally understanding the global insights of the process, methodically determining its attainable performance targets, and systematically making decisions leading to the realization of these targets.

Process integration involves the following activities:

1. Task Identification: The first step in synthesis is to explicitly express the goal we are aiming to achieve and describe it as an *actionable* task. The actionable task should be defined in such a way so as to capture the essence of the original goal. For instance, quality enhancement may be described as a task to reach a specific

composition or certain properties of a product. Another example is in the AN case study, the debottlenecking objective may be expressed as a wastewater-reduction task which entails water integration. In the case of the coal pyrolysis example, the task of stabilizing the middle and heavy distillates can be expressed as a hydrodesulfurization/hydrotreating task which involves hydrogen integration. In characterizing the task, we should describe the salient information and constraints. Additionally, the task should be characterized by some quantifiable metrics. For instance, the task may be quantified with an extreme performance (e.g., minimum wastewater discharge), a specific value (e.g., 50% reduction in wastewater), or as a multivariable function (e.g., relationship between extent of wastewater reduction and pollutant content).

2. Targeting: The concept of targeting is one of the most powerful contributions of process integration. *Targeting refers to the identification of performance benchmarks ahead of detailed design.* In a way, you can find the ultimate answer without having to specify how it may be reached! For instance, in the AN example what is the target for minimum wastewater discharge. As shown in Chapter Five, this target is 4.8 kg/s and it can be determined without discussing how it can be reached. Similarly, in the pharmaceutical process, the targets for minimum heating and cooling utility requirements are 2620 and 50 kW, respectively. Again, these targets can be rigorously determined without conjecturing how the implementation of the heat-integration scheme looks like. Targeting allows us to determine how far we can push the process performance and sheds useful insights on the exact potential and realizable opportunities for the process. Even if we elect not to reach the target, it is still useful to benchmark current performance versus the ultimate performance.

3. Generation of Alternatives (Synthesis): Given the enormous number of possible solutions to reach the target (or the defined task), it is necessary to use a framework that is rich enough to embed all configurations of interest and represent alternatives that aid in answering questions such as: How should streams be rerouted? What are the needed transformations (e.g., separation, reaction, heating, etc.)? For example, should we use separations to clean up wastewater for reuse? To remove what? How much? From which streams? What technologies should be employed? For instance, should we use extraction, stripping, ion exchange, or a combination? Where should they be used? Which solvents? What type of columns? Should we change operating conditions of some units? Which units and which operating conditions? The right level of representation for generating alternatives is critically needed to set capture the appropriate design space. Westerberg (2004) underscores this point by stating that "It is crucial to get the representation right. The right representation can enhance insights. It can aid innovation".

4. Selection of Alternative(s) (Synthesis): Once the search space has been generated to embed the appropriate alternatives, it is necessary to extract the optimum solution from among the possible alternatives. This step is typically guided by some performance metrics that assist in ranking and selecting the optimum alternative. Graphical, algebraic, and mathematical optimization techniques may be used to select the optimum alternative(s). It is worth noting that the generation and selection of alternatives are *process synthesis* activities.

5. Analysis of Selected Alternative(s): Process analysis techniques can be employed to evaluate the selected alternative. This evaluation may include prediction of performance, techno-economic assessment, safety review, environmental impact assessment, etc.

It is instructive to reiterate the difference between targeting and the generation/selection of alternatives. Targeting is a structure-independent approach while the generation and selection of alternative configurations is structure based (El-Halwagi and Spriggs 1998; El-Halwagi 1997). The structure-independent (or targeting) approach is based on tackling the task via a sequence of stages. Within each stage, a design target can be identified and employed in subsequent stages. Such targets are determined ahead of detailed design and without commitment to the final system configuration. The targeting approach offers two main advantages. First, within each stage, the problem dimensionality is reduced to a manageable size, avoiding the combinatorial problems. Second, this approach offers valuable insights into the system performance and characteristics.

The structure-dependent approach to the generation and selection of alternatives involves the development of a framework that embeds all potential configurations of interest. Examples of these frameworks include process graphs (e.g. Brendel et al., 2000; Kovacs et al., 2000; Friedler et al., 1995), state-space representation (e.g., Martin and Manousiouthakis 2001; Bagajewicz and Manousiouthakis 1992), and superstructures (e.g., Biegler et al., 1997; Floudas et al., 1986). The mathematical representation used in this approach is typically in the form of mixed-integer non-linear programs (MINLPs). The objective of these programs is to identify two types of variables; integer and continuous. The integer variables correspond to the existence or absence of certain technologies and pieces of equipment in the solution. For instance, a binary integer variable can assume a value of one when a unit is selected and zero when it is not chosen as part of the solution. On the other hand, the continuous variables determine the optimal values of nondiscrete design and operating parameters such as flowrates, temperatures, pressures, and unit sizes. Although this approach is potentially more robust than the structure-independent strategies, its success depends strongly on three challenging factors. First, the system representation should embed as many potential alternatives as possible. Failure to incorporate certain configurations may result in suboptimal solutions. Second, the non-linearity properties of the mathematical formulations mean that obtaining a global solution to these optimization programs can sometimes be an illusive goal. Finally, once the synthesis task is formulated as an MINLP, the engineer's input, preference, judgment, and insights are set aside. Therefore, it is important to incorporate these insights as part of the problem formulation. This can be a tedious task.

1.7 CATEGORIES OF PROCESS INTEGRATION

Over the past two decades, numerous contributions have been made in the field of process integration. These contributions may be classified in different ways. One method of classification is based on the two main commodities consumed and processed in a typical facility: energy and mass. Therefore, from the perspective of resource integration, process integration may be classified into energy integration and mass integration. *Energy integration* is a systematic methodology that provides a fundamental understanding of energy utilization within the process and employs this understanding in identifying energy targets and optimizing heat-recovery and energy-utility systems. On the other hand, *mass integration* is a systematic methodology that provides a fundamental understanding of the global flow of mass within the process and employs this understanding in identifying performance targets and optimizing the generation and routing of species throughout the process. The fundamentals and applications of energy and mass integration have been reviewed in literature (e.g., Rossiter 2004; Dunn and El-Halwagi 2003; Hallale 2001; Smith 2000; El-Halwagi and Spriggs 1998; El-Halwagi 1997; Shenoy 1995; Linnhoff 1994; Gundersen and Naess 1988; Douglas 1988). More recently, a new category of process integration has been introduced. It is referred to as "property integration". Chapter Eight provides on overview of property integration and how it can be used to optimize the process and conserve resources by tracking properties and functionalities.

1.8 STRUCTURE OF THE BOOK

This book presents the fundamentals and applications of process integration. Holistic approaches, methodical techniques, and step-by-step procedures are presented and illustrated by a wide variety of case studies. Visualization, algebraic, and mathematical programming techniques are used to explain and address process integration problems. The first five chapters of the book focus on graphical approaches. Chapters six and seven illustrate the use of algebraic tools. The rest of the book introduces mathematical programming techniques in conjunction with graphical and algebraic methods. The covered topics include mass integration, energy integration, and property integration. The scope of problems ranges from identification of overall performance targets to integration of separation systems, recycle networks, and heat exchange networks. Numerous case studies are used to illustrate the theories and concepts. It is hoped that this book introduces you to the fascinating world of process integration, concepts, tools, and applications. Welcome aboard!

1.9 REFERENCES

Bagajewicz, M.J. and Manousiouthakis, V. 1992, 'Mass-heat exchange network representation of distillation networks,' *AIChE J.*, vol. 38, no. 11, pp. 1769-1800.

Biegler, L.T., Grossmann, I.E., and Westerberg, A.W. 1997, *Systematic Methods of Chemical Process Design*, Prentice Hall, New Jersey.

Brendel, M.H., Friedler, F., and Fan, L.T. 2000, 'Combinatorial foundation for logical formulation in process network synthesis,' *Computers Chem. Engng.*, vol. 24, pp. 1859-1864.

Douglas, J.M. 1988, *Conceptual Design of Chemical Processes*, McGraw Hill, New York.

Dunn, R.F. and El-Halwagi, M.M. 2003, 'Process integration technology review: Background and applications in the chemical process industry,' *J. Chem. Tech. and Biotech.*, vol. 78, pp. 1011-1121.

El-Halwagi, M.M. 1997, *Pollution Prevention Through Process Integration*, Academic Press, San Diego.

El-Halwagi, M.M. and Spriggs, H.D. 1998, 'Solve design puzzles with mass integration,' *Chem. Eng. Prog.*, vol. 94, pp. 25-44.

Floudas, C.A., Ciric, A.R., and Grossmann, I.E. 1986, 'Automatic synthesis of optimum heat exchange network configurations,' *AIChE J.*, vol. 32, no. 2, pp. 276-290.

Friedler, F., Varga, J.B., and Fan, L.T. 1995, 'Algorithmic approach to the integration of total flowsheet synthesis and waste minimization,' *AIChE Symp. Ser. AIChE, NY*, vol. 90, no. 303, pp. 86-97.

Gundersen, T. and Naess, L. 1988, 'The synthesis of cost optimal heat exchanger networks: An industrial review of the state of the art,' *Comp. Chem. Eng.*, vol. 12, no. 6, pp. 503-530.

Hallale, N. 2001, 'Burning bright: Trends in process integration,' *Chem. Eng. Prog.*, vol. 97, no. 7, pp. 30-41.

Harmsen, G.J. 2004, 'Industrial best practices of conceptual process design,' *Chem. Eng. & Processing*, vol. 43, pp. 677-681.

Johns, W.R. 2001, 'Process synthesis poised for a wider role,' *Chem. Eng. Prog.*, vol. 97, no. 4, pp. 59-65.

Keller, G.E. II and Bryan, P.F. 2000, 'Process engineering: Moving in new directions,' *Comp. Chem. Eng.*, vol. 96, no. 1, pp. 41-50.

Kovacs, Z., Ercsey, Z., Friedler, F., and Fan, L.T. 2000, 'Separation-network synthesis: Global optimum through rigorous super-structure,' *Computers Chem. Engg.*, vol. 24, pp. 1881-1900.

Linnhoff, B. 1994, 'Use pinch analysis to knock down capital costs and emissions,' *Chem. Eng. Prog.*, vol. 90, no. 8, pp. 33-57.

Martin, L.L. and Manousiouthakis, V. 2001, 'Total annualized cost optimality properties of state space models for mass and heat exchanger networks,' *Chem. Eng. Sci.*, vol. 56, no. 20, pp. 5835-5851.

Rossiter, A.P. 2004, 'Success at process integration,' *Chem. Eng. Prog.*, vol. 100, no. 1, pp. 58-62.

Rudd, D.F., Powers, G.J., and Siirola, J.J. 1973, *Process Synthesis*, Prentice Hall, New Jersey.

Seider, W.D., Seader, J.D., and Lewin, D.R. 2003, *Product and Process Design Principles*, Wiley, New York.

Shenoy, U.V. 1995, *Heat Exchange Network Synthesis: Process Optimization by Energy and Resource Analysis*, Gulf Pub. Co., Houston, TX.

Sikdar, S.K. and El-Halwagi, M.M eds. 2001, *Process Design Tools for the Environment*, Taylor and Francis.

Smith, R. 1995, *Chemical Process Design*, McGraw Hill, New York.

Smith, R. 2000, 'State of the art in process integration,' *Applied Thermal Eng.*, vol. 20, pp. 1337-1345.

Stephanopoulos, G. and Townsend, D. 1986, 'Synthesis in process development,' *Chem. Eng. Res. Des.*, vol. 64, pp. 160-174.

Westerberg, A.W. 1987, 'Process synthesis: A morphological view,' in *Recent Developments in Chemical Process and Plant Design*, eds. Y.A. Liu, H.A. McGee Jr. and W.R. Epperly, pp. 127-145, Wiley, New York.

Westerberg, A.W. 2004, 'A Retrospective on Design and Process Synthesis,' *Comp. Chem. Eng.*, vol. 28, pp. 447-458.

2

OVERALL MASS TARGETING

In many cases, it is useful to determine the potential improvement in the performance of a whole process or sections of the process without actually developing the details of the solution. In this context, the concept of targeting is very useful. Targeting is aimed at benchmarking the performance of a process ahead of detailed design and without commitment to the specific details of the strategies leading to improvement. This *a priori* approach ensures that the process capabilities are explored without exhausting the designer's time and effort. Examples of important mass targets include minimum consumption of material utilities (solvents, water, etc.), minimum discharge of wastes, minimum purchase of fresh raw materials, minimum production of undesirable byproducts, and maximum sales of desirable products.

To illustrate the concept of overall mass targeting, let us start by considering the objective of minimizing the discharge of a waste. Later, other objectives will be discussed.

2.1 TARGETING FOR MINIMUM DISCHARGE OF WASTE

Consider the case when it is desired to determine the target for minimizing the load of a discharged species (e.g., pollutants in effluents). Three sets of data for that species are first collected: fresh usage, generation/depletion, and terminal discharge. The fresh usage (F) corresponds to the quantity of the targeted species in streams entering the process (the waste stream may have entered the process as a fresh feedstock or a material utility). Within any process, several phenomena contribute to the net balance of a species. Generation (G) refers to the net amount of the targeted species which is produced through chemical reaction. Depletion (D) may take place through chemical reactions but it may also be attributed to leaks, fugitive emissions, and other losses that are not explicitly accounted for. The net generation (Net_G) of a targeted species is defined as the difference between generation and depletion. Therefore,

$$Net\_G = G - D \qquad (2.1)$$

Finally, the terminal discharge (T) is used to refer to the load of the targeted species in streams designated as waste streams or point sources for pollution.

An overall material balance on the targeted species before mass integration (BMI) is shown in Figure 2-1 and can be expressed as:

$$T^{BMI} = F^{BMI} + Net\_G^{BMI} \qquad (2.2)$$

where T^{BMI}, F^{BMI}, and Net_G^{BMI}, respectively refer to the terminal load, fresh load, and net generation of the targeted species before any mass integration changes.

In order to reduce the terminal discharge of the targeted species, the two terms on the right hand side of equation (2.1) are to be reduced. The general solution strategy is described by Noureldin and El-Halwagi (1999, 2000). However, there are several special cases for which simplified targeting methods can be developed. For instance, when the net generation is independent of stream recycle and adjustments in fresh feed, we can use the following two-step shortcut method to identify the minimum-discharge target. In the first step, the net generation of the targeted species is minimized. The net generation can be described in terms of the process design and operating variables that are allowed to be modified. The value of the minimum net generation of the targeted species is referred to as Net_G^{MIN}.

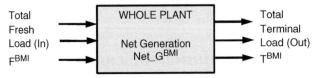

FIGURE 2-1 OVERALL MATERIAL BALANCE FOR THE TARGETED SPECIES BEFORE MASS INTEGRATION

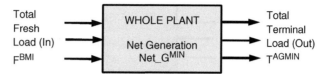

FIGURE 2-2 OVERALL MATERIAL BALANCE FOR THE TARGETED SPECIES AFTER MINIMIZING THE NET GENERATION

As can be seen from Figure 2-2, once the net generation has been minimized, we can calculate the terminal discharge load after generation minimization (T^{AGMIN}) through an overall material balance:

$$T^{AGMIN} = F^{BMI} + Net\_G^{MIN} \qquad (2.3)$$

In the second step, the fresh usage in the process is minimized. First, the requirement for a fresh supply of the targeted species is reduced to the extent feasible by adjusting design and operating variables of the units that employ the targeted species in fresh streams. This entails answering the following questions:

- What are the design and operating variables that influence fresh usage?
- Which of these variables are allowed to be changed (manipulated variables)?
- What is the functional relationship between these variables and fresh consumption? This relationship may be expressed as:

$$F = f(\text{manipulated design variables, manipulated operating variables})$$
$$(2.3a)$$

Equation (2.3a) can be minimized to identify the minimum load of the targeted species in the fresh feeds after reduction (referred to as F^{AFR}). As can be seen from Figure 2-3, the terminal load after minimization of net generation and reduction in fresh feed is expressed as:

$$T^{AGMIN, AFR} = F^{AFR} + Net\_G^{MIN} \qquad (2.3b)$$

Finally, it is possible to recover almost all of the targeted species from the terminal streams or from paths leading to the terminal streams and render the recovered species in a condition which enables its use in lieu of the targeted species in the fresh feed. Clearly, the higher the recovery and recycle to replace the fresh load, the lower the net fresh use and the lower the terminal discharge. Consequently, in order to minimize the terminal discharge of the targeted species, we should recycle the maximum amount from terminal streams (or paths leading to terminal streams) to replace fresh feed.

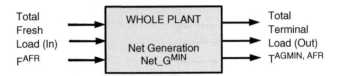

FIGURE 2-3   OVERALL MATERIAL BALANCE OF THE TARGETED SPECIES AFTER FRESH REDUCTION AND NET-GENERATION MINIMIZATION

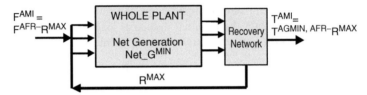

FIGURE 2-4   TARGETING FOR MINIMUM DISCHARGE OF TARGETED SPECIES

The maximum recyclable load of the targeted species is the lower of the two loads: the fresh and the terminal, i.e.,

$$R^{MAX} = \text{argmin} \{F^{AFR}, T^{AGMIN, AFR}\} \qquad (2.4)$$

where argmin refers to the lowest value in the set of loads (in this case the fresh and terminal loads).

As a result of recycle, the fresh load after mass integration becomes

$$F^{AMI} = F^{AFR} - R^{MAX} \qquad (2.5)$$

Therefore, the target for minimum discharge of the targeted species (Figure 2-4) after mass integration (T^{AMI}) can be calculated through the following overall material balance after mass integration:

$$T^{AMI} = F^{AMI} + Net\_G^{MIN} \qquad (2.6)$$

Alternatively, the target for minimum discharge can be calculated from a material balance around the recovery and recycle system:

$$T^{AMI} = T^{AGMIN, AFR} - R^{MAX} \qquad (2.7)$$

It is worth mentioning that in the spirit of targeting, it is not necessary to determine the type or techno-economic details of the recovery system. The objective is to determine the target for minimum load to be discharged from the plant. Later, we discuss how the various alternatives are systematically screened with the objective of meeting this target at minimum cost.

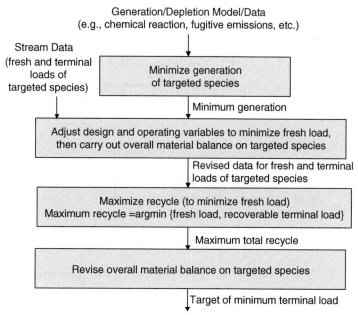

FIGURE 2-5 PROCEDURE FOR IDENTIFYING MINIMUM WASTE DISCHARGE

The foregoing procedure can be summarized by the flowchart shown in Figure 2-5. First, generation and depletion information are gathered as data or models. These include information on the depletion or generation of the targeted species through chemical reactions, nonpoint loss of the species through fugitive emissions or leaks, etc. Data are also collected on the amount of the targeted species in fresh feeds entering the process or terminal streams leaving the process. The net generation is first minimized. Then, the fresh feed of the targeted species is minimized by adjusting design and operating variables. Next, a recovery system is placed to recover maximum recyclable load (the lower of fresh and terminal loads). Hence, the fresh feed is minimized as maximum recycle is used to replace fresh feed. Finally, an overall material balance is used to calculate the target for minimum waste discharge.

2.2 TARGETING FOR MINIMUM PURCHASE OF FRESH MATERIAL UTILITIES

We now move to the case of targeting minimum fresh usage of material utilities (e.g, water, solvents, additives, etc.). The overall material balance can be written as:

$$F^{BMI} = T^{BMI} - Net\_G^{BMI} \tag{2.8}$$

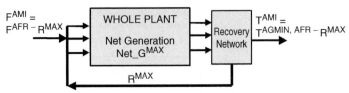

FIGURE 2-6   TARGETING FOR MINIMUM USAGE OF MATERIAL UTILITIES

For the special case, when the net generation is independent of stream recycle and adjustments in fresh feed, we can use a procedure similar to the one described for minimum waste discharge. The key difference is that the net generation of the targeted species is maximized (as opposed to minimized in the case of waste discharge). This can be deduced from the overall material balance given by equation (2.8) which entails maximizing the net generation in order to minimize the fresh load. The targeting scheme is shown in Figure 2-6.

It is worth pointing out that when the net generation is not to be altered, the target procedures for minimizing waste discharge and fresh usage become identical. Other objectives such as minimizing raw materials and maximizing yield can be similarly developed (e.g., Al-Otaibi and El-Halwagi 2004).

2.3 MASS-INTEGRATION STRATEGIES FOR ATTAINING TARGETS

Once a target is determined, it is necessary to develop cost-effective strategies to reach the target. In general, these strategies include stream segregation/mixing, recycle, interception using separation devices, changes in design and operating conditions of units, materials substitution, and technology changes including the use of alternate chemical pathways. These strategies can be classified into a hierarchy of three categories (El-Halwagi 1999; Noureldin and El-Halwagi 2000):

- No/low cost changes
- Moderate cost modifications, and
- New technologies

Three main factors can be used in describing these strategies, economics, impact, and acceptability. The economic dimension can be assessed by a variety of criteria such as capital cost, return on investment, net present worth, and payback period. Impact is a measure of the effectiveness of the proposed solution in reducing negative ecological and hazard consequences of the process such as reduction in emissions and effluents from the plant. Acceptability is a measure of the likelihood of a proposed strategy to be accepted and implemented by the plant. In addition to cost, acceptability depends upon several factors including corporate culture, dependability, safety, and operability. Figure 2-7 is a schematic representation of the typical hierarchy of mass-integration strategies. These strategies are typically

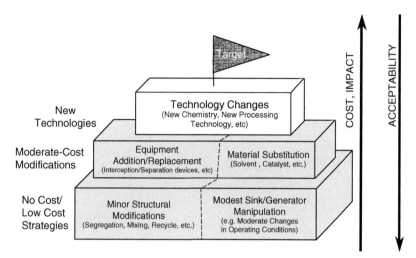

FIGURE 2-7 HIERARCHY OF MASS-INTEGRATION STRATEGIES (EL-HALWAGI 1999)

in ascending order of cost and impact and in descending order of acceptability.

The development of process-related mass-integration strategies will be covered throughout this book.

EXAMPLE 2-1 REDUCTION OF ACETIC ACID FRESH USAGE AND TERMINAL LOSSES IN A VINYL ACETATE PLANT

Vinyl acetate monomer "VAM" is manufactured by reacting acetic acid with oxygen and ethylene according to the following chemical reaction:

$$C_2H_4 \quad + \quad 0.5\,O_2 + \quad C_2H_4O_2 \quad = \quad C_4H_6O_2 \quad + \quad H_2O$$

Ethylene Oxygen Acetic Acid   VAM   Water

Consider the process shown in Figure 2-8. A fresh feed of 10,000 kg/h of acetic acid "AA" along with 200 kg/h of water are evaporated in an acid tower. The vapor is fed with oxygen and ethylene to the reactor where 7000 kg/h of acetic acid are reacted and 10,000 kg/h of VAM are formed. The reactor off-gas is cooled and fed to the first absorber where AA (5100 kg/h) is used as a solvent. Almost all the gases leave from the top of the first absorption column together with 1200 kg/h of AA. This stream is fed to the second absorption column where water (200 kg/h) is used to scrub acetic acid. The bottom product of the first absorption column is fed to the primary distillation tower where VAM is recovered as a top product (10,000 kg/h) along with 200 kg/h of water and a small amount of AA (100 kg/h) which is not economically justifiable to recover. This stream is sent to final finishing. The bottom product of the primary tower (6800 kg/h of AA and 2300 kg/h of water) is mixed with the bottom product of the second absorption column

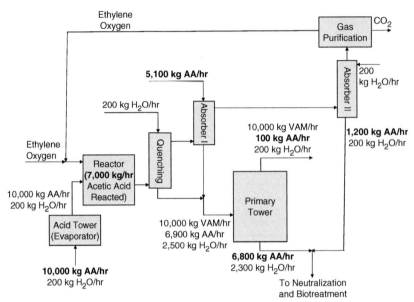

FIGURE 2-8   VAM PRODUCTION PROCESS

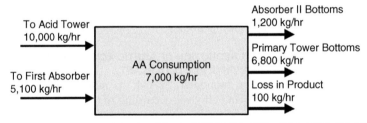

FIGURE 2-9   OVERALL AA MATERIAL BALANCE FOR THE VAM PROCESS BEFORE MASS INTEGRATION

(1200 kg/h of AA and 200 kg/h of water). The mixed waste is fed to a neutralization system followed by biotreatment. In this example, let us consider the case when no changes are made to the consumption by chemical reaction and there are no adjustments in design or operating conditions to reduce fresh AA consumption. What is the target for minimum fresh usage and minimum terminal losses of AA?

SOLUTION

First, we extract the pertinent information regarding the fresh and terminal loads as well as consumption by chemical reaction. The data are presented in Figure 2-9. Clearly, a net consumption of 7000 kg AA/h is the equivalent of −7000 kgAA/h of net generation. Since there are no allowable changes in reaction or design and operating conditions affecting fresh consumption, we resort to recovery from terminal streams, recycle, and

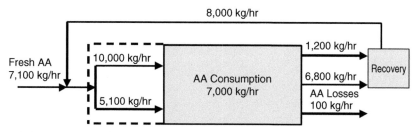

FIGURE 2-10 OVERALL AA MATERIAL BALANCE FOR THE VAM PROCESS AFTER MASS INTEGRATION

replacement of fresh AA. It is worth noting that the recoverable load from the terminal streams (8000 kg AA/h) is less than the fresh load (15,100 kg AA/h). Hence, R^{MAX} is 8000 kg AA/h. Therefore, according to equation (2.5), we get a target for minimum fresh usage to be:

$$F^{AMI} = 15,100 - 8000 = 7100 \, \text{kg AA/h} \qquad (2.9)$$

Consequently, as can be seen from Figure 2-10 the target for minimum terminal losses of AA is 100 kg /h.

EXAMPLE 2-2 REDUCTION OF DISCHARGE IN A TIRE-TO-FUEL PLANT

This case study is adapted from El-Halwagi (1997) and Noureldin and El-Halwagi (1999). It involves a processing facility that converts scrap tires into fuel via pyrolysis. Figure 2-11 is a simplified flowsheet of the process. The discarded tires are fed to a high-temperature reactor where hydrocarbon content of the tires are broken down into oils and gaseous fuels. The oils are further processed and separated to yield transportation fuels. As a result of the pyrolysis reactions, water is formed. The amount of generated water is a function of the reaction temperature, T_{rxn}, through the following correlation:

$$W_{rxn} = 0.152 + (5.37 - 7.84 \times 10^{-3} T_{rxn}) e^{(27.4 - 0.04 T_{rxn})} \qquad (2.10)$$

where W_{rxn} is in kg/s and T_{rxn} is in K. At present, the reactor is operated at 690 K which leads to the generation of 0.12 kg water/s. In order to maintain acceptable product quality, the reaction temperature should be maintained within the following range:

$$690 \leq T_{rxn}(\text{K}) \leq 740 \qquad (2.11)$$

The gases leaving the reactor are passed through a cooling/condensation system to recover some of the light oils. In order to separate the oils, a decanter is used to separate the mixture into two layers: aqueous and organic. The aqueous layer is a wastewater stream whose flowrate is designated as W_1

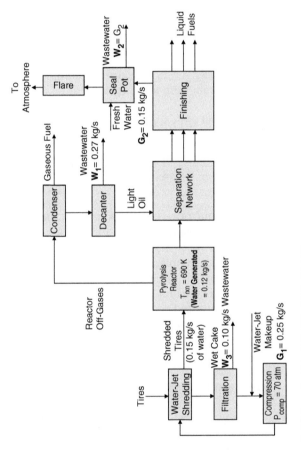

FIGURE 2-11   SIMPLIFIED FLOWSHEET OF TIRE-TO-FUEL PLANT

and it contains phenol as the primary pollutant. The organic layer is mixed with the liquid products of the reactor and fed to finishing. As a result of fuel finishing, a gaseous waste is produced and flared. As a safety precaution to prevent the back-propagation of fire from the flare, a seal pot (or a water valve) is placed before the flare to provide a buffer zone between the fire and the flare gas. The flowrate of the water stream passing through the seal pot is referred to as G_2 and an equivalent flowrate of wastewater stream, $W_2 = G_2$, is withdrawn from the seal pot.

Tire shredding is achieved by using high-pressure water jets. The shredded tires are fed to the process while the spent water is filtered. The wet cake collected from the filtration system is forwarded to solid waste handling. The filtrate is mixed with fresh water-jet makeup "G_1" to compensate for water losses with the wet cake "W_3" and the shredded tires. The mixture of the filtrate and the water makeup is fed to a high-pressure compression station for recycle to the shredding unit. The flowrate of water-jet makeup depends on the applied pressure coming out of the compression stage "P_{comp}" via the following expression:

$$G_1 = 0.47 \, e^{-0.009 P_{comp}} \tag{2.12}$$

where G_1 is in kg/s and P_{comp} is in atm. In order to achieve acceptable shredding, the jet pressure may be varied within the following range:

$$70 \le P_{comp}(\text{atm}) \le 95 \tag{2.13}$$

At present, P_{comp} is 70 atm which requires a water-jet makeup flowrate of 0.25 kg/s.

The water lost in the cake is related to the mass flowrate of the water-jet makeup through:

$$W_3 = 0.4 \, G_1 \tag{2.14}$$

In addition to the water in the wet cake, the plant has two primary sources for wastewater; from the decanter (W_1) and from the seal pot (W_2). At present, the values of W_1, W_2, and W_3 are 0.27, 0.15, and 0.10 kg/s, respectively. The wastewater from the decanter contains about 500 ppm of phenol. Within the range of allowable operating changes, this concentration can be assumed to remain constant. At present, the wastewater from the seal pot contains no phenol. The plant has been shipping the wastewater streams W_1 and W_2 for off-site treatment. The cost of wastewater transportation and treatment is $0.10/kg leading to a wastewater treatment cost of approximately $1.33 million/year. W_3 has been processed on site. Because of the characteristics of W_3, the plant does not allow its recycle back to the process even after waste-handling processing. The plant wishes to reduce off-site treatment of wastewater streams W_1 and W_2 to avoid cost of off-site treatment and alleviate legal-liability concerns in case of transportation accidents or inadequate treatment of the wastewater. The objective of this problem is to determine a target for reduction in flowrate of terminal discharges W_1 and W_2.

SOLUTION

Figure 2-12 shows an overall water balance for the process before mass integration.

The first step in the analysis is to reduce the terminal discharge by minimizing the net generation of water. Figure 2-13 is a graphical representation of equation (2.10) illustrating the net generation of water through chemical reaction as a function of the reaction temperature. As can be seen from this graph, the minimum generation of water is 0.08 kg/s and is attained at a reaction temperature of 710 K.

Next, we adjust design and operating parameters so as to minimize fresh water consumption. As mentioned earlier, the fresh water used in shredding is a function of pressure as given by equation (2.12):

$$G_1 = 0.47 \, e^{-0.009 P_{comp}}$$

where P_{comp} should be maintained within a permissible interval of [70 atm, 95 atm]. Therefore, in order to minimize the fresh water needed for shredding, the value of P_{comp} should be set to its maximum limit of 95 atm.

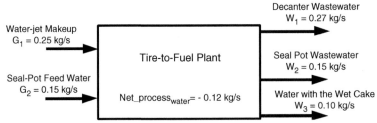

FIGURE 2-12   OVERALL WATER BALANCE FOR THE TIRE-TO-FUEL PROCESS

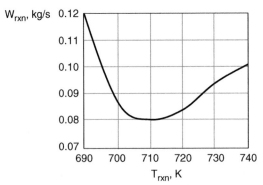

FIGURE 2-13   RATE OF WATER CONSUMPTION AS A FUNCTION OF REACTION TEMPERATURE

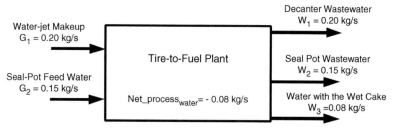

FIGURE 2-14 OVERALL WATER BALANCE AFTER SINK/GENERATOR MANIPULATION

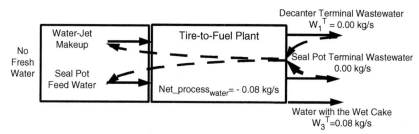

FIGURE 2-15   TARGETING FOR MINIMUM WATER USAGE AND DISCHARGE

Consequently, G_1 is reduced to $0.20\,\text{kg/s}$. According to equation (2.14), the new value of W_3 is given by:

$$W_3 = 0.4 \times 0.20 = 0.08 \text{ kg/s} \qquad (2.15)$$

With the new values of G_1 and W_3 and with the water generation minimized to $0.08\,\text{kg/s}$, an overall water balance provides the value of W_2 to be $0.15\,\text{kg/s}$. These results are shown in Figure 2-14 and represent the overall water balance after sink/generator manipulation with existing units and current process configuration. Next, we calculate the target for water usage and discharge using interception (cleaning up of recycled water) and recycle. This targeting analysis is shown in Figure 2-15 and it yields a target of zero fresh water and $0.08\,\text{kg/s}$ for wastewater discharge.

2.4 PROBLEMS

2.1 Consider the VAM process described in Example 2-1. A new reaction pathway has been developed and will be used for the production of VAM. This new reaction does not involve acetic acid. The rest of the process remains virtually unchanged and the AA losses with the product are $100\,\text{kg/h}$. What are the targets for minimum fresh usage and discharge/losses of AA?

2.2 Consider the magnetic-tape manufacturing process shown by Figure 2-16. In this process (Dunn et al., 1995; El-Halwagi 1997), coating ingredients

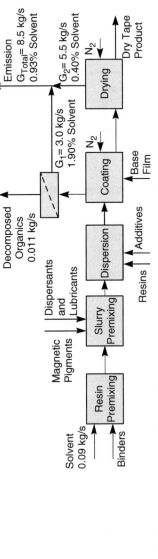

FIGURE 2-16   MAGNETIC TAPE PLANT (DUNN ET AL., 1995, EL-HALWAGI 1997)

are dissolved in 0.09 kg/s of organic solvent and mixed to form a slurry. The slurry is suspended with resin binders and special additives. Next, the coating slurry is deposited on a base film. Nitrogen gas is used to induce evaporation rate of solvent that is proper for deposition. In the coating chamber, 0.011 kg/s of solvent are decomposed into other organic species. The decomposed organics are separated from the exhaust gas in a membrane unit. The retentate stream leaving the membrane unit has a flowrate of 3.0 kg/s and is primarily composed of nitrogen that is laden with 1.9 wt/wt% of the organic solvent. The coated film is passed to a dryer where nitrogen gas is employed to evaporate the remaining solvent. The exhaust gas leaving the dryer has a flowrate of 5.5 kg/s and contains 0.4 wt/wt% solvent. The two exhaust gases are mixed and disposed off.

In addition to the environmental problem, the facility is concerned about the waste of resources, primarily in the form of used solvent (0.09 kg/s) that costs about $2.3 million/year. It is desired to undertake a mass-integration analysis to optimize solvent usage, recovery, and losses. Determine the target for minimizing fresh solvent usage in the process.

2.3 Every year, significant quantities of plastic wastes are disposed off to landfills. An emerging processing technique for reclaiming plastic waste is to convert it into liquid fuels. A schematic process flowsheet is given by Figure 2-17. The data in this problem are taken from Hamad and El-Halwagi (1998).

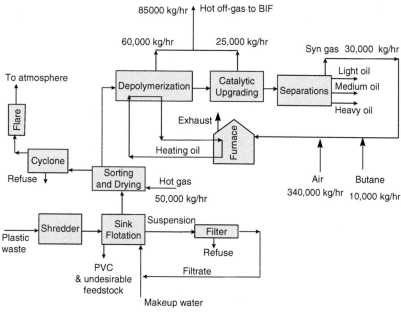

FIGURE 2-17 A SIMPLIFIED PLANT FOR OIL PRODUCTION FROM WASTE PLASTIC (HAMAD AND EL-HALWAGI 1998)

Plastic waste (feedstock) is first shredded then sorted in a sink/flotation unit to remove polyvinyl chloride (PVC) and undesirable feedstock. The suspension from the sink floatation unit is filtered. The refuse from filtration is rejected and the filtrate is recycled to the sink flotation unit. The remaining feedstock is sorted and dried using 50,000 kg/h of hot gas. At present, this hot gas contains no butane. Also, no butane is formed in sorting and drying. The hot gas leaving the sorting and drying process is passed through a cyclone to remove suspended solids as refuse. The cyclone is followed by a flare where any organics are burned.

The sorted/dried feedstock enters a depolymerization unit where a butane-laden gaseous stream (total flowrate of gas is 60,000 kg/h and it contains 7200 ppmw butane) is generated. The slurry leaving the depoly-merization unit is upgraded in a catalytic unit then separated into various hydrocarbon cuts. The off-gas leaving the catalytic upgrading unit has a flowrate of 25,000 kg/h and contains 80,000 ppmw of butane.

The depolymerization unit is heated using a recirculating heating oil coming from a furnace. The feed to the furnace consists of 30,000 kg/h of syngas (composed primarily of butane), 10,000 kg/h of butane, and 340,000 kg/h of air. The exhaust from the furnace contains almost no butane.

Until recently, the gaseous streams from the depolymerization and catalytic units were mixed and fed to a boiler/industrial furnace (BIF). Due to economic, safety, and environmental concerns the BIF operation is to be discontinued and the mixed off-gas is to be rerouted in the plant.

a. What is current flowrate (kg/h) of purchasd fresh butane?
b. What is the total amount (kg/h) of butane generated by chemical reaction in the process? *Hint*: butane generation by chemical reaction takes place in the depolymerization and catalytic upgrading units.
c. How much (kg/h) butane is depleted by chemical reaction in the process? *Hint*: butane depletion by chemical reaction takes place in the furnace.
d. What is the target (kg/h) for minimum purchase of fresh butane?

2.5 REFERENCES

Al-Otaibi, M. and El-Halwagi, M.M. 2004, 'Inclusion Techniques for Integrated Yield Enhancement', paper 110d, *AIChE Spring Meeting*, New Orleans, April.

Dunn, R.F., El-Halwagi, M.M., Lakin, J., and Serageldin, M. 1995, Selection of organic solvent blends for environmental compliance in the coating industries, *Proceedings of the First International Plant Operations and Design Conference*, E.D. Griffith, H. Kahn, and M.C. Cousinseds., AIChE, New York. Vol. III. pp. 83-107.

El-Halwagi, M.M. 1997, *Pollution Prevention through Process Integration*, Academic Press, San Diego.

El-Halwagi, M.M. 1999, Sustainable pollution prevention through mass integration, in *Tools and Methods for Pollution Prevention*. S. Sikdar, and U. Diwekar eds., UK. pp. 233-275.

Hamad, A.A. and El-Halwagi, M.M. 1998, 'Simultaneous synthesis of mass separating agents and interception networks,' *Chem. Eng. Res. Des., Trans. Inst. Chem. Eng.,* vol. 76, pp. 376-388.

Noureldin, M.B. and El-Halwagi, M.M. 2000, 'Pollution-Prevention targets through integrated design and operation,' *Comp. Chem. Eng.,* vol. 24, pp. 1445-1453.

Noureldin, M.B. and El-Halwagi, M.M. 1999, 'Interval-Based Targeting for Pollution-Prevention via Mass Integration,' *Comp. Chem. Eng.,* vol. 23, pp. 1527-1543.

3

GRAPHICAL TECHNIQUES FOR DIRECT-RECYCLE STRATEGIES

Processing facilities are characterized by the use of tremendous amounts of material resources. If not managed properly, such enormous usage can lead to the depletion of natural resources poses many economic, social, and ecological challenges. Therefore, the process industries have embarked on major material–conservation initiatives to enhance market competitiveness and sustainability. Several strategies can lead to material conservation including segregation, mixing, recycle/reuse, material substitution, reaction alteration, and process modification. In this chapter, we focus on stream rerouting including segregation, mixing, and recycle/reuse. *Segregation* refers to avoiding the mixing of streams. Segregation of streams with different compositions avoids unnecessary loss of driving force of streams. Such management of driving forces enhances the performance of process units (e.g., separators) and can also provide composition levels that allow the streams to be recycled directly to process units. *Recycle* refers to the utilization of a process stream (e.g., a waste or a low-value stream) in a process unit (a sink). While reuse is distinguished from recycle by emphasis that reuse corresponds to the reapplication of the stream for the original intent, we will use the term recycle in a general sense that includes reuse. In particular, we will start with *direct recycle* where the streams are rerouted without the installation of new devices. Direct-recycle is typically classified as low cost strategies, since it basically involves pumping and piping and in some cases may even be achieved without the need for additional pumping or piping.

First, a targeting technique will be described to identify bounds on minimum usage of fresh resources and minimum discharge of wastes through recycle/reuse. Next, a systematic procedure is presented to implement the specific stream rerouting that attains the identified target.

3.1 PROBLEM STATEMENT

Consider a process with a number of process sources (e.g., process streams, wastes) that can be considered for possible recycle and replacement of the fresh material and/or reduction of waste discharge. Each source, i, has a given flowrate, W_i, and a given composition of a targeted species, y_i. Available for service is a fresh (external) resource that can be purchased to supplement the use of process sources in sinks. The sinks are process units such as reactors, separators, etc. Each sink, j, requires a feed whose flow rate, G_j^{in}, and an inlet composition of a targeted species, z_j^{in}, must satisfy certain bounds on their values.

The objective is to develop a graphical procedure that determines the target for minimum usage of the fresh resource, maximum material reuse, and minimum discharge to waste. Therefore, the design questions to be answered include:

- Should a stream (source) be segregated and split? To how many fractions? What should be the flowrate of each split?
- Should streams or splits of streams be mixed? To what extent?
- What should be the optimum feed entering each sink? What should be its composition?
- What is the minimum amount of fresh resource to be used?
- What is the minimum discharge of unused process sources?

3.2 SOURCE–SINK MAPPING DIAGRAM AND LEVER-ARM RULES

The source–sink mapping diagram (El-Halwagi and Spriggs 1996; El-Halwagi 1997) is a visualization tool that can be used to derive useful recycle rules. As mentioned in the problem statement, there are bounds on flowrate and composition entering each sink. These bounds are described by the following constraints:

$$G_j^{min} \leq G_j^{in} \leq G_j^{max} \quad \text{where } j = 1, 2, \ldots, N_{sinks} \quad (3.1)$$

where $G_j^{\min}$ and $G_j^{\max}$ are given lower and upper bounds on admissible flowrate to unit j.

$$z_j^{\min} \leq z_j^{\text{in}} \leq z_j^{\max} \quad \text{where } j = 1, 2, ..., N_{\text{sinks}} \tag{3.2}$$

where $z_j^{\min}$ and $z_j^{\max}$ are given lower and upper bounds on admissible composition to unit j.

The flowrate and composition bounds for each sink can be determined based on several considerations such as:

- Technical considerations (e.g., manufacturer's specifications operable composition ranges to avoid scaling, corrosion buildup, etc. Operable flowrate ranges such as weeping/flooding flowrates).
- Safety (e.g., to stay away from explosion regions).
- Physical (e.g., saturation limits).
- Monitoring: These bounds can also be determined from historical data of operating the unit which are typically available through the process information monitoring system. Figures 3-1a, b illustrate the bounding of feed flowrate and composition for sink j based on monitored data for which the sink has performed acceptably.
- Constraint propagation: In some cases (Figure 3-2), the constraints on a sink (j) are based on critical constraints for another unit ($j+1$). Using a process model to relate the inlets of units j and $j+1$, we can derive the constraints for unit j based on those for unit $j+1$. For instance, suppose that the constraints for unit $j+1$ are given by:

$$0.06 \leq z_{j+1}^{\text{in}} \leq 0.08 \tag{3.3a}$$

and the process model relating the inlet compositions to units j and $j+1$ can be expressed as:

$$z_{j+1}^{\text{in}} = 2\, z_j^{\text{in}} \tag{3.3b}$$

Therefore, the bounds for unit j are calculated to be:

$$0.03 \leq z_j^{\text{in}} \leq 0.04 \tag{3.3c}$$

For each targeted species, a diagram called the source–sink mapping diagram (Figure 3-3) is constructed by plotting the flowrate versus composition. On the source–sink mapping diagram, sources are represented by shaded circles and sinks are represented by hollow circles. The constraints on flowrate and composition are respectively represented by horizontal and vertical bands. The intersection of these two bands provides a zone of acceptable load and composition for recycle. If a source (e.g., source a) lies within this zone, it can be directly recycled to the sink (e.g., sink S).

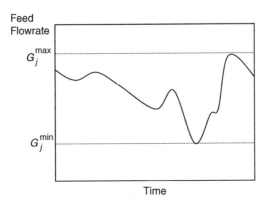

FIGURE 3-1a　BOUNDING FEED FLOWRATE BASED ON MONITORED DATA

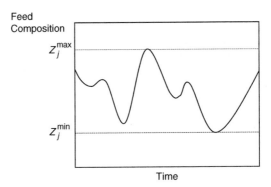

FIGURE 3-1b　BOUNDING FEED COMPOSITION BASED ON MONITORED DATA

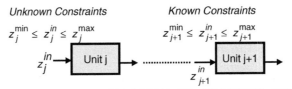

FIGURE 3-2 CONSTRAINT PROPAGATION TO DETERMINE COMPOSITION BOUNDS

Moreover, sources b and c can be mixed using the lever-arm principle to create a mixed stream that can be recycled to sink S.

The lever arms can be determined based on material balance. Consider the mixing of two sources, a and b shown in Figure 3-4. The flowrates of the two sources are W_a and W_b and their compositions are y_a and y_b. The mixture resulting from the two sources has a flowrate $W_a + W_b$ and a

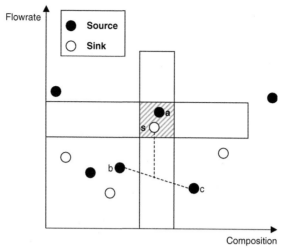

FIGURE 3-3   SOURCE–SINK MAPPING DIAGRAM

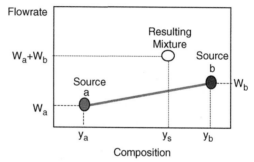

FIGURE 3-4   MIXING OF SOURCES *a* AND *b*

composition, y_s. The resulting flowrate and composition of the mixture satisfy the flowrate and composition constraints for sink S.

A material balance for the targeted species around the mixing operation can be expressed as:

$$y_s(W_a + W_b) = y_a W_a + y_b W_b \tag{3.4a}$$

Rearranging terms, we get

$$\frac{W_a}{W_b} = \frac{y_b - y_s}{y_s - y_a} \tag{3.4b}$$

or

$$\frac{W_a}{W_b} = \frac{\text{Arm for } a}{\text{Arm for } b} \tag{3.4c}$$

where

$$\text{Arm for } a = y_b - y_s \qquad (3.4d)$$

and

$$\text{Arm for } b = y_s - y_a \qquad (3.4e)$$

Similarly,

$$\frac{W_a}{W_a + W_b} = \frac{\text{Arm for } a}{\text{Total arm}} \qquad (3.5a)$$

where the total arm is the sum of arm for a and arm for b. Hence,

$$\text{Total arm} = y_b - y_a \qquad (3.5b)$$

The lever arms for the individual sources as well as the resulting mixture are shown in Figure 3-5. It is worth noting that these are the horizontal lever arms. Those arms can also be shown on the tilted line connecting sources a and b. In such cases, the ratio of the arms will be exactly the same as the ratio of the horizontal arms because of angle similitude.

We are now in a position to derive key recycle rules based on lever arms. In particular, it is possible to identify useful rules for composition of feed entering a sink and prioritization of sources to be recycled. Consider the sink j shown in Figure 3-6. The composition bounds are given by constraint (3.2). Currently, a fresh resource is used to satisfy the sink requirements. A process source a (e.g., waste stream) may be rerouted to sink j where a portion of its flowrate can be mixed with the fresh resource

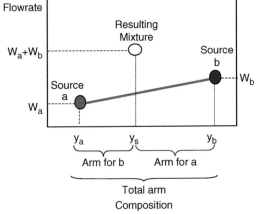

FIGURE 3-5   LEVER-ARM RULE FOR MIXING

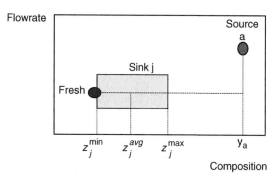

Flowrate

Source
a

Sink j

Fresh

z_j^{min} z_j^{avg} z_j^{max} y_a

Composition

FIGURE 3-6   SELECTION OF FEED COMPOSITION TO SINK

to reduce the consumption of the fresh resource in the sink. In order to minimize the usage of the fresh flowrate while satisfy the composition constraints for the sink, what should be the composition of the feed entering the sink? Should it be z_j^{min}, z_j^{max} or some intermediate value, z_j^{avg}? According to the lever-arm rule described by equation (3.5a), we get:

$$\frac{\text{Fresh flowrate used in sink}}{\text{Total flowrate fed to sink}} = \frac{\text{Fresh arm}}{\text{Total arm}} \qquad (3.6a)$$

i.e.,

$$\frac{\text{Fresh flowrate used in sink}}{\text{Total flowrate fed to sink}} = \frac{y_a - z_{\text{Feed to sink}}}{y_a - y_F} \qquad (3.6b)$$

The right hand side of equation (3.6a) or (3.6b) is referred to as the relative fresh arm. For a given requirement of the total flowrate fed to the sink, the flowrate of the fresh is minimized when $z_{\text{Feed to sink}}$ is maximized. Hence, the composition of the feed entering the sink should be set to z_j^{max}. This analysis leads to the following rule:

Sink Composition Rule: When a fresh resource is mixed with process source(s), the composition of the mixture entering the sink should be set to a value that minimizes the fresh arm. For instance, when the fresh resource is a pure substance that can be mixed with pollutant-laden process sources, the composition of the mixture should be set to the maximum admissible value.

Another important aspect is the prioritization of sources. Consider the system shown in Figure 3-7. A process sink currently uses a fresh resource. In order to reduce the fresh usage, two process sources are considered for recycle. Both sources have sufficient flowrate to satisfy the sink but their compositions exceed the maximum admissible composition to the sink. Nonetheless, upon mixing with the fresh resource in proper proportions, both process sources can be fed to the sink. The question is which source should be used: *a* or *b*?

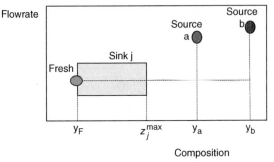

FIGURE 3-7   SOURCE PRIORITIZATION RULE

According to Figure 3-7 and equation 3.5b:

$$\text{Fresh arm when source } a \text{ is used} = y_a - z_j^{\max} \qquad (3.7)$$

where $z_j^{\max}$ is set as the composition of the mixture as described in the sink composition rule. Similarly,

$$\text{Fresh arm when source } b \text{ is used} = y_b - z_j^{\max} \qquad (3.8)$$

The arms given by equations (3.7) and (3.8) are referred to as the absolute fresh arms when sources a and b are used. Clearly, the absolute fresh arm for a is shorter than the fresh arm for b. However, equation (3.6) indicates that in order to minimize the usage of the fresh resource, the relative fresh arm (which is the ratio of the absolute fresh arm to the total arm of the mixture) should be minimized. It can be shown that minimizing the absolute fresh arm for a source corresponds to minimizing the relative fresh arm for the source. To prove this observation, we compare the relative fresh arms for sources a and b:

$$\text{Relative fresh arm for } a = \frac{y_a - z_j^{\max}}{y_a - y_F} \qquad (3.9a)$$

and

$$\text{Relative fresh arm for } b = \frac{y_b - z_j^{\max}}{y_b - y_F} \qquad (3.9b)$$

Let us now evaluate the ratio of the two relative fresh arms:

$$\frac{\text{Relative fresh arm for } a}{\text{Relative fresh arm for } b} = \frac{(y_a - z_j^{\max})/(y_a - y_F)}{(y_b - z_j^{\max})/(y_b - y_F)}$$

$$= \frac{(y_a - z_j^{\max})/(y_a - y_F)}{[(y_b - y_a) + (y_a - z_j^{\max})]/[(y_b - y_a) - (y_a - y_F)]} \qquad (3.10)$$

$$= \frac{[(y_b - y_a)/(y_a - y_F)] + 1}{[(y_b - y_a)/(y_a - z_j^{\max})] + 1}$$

But, as can be seen from Figure 3-7

$$y_a - y_F > y_a - z_j^{\max} \tag{3.11a}$$

Hence,

$$\frac{y_b - y_a}{y_a - y_F} < \frac{y_b - y_a}{y_a - z_j^{\max}} \tag{3.11b}$$

Combining inequalities (3.10) and (3.11b), we get

$$\text{Relative fresh arm for } a \ < \ \text{Relative fresh arm for } b \tag{3.12}$$

Inequality (3.12) allows us to look at the absolute fresh arms instead of calculating the relative fresh arms. We can now state the following important observation:

Source Prioritization Rule: In order to minimize the usage of the fresh resource, recycle of the process sources should be prioritized in order of their fresh arms starting with the source having the shortest fresh arm. For instance, in Figure 3-7, source a should be recycled before source b is considered. In other words, the recyclable source with the shorted relative fresh arm should be used first until it is completely recycled[1] then the source with next-to-shortest fresh arm is recycled, and so on.

3.3 SELECTION OF SOURCES, SINKS, AND RECYCLE ROUTES

In principle, it is possible to replace any fresh source of the targeted species with an equivalent amount of recycle from a terminal or an in-process stream. If the flowrate and composition of the recycled stream meet constraints (3.1) and (3.2) for units employing fresh sources, then we can undertake direct-recycle from those terminal streams to those units employing fresh sources. On the other hand, if flowrate and/or composition constraints are not met, then the terminal streams must be intercepted to render them in a condition that allows replacement of fresh sources. Here, we focus on direct recycle opportunities particularly for cases when the net generation of the targeted species is independent of the recycle strategy. In such cases, it is important to note that these *recycle activities should be directed to rerouting*

---

[1]"Completely recycled" can be to more than one unit. Hence, it is possible to recycle a portion of this source and mix it with another source, then use the rest of the source in another sink. For instance, if using the shortest-arm source in a sink will not satisfy the sink composition rule (e.g., the use of the shortest-arm source will not maximize the inlet composition). In such cases, a portion of the shortest-arm source is mixed with the next source to maximize the inlet composition of the sink and the rest of this source is used in another sink. Therefore, the shortest-arm source will still be completely used or committed (in more than one unit) before the next source is used.

process streams to units that employ fresh resources so as to replace the fresh resources with recycled process streams. In order to illustrate this observation, let us consider the process shown in Figure 3-8a. In this process, three fresh streams ($j = 1$–3) carry the targeted species. The required input load of the kth targeted species in these streams is denoted by Fresh_Load$_{k,j}$. The targeted species leave the process in four terminal streams; two of which ($i = 1,2$) are recyclable (with or without interception) and the other two ($i = 3,4$) are forbidden from being recycled. The total load from the four terminal streams is given by Terminal_Load$_{k,1}$ + Terminal_Load$_{k,2}$ + Terminal_Load$_{k,3}$ + Terminal_Load$_{k,4}$.

Let us first consider recycle from terminal streams to units that do not employ fresh resources. For instance, as shown by Figure 3-8b, let us recycle a load of R$_{k,1}$ from $i = 1$ to the inlet of unit #5 and a load of R$_{k,2}$ from $i = 2$ to the inlet of unit #4. Since we are dealing with the case where recycle activities have no effect on Net_process$_k$, the loads in the individual terminal streams are simply redistributed with the total terminal load remaining the same (Terminal_Load$_{k,1}$ + Terminal_Load$_{k,2}$ + Terminal_Load$_{k,3}$ + Terminal_ Load$_{k,4}$). Therefore, in this case, sinks that do not employ fresh sources of the targeted species are poor destinations for recycle.

Next, we consider recycles that reduce fresh loads. For instance, let us examine the effect of recycling a load of R$_{k,1}$ from $i = 1$ to the inlet of unit #2 and a load of R$_{k,2}$ from $i = 2$ to the inlet of unit #1. This is shown by Figure 3-8c. The result of the fresh source replacement is a net reduction of R$_{k,1}$ + R$_{k,2}$ from Fresh_Load$_k$ and consequently the total terminal loads are reduced by R$_{k,1}$ + R$_{k,2}$. It is worth noting that these appropriate

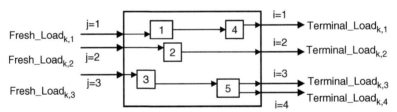

FIGURE 3-8a A GENERIC PROCESS BEFORE RECYCLE (NOURELDIN AND EL-HALWAGI 2000)

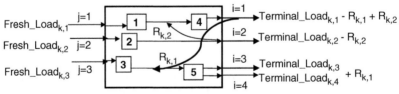

FIGURE 3-8b THE PROCESS AFTER RECYCLE TO POOR SINKS (NOURELDIN AND EL-HALWAGI 2000) (TOTAL TERMINAL LOAD OF TARGETED SPECIES REMAINS UNCHANGED)

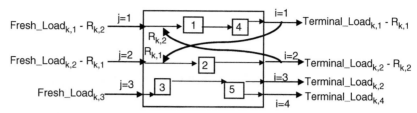

FIGURE 3-8c **THE PROCESS AFTER RECYCLE TO PROPER SINKS TO REPLACE FRESH LOADS OF TARGETED SPECIES (NOURELDIN AND EL-HALWAGI 2000) (TOTAL TERMINAL LOAD OF TARGETED SPECIES IS REDUCED)**

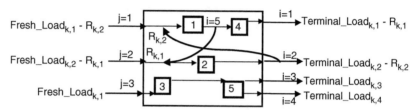

FIGURE 3-8d **RECYCLE TO REPLACE FRESH SOURCES USING IN-PLANT AND TERMINAL STREAMS (NOURELDIN AND EL-HALWAGI 2000)**

recycles are not limited to terminal streams. Instead, what is needed is the replacement of fresh loads with recycled loads from an in-plant or a terminal source. For example, the same effect shown in Figure 3-8c can be accomplished by recycling (with or without interception) from in-plant sources (e.g., $i = 5$) as shown in Figure 3-8d.

The foregoing discussion illustrates that in the case where the net generation/depletion of the targeted species is independent of stream rerouting activities, recycle should be allocated to sinks that consume the fresh resource. The recyclable sources may be terminal streams or in-plant streams on the path to terminal streams. The result of such selection of sources and sinks leads to a reduction in both fresh consumption and waste discharge.

EXAMPLE 3-1 MINIMIZATION OF FRESH REACTANT IN A CHEMICAL PLANT

Consider the chemical process shown in Figure 3-9. The reactor consumes 100 kg/s of fresh feed. In order to reduce the consumption of the fresh reactant, two process sources are considered for recycle to the reactor: the top and bottom products of a separation system. The flowrates of the top product is 60 kg/s and it has 10% (mass basis) of impurities. The bottom product has a flowrate of 216 kg/s and its content of impurities is 75% (mass basis). Determine the recycle strategies that minimize the usage of the fresh reactant.

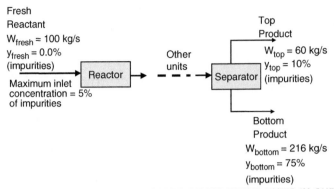

FIGURE 3-9   SIMPLIFIED PROCESS SHEET OF RELEVANT UNITS IN CHEMICAL PLANT OF EXAMPLE 3-1

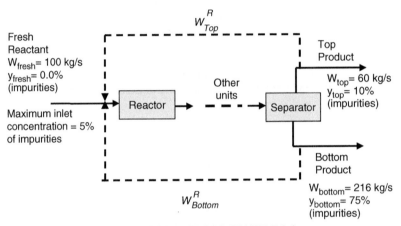

FIGURE 3-10   RECYCLE ALTERNATIVES FOR EXAMPLE 3-1

SOLUTION

The problem entails two candidate recycle alternatives as shown in Figure 3-10. The flowrates to be recycled from the top and bottom products are designated by W_{Top}^R and W_{Bottom}^R. In order to determine the recycle strategies, we represent the sources and the sink on the source–sink mapping diagram (Figure 3-11).

Since the top product has the shortest fresh arm, we start with recycle from top product. According to lever-arm equation (3.6b),

$$\frac{\text{Fresh}}{100} = \frac{0.10 - 0.05}{0.10 - 0.00} \qquad (3.13a)$$

or

$$\text{Fresh} = 50.0\,\text{kg/s} \qquad (3.13b)$$

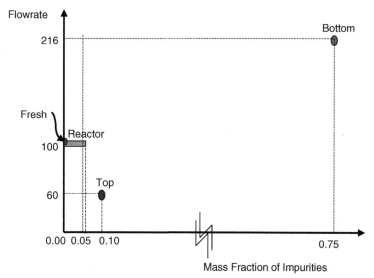

FIGURE 3-11   SOURCE–SINK MAPPING DIAGRAM FOR EXAMPLE 3-1

The flowrate of the top product recycled to the reactor can be calculated by material balance:

$$\text{Fresh} + W^R_{\text{Top}} = 100.0 \tag{3.14a}$$

i.e.,

$$W^R_{\text{Top}} = 50.0\,\text{kg/s} \tag{3.14b}$$

It is worth noting that since not all the top product has been recycled, there is no need to consider the bottom product for recycle (i.e., $W^R_{\text{Bottom}} = 0.0\,\text{kg/s}$). This is attributed to the source prioritization rule which indicates that the source with the shortest fresh arm should be completely used before the next-to-shortest fresh armed source be considered.

Alternatively, the same result can be obtained from a component material balance including recycle from the top stream (shortest fresh arm) and the fresh reactant:

$$\text{Fresh} \times 0.00 + W^R_{\text{Top}} \times 0.10 = 100.0 \times 0.05 \tag{3.15}$$

Along with equation (3.14a), we get $W^R_{\text{Top}} = 50\,\text{kg/s}$ and Fresh = 50.0 kg/s. Therefore, a 50% reduction in fresh consumption is accomplished through direct-recycle.

It is instructive to examine the case if the shortest arm rule was not followed and recycle from the bottom product was considered. In this case, the overall and component material balances can be expressed as:

$$\text{Fresh} + W^R_{\text{Bottom}} = 100.0 \qquad (3.16)$$

$$\text{Fresh} \times 0.0 + W^R_{\text{Bottom}} \times 0.75 = 100 \times 0.05 \qquad (3.17)$$

Therefore,

$$W^R_{\text{Bottom}} = 6.7 \, \text{kg/s} \qquad (3.18)$$

and

$$\text{Fresh} = 100.0 - 6.7 = 92.3 \, \text{kg/s} \qquad (3.19)$$

which corresponds to 6.7% reduction in fresh consumption.

The previous analysis illustrates the merit of the source prioritization rule by ordering the sequence of recycles accoring to the length of the fresh arms.

■ ■ ■

EXAMPLE 3-2 NITROGEN RECYCLE IN A MAGNETIC TAPE PLANT

Consider the magnetic tape manufacturing process shown in Figure 3-12. In this process (Dunn et al., 1995; El-Halwagi 1997), coating ingredients are dissolved in 0.09 kg/s of organic solvent and mixed to form a slurry. The slurry is suspended with resin binders and special additives. Next, the coating slurry is deposited on a base film. Nitrogen gas is used to induce evaporation rate of solvent that is proper for deposition. In the coating chamber, 0.011 kg/s of solvent are decomposed into other organic species. The decomposed organics are separated from the exhaust gas in a membrane

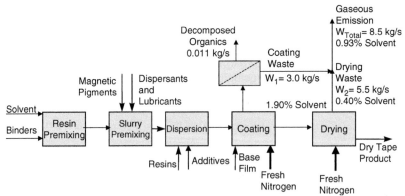

FIGURE 3-12 SCHEMATIC REPRESENTATION OF A MAGNETIC TAPE MANUFACTURING PROCESS (EL-HALWAGI 1997)

unit. The retentate stream leaving the membrane unit has a flowrate of 3.0 kg/s and is primarily composed of nitrogen that is laden with 1.9 wt/wt% of the organic solvent. The coated film is passed to a dryer where nitrogen gas is employed to evaporate the remaining solvent. The exhaust gas leaving the dryer has a flowrate of 5.5 kg/s and contains 0.4 wt/wt% solvent. The two exhaust gases are mixed and disposed off.

It is desired to undertake a direct-recycle initiative to use solvent-laden nitrogen (gaseous wastes) in lieu of fresh nitrogen gas in the coating and drying chambers. The following constraints on the gaseous feed to these two units should be observed:

Coating

$$3.0 \leq \text{flowrate of gaseous feed (kg/s)} \leq 3.2 \qquad (3.20)$$

$$0.0 \leq \text{wt\% of solvent} \leq 0.2 \qquad (3.21)$$

Dryer

$$5.5 \leq \text{flowrate of gaseous feed (kg/s)} \leq 6.0 \qquad (3.22)$$

$$0.0 \leq \text{wt \% of solvent} \leq 0.1 \qquad (3.23)$$

It may be assumed that the outlet gas compositions from the coating and the dryer chambers are independent of the entering gas compositions.

Using segregation, mixing and direct-recycle, what is the minimum consumption of nitrogen gas that should be used in the process? What are the strategies leading to the target?

SOLUTION

The first step in solving the problem is to determine the sources and sinks to be included in the analysis. The exhaust gas emission is segregated to its two sources: coating waste and drying waste. We will consider the recycle of the segregated sources and not the mixed source. Since mixing of sources is one of the solution strategies, there is no loss of generality and if the mixed source is to be recycled in its mixed form, it will appear in the analysis by showing the mixing of the segregated sources. Regarding the selection of the sinks, we select the coating and the drying chambers since they employ fresh nitrogen (as described in Figures 3-8a–d).

The source–sink mapping diagram for the problem is shown in Figure 3-13. The admissible recycle regions for the two sinks are shown as the boxes resulting from the intersection of the flowrate constraints with the composition constraints (inequalities (3.20) – (3.23)).

For the coating unit, the shortest fresh arm of process sources is that of drying waste. As can be seen from Figure 3-13, the flowrate and composition of the drying waste exceed the maximum allowable for coating. The excess flowrate is an easily resolvable problem since a portion of the drying waste

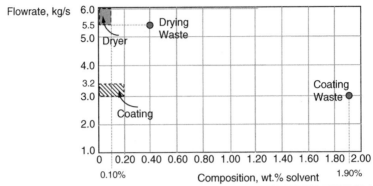

FIGURE 3-13   SOURCE–SINK MAPPING DIAGRAM FOR MAGNETIC TAPE PLANT

flowrate is to be recycled and the rest is to be bypassed. However, the unacceptably high composition of the solvent in the drying waste must be reduced. One way of reducing the solvent content is to use an interception device (e.g., separator). This will be covered in later chapters. However, since our focus here is on direct-recycle where no new equipment is to be added, we must consider mixing a portion of the drying waste with fresh nitrogen, thereby adjusting the composition of the mixture to lie inside the coating box. According to the sink composition rule, in order to minimize consumption of fresh nitrogen, the composition of the mixture entering the coating should be selected so as to minimize the fresh arm of the drying waste. Hence, the composition of the mixture is set to its maximum admissible value which is 0.002 (mass fraction of solvent). As for the flowrate of the feed to the coating unit, since the range is 3.0 to 3.2 kg/s, we choose 3.0 kg/s to reduce fresh nitrogen requirements. Let us designate the flowrate of the drying waste that is to be recycled to the coating as $W_{\text{Drying Waste}\rightarrow\text{Coating}}$. This recycled flowrate is to be calculated along with the fresh requirement of the coating unit according to the scheme shown in Figure 3-14.

According to lever-arm equation (3.6b),

$$\frac{\text{Fresh nitrogen in coating}}{3.0} = \frac{0.004 - 0.002}{0.004 - 0.00} \tag{3.24a}$$

or

$$\text{Fresh nitrogen used in coating} = 3.0 - 1.5 = 1.5\,\text{kg/s} \tag{3.24b}$$

and

$$W_{\text{DryingWaste}\rightarrow\text{Coating}} = 3.0 - 1.5 = 1.5\,\text{kg/s} \tag{3.25}$$

Alternatively, component material balance for the feed of the coating yields

$$(W_{\text{DryingWaste}\rightarrow\text{Coating}} \times 0.004 + \text{Fresh nitrogen to Coating} \times 0.000)$$
$$= 3.0 \times 0.002$$

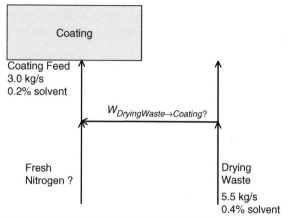

FIGURE 3-14   MIXING OF DRYING WASTE WITH FRESH NITROGEN TO PROVIDE FEED TO COATING

i.e.,

$$W_{\text{DryingWaste}\to\text{Coating}} = 1.5\,\text{kg/s} \tag{3.26}$$

Since the drying waste has not been fully utilized, its remaining flowrate should be considered for recycle to the drying unit as it provides a shorter fresh arm than the coating waste. For the drying unit, equation (3.6b) can be written as:

$$\frac{\text{Fresh nitrogen used in drying}}{5.5} = \frac{0.004 - 0.001}{0.004 - 0.00} \tag{3.27a}$$

or

$$\text{Fresh nitrogen used in drying unit} = 4.125\,\text{kg/s} \tag{3.27b}$$

Hence,

$$W_{\text{DryingWaste}\to\text{Drying}} = 5.500 - 4.125 = 1.375\,\text{kg/s} \tag{3.28a}$$

Alternatively, component material balance for the feed of the dryer yields

$$(W_{\text{DryingWaste}\to\text{Coating}} \times 0.004 + \text{Fresh nitrogen to Coating} \times 0.000)$$
$$= 5.50 \times 0.001 \tag{3.28b}$$
$$W_{\text{DryingWaste}\to\text{Drying}} = 1.375\,\text{kg/s}$$

Based on equations (3.24b) and (3.27b), we have

$$\text{Minimum fresh nitrogen consumption in coating and drying after}$$
$$\text{direct recycle} = 1.500 + 4.125 = 5.625\,\text{kg/s} \tag{3.29}$$

Compared to the original fresh nitrogen consumption of 8.500 kg/s, direct-recycle saves 34% of fresh nitrogen purchase.

■ ■ ■

EXAMPLE 3-3 MULTIPLICITY OF RECYCLE IMPLEMENTATIONS: VINYL ACETATE CASE STUDY

An important question is whether, or not, it is possible to develop more than one direct-recycle strategy that can achieve the same target (e.g., of fresh consumption or waste discharge). It is indeed possible in some cases to have multiple (sometimes infinite) direct-recycle strategies that can be implemented to attain the same target. This is the case of degenerate solutions. *Degeneracy* in recycle strategies refers to the ability of different recycle structures to achieve the same target. To illustrate this concept, let us consider the vinyl acetate monomer (VAM) plant (Figure 3-15) presented in Example 2-1 with more data and process details to enable the consideration of recycle strategies.

Vinyl acetate monomer "VAM" is manufactured by reacting acetic acid with oxygen and ethylene according to the following chemical reaction:

$$C_2H_4 \quad + \quad 0.5\,O_2 + \quad C_2H_4O_2 \quad = \quad C_4H_6O_2 \quad + \quad H_2O$$

Ethylene   Oxygen   Acetic Acid   VAM   Water

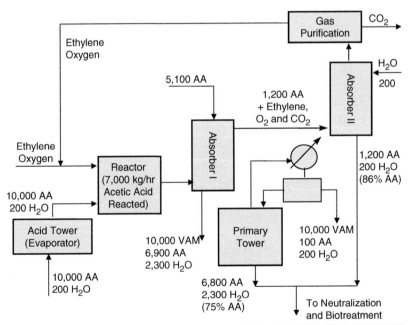

FIGURE 3-15 SCHEMATIC FLOWSHEET OF VAM PROCESS (ALL NUMBERS ARE IN kg/h)

Consider the process shown in Figure 3-15. 10,000 kg/h of acetic acid "AA" along with 200 kg/h of water are evaporated in an acid tower. The vapor is fed with oxygen and ethylene to the reactor where 7000 kg/h of acetic acid are reacted and 10,000 kg/h of VAM are formed. The reactor off-gas is cooled and fed to the first absorber where AA (5100 kg/h) is used as a solvent. Almost all the gases leave from the top of the first absorption column together with 1200 kg/h of AA. This stream is fed to the second absorption column where water (200 kg/h) is used to scrub acetic acid. The bottom product of the first absorption column is fed to the primary distillation tower where VAM is recovered as a top product (10,000 kg/h) along with small amount of AA which are not worth recovering (100 kg/h) and water (200 kg/h). This stream is sent to final finishing. The bottom product of the primary tower (6800 kg/h of AA and 2300 kg/h of water) is mixed with the bottom product of the second absorption column (1200 kg/h of AA and 200 kg/h of water). The mixed waste is fed to a neutralization system followed by biotreatment.

The following technical constraints should be observed in any proposed solution:

Neutralization

$$0 \leq \text{Flowrate of Feed to Neutralization (kg/h)} \leq 11{,}000 \qquad (3.30)$$

$$0 \leq \text{AA in Feed to Neutralization (wt\%)} \leq 85\% \qquad (3.31)$$

Acid Tower

$$10{,}200 \leq \text{Flowrate of Feed to Acid Tower (kg/h)} \leq 11{,}200 \qquad (3.32)$$

$$0.0 \leq \text{Water in Feed to Acid Tower (wt\%)} \leq 0.0 \qquad (3.33)$$

First Absorber

$$5{,}100 \leq \text{Flowrate of Feed to Absorber I (kg/h)} \leq 6000 \qquad (3.34)$$

$$0.0 \leq \text{Water in Feed to Absorber I (wt\%)} \leq 5.0 \qquad (3.35)$$

It can be assumed that the process performance will not significantly change as a result of direct-recycle activities.

What is the target for minimum usage (kg/h) of fresh acetic acid in the process if segregation, mixing, direct-recycle are used? Can you find more than one strategy to reach the target?

SOLUTION

We start by selecting the sources and sinks to be included in the analysis. The wastewater fed to neutralization and biotreatment is segregated to its two sources: bottoms of absorber II (referred to as R_1) and bottoms of primary tower (designated by R_2). As for the choice of

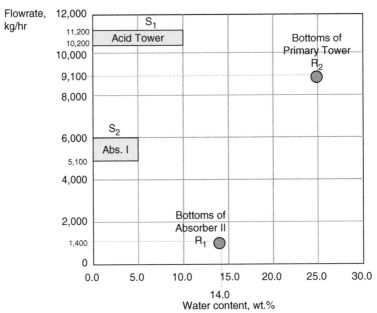

FIGURE 3-16   SOURCE–SINK MAPPING DIAGRAM FOR THE VAM CASE STUDY

the sinks, we select the acid tower and absorber I (referred to as S_1 and S_2, respectively) since they employ fresh acetic acid (as described in Figures 3-8a–d). The source–sink mapping diagram for the problem is shown in Figure 3-16.

For the acid tower (sink S_1), R_1 shows the shortest fresh (acetic acid) arm among process sources. Let us determine what flowrate of R_1 should be recycled to S_1. Assuming that only R_1 and fresh AA are mixed and fed to S_1, equation (3.6b), we get:

$$\frac{\text{Fresh AA used in } S_1}{10,200} = \frac{0.14 - 0.10}{0.14 - 0.00} \tag{3.36}$$

or

$$\text{Fresh AA in } S_1 = 2914 \, \text{kg/h} \tag{3.37}$$

Hence, the flowrate to be recycled from R_1 to S_1 would be

$$W_{R_1 \to S_1} = 10,200 - 2914 = 7286 \, \text{kg/s} \tag{3.38}$$

However, the maximum recyclable flowrate of R_1 is 1400 kg/h. Therefore, all of R_1 will be recycled to S_1. Now that all of R_1 is completely recycled, we move to the next-to-shortest arm which corresponds to R_2. The question is what flowrate of R_2 is recyclable to S_1 (referred to as $W_{R_2 \to S_1}$) to minimize the usage of fresh AA? A water balance for S_1 is expressed as:

$$1400 \times 0.14 + W_{R_2 \to S_1} \times 0.25 + \text{Fresh AA in } S_1 \times 0.0 = 10,200 \times 0.10$$
$$(3.39a)$$

Therefore,

$$W_{R_2 \to S_1} = 3296 \, \text{kg/h} \qquad (3.39b)$$

and

$$\text{Fresh AA in } S_1 = 10200 - 1400 - 3296 = 5504 \, \text{kg/h} \qquad (3.40)$$

Next, we recycle from R_2 to S_2 and mix with fresh AA. Hence,

$$\frac{\text{Fresh AA used in } S_2}{5100} = \frac{0.25 - 0.05}{0.25 - 0.00} \qquad (3.41a)$$

or

$$\text{Fresh AA in } S_2 = 4080 \, \text{kg/h} \qquad (3.41b)$$

Hence, the flowrate to be recycled from R_2 to S_2 would be

$$W_{R_2 \to S_2} = 5100 - 4080 = 1020 \, \text{kg/h} \qquad (3.42)$$

Based on equations (3.40) and (3.41b), we get

Target for minimum fresh usage when direct recycle is implemented
$$= 5504 + 4080 = 9584 \, \text{kg/h} \qquad (3.43)$$

A schematic representation of this solution is shown in Figure 3-17a.

Next, it is required to determine if there are other direct-recycle strategies leading to the same AA target. One alternative is to start the recycle to sink S_2. The shortest fresh arm is provided by R_1. It can be shown that all of R_1 is recyclable to S_2 with the rest of the feed to S_2 coming from R_2 and fresh AA. A water balance for S_2 is expressed as:

$$1400 \times 0.14 + W_{R_2 \to S_2} \times 0.25 + \text{Fresh AA in } S_2 \times 0.0$$
$$= 5100 \times 0.05 \qquad (3.44a)$$

Therefore,

$$W_{R_2 \to S_2} = 236 \, \text{kg/h} \qquad (3.44b)$$

and

$$\text{Fresh AA in } S_2 = 5100 - 1400 - 236 = 3464 \, \text{kg/h} \qquad (3.45)$$

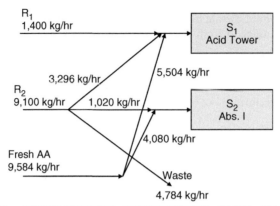

FIGURE 3-17a DIRECT-RECYCLE CONFIGURATION WHEN ALL OF R_1 IS FED TO S_1

Next, we recycle from R_2 to S_1 and mix with fresh AA. Hence,

$$\frac{\text{Fresh AA used in } S_1}{10,200} = \frac{0.25 - 0.10}{0.25 - 0.00} \qquad (3.46a)$$

or

$$\text{Fresh AA in } S_1 = 6120 \, \text{kg/h} \qquad (3.46b)$$

Hence, the flowrate to be recycled from R_2 to S_1 would be

$$W_{R_2 \rightarrow S_1} = 10,200 - 6120 = 4080 \, \text{kg/h} \qquad (3.47)$$

Based on equations (3.45) and (3.46b), we get

$$\text{Target for minimum fresh usage when direct recycle is implemented}$$
$$= 3464 + 6120 = 9584 \, \text{kg/h} \qquad (3.48)$$

which is the same target found through an alternative recycle strategy. A schematic representation of this solution is shown in Figure 3-17b.

Another alternative can be found by splitting R_1 between S_1 and S_2 in any proportion. For instance, let us split R_1 equally between S_1 and S_2. In this case, a water balance for S_1 is expressed as:

$$700 \times 0.14 + W_{R_2 \rightarrow S_1} \times 0.25 + \text{Fresh AA in } S_1 \times 0.0$$
$$= 10,200 \times 0.10 \qquad (3.49)$$

Therefore,

$$W_{R_2 \rightarrow S_1} = 3688 \, \text{kg/h} \qquad (3.50)$$

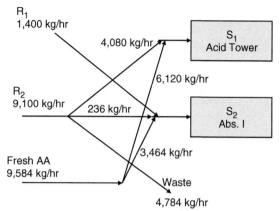

FIGURE 3-17b   DIRECT-RECYCLE CONFIGURATION WHEN ALL OF R_1 IS FED TO S_2

and

$$\text{Fresh AA in } S_1 = 10{,}200 - 700 - 3688 = 5812 \, \text{kg/h} \qquad (3.51)$$

Similarly, a water balance for S_2 is expressed as:

$$700 \times 0.14 + W_{R_2 \to S_2} \times 0.25 + \text{Fresh AA in } S_2 \times 0.0 = 5100 \times 0.05 \quad (3.52\text{a})$$

Therefore,

$$W_{R_2 \to S_2} = 628 \, \text{kg/h} \qquad (3.52\text{b})$$

and

$$\text{Fresh AA in } S_2 = 5100 - 700 - 628 = 3772 \, \text{kg/h} \qquad (3.53)$$

Based on equations (3.51) and (3.53), we get

$$\begin{aligned}\text{Target for minimum fresh usage when direct recycle is implemented} \\ = 5812 + 3772 = 9584 \, \text{kg/h}\end{aligned} \qquad (3.54)$$

which is the same target found through an alternative recycle strategy. A schematic representation of this solution is shown in Figure 3-17c.

Along the same lines, R_1 can be split between S_1 and S_2 in any proportion (while completely recycling all its flowrate), and the target for minimum usage of fresh AA will be the same. Hence, there are infinite number of recycle strategies that attain the same target for fresh AA consumption. Consequently, it is critical to have the ability to identify the target without detailing the recycle strategy. This is the objective of the next section.

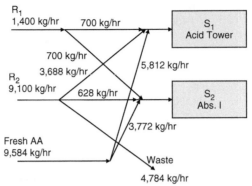

FIGURE 3-17c   DIRECT-RECYCLE CONFIGURATION WHEN R_1 IS SPLIT EQUALLY BETWEEN S_1 AND S_2

3.4 DIRECT-RECYCLE TARGETS THROUGH MATERIAL RECYCLE PINCH DIAGRAM

In many cases, it is useful to identify performance targets as a result of direct-recycle strategies without detailing those strategies. Targeting is an important activity since it spares the designer from the detailed computations particularly when there are numerous solutions or when there are many sources and sinks. Additionally, targeting for the whole system emphasizes the integrated nature without getting entangled in the detailed analysis. It also sheds useful insights on the key characteristics of the system. In the following, a rigorous targeting method will be presented to benchmark performance of the process when direct-recycle strategies are considered.

To illustrate the targeting procedure, let us first consider the case of a *pure fresh* resource that is to be replaced by process sources. Later, this restriction will be relaxed. Therefore, constraint (3.2) can be rewritten as:

$$0 \leq z_j^{\text{in}} \leq z_j^{\text{max}} \quad \text{where } j = 1, 2, \ldots N_{\text{sinks}} \tag{3.55}$$

Additionally, we consider the case when the flowrate to be fed to each sink, G_j, is given. The load of impurities entering the jth sink is given by:

$$M_j^{\text{sink}} = G_j z_j^{\text{in}} \tag{3.56}$$

Therefore, the composition constraint of the sink (3.55) may be replaced by the following constraint on load:

$$0 \leq M_j^{\text{sink}} \leq M_j^{\text{max}} \quad \text{where } j = 1, 2, \ldots, N_{\text{sinks}} \tag{3.57}$$

where

$$M_j^{\text{max}} = G_j z_j^{\text{max}} \tag{3.58}$$

Therefore, we can restate the sink-composition rule as the following *sink-load rule*: *If the sink requires the use of fresh source, its inlet impurities load should be maximized, i.e.,*

$$M_j^{\text{in,optimum}} = M_j^{\text{max}} \quad \text{where } j = 1, 2, \ldots, N_{\text{sinks}} \tag{3.59}$$

unless no fresh resource is to be used in this sink (in which case, the inlet load of the sink is that of the recycled/reused sources).

The sink-load rule coupled with the source prioritization rule constitute the basis for the following graphical procedure referred to as the *material recycle pinch diagram* (El-Halwagi et al., 2003):

1. Rank the sinks in ascending order of maximum admissible composition of impurities,

$$z_1^{\text{max}} \le z_2^{\text{max}} \le \ldots z_j^{\text{max}} \ldots \le z_{N_{\text{sinks}}}^{\text{max}}$$

2. Rank sources in ascending order of impurities composition, i.e.,

$$y_1 < y_2 < \ldots y_i \ldots < y_{N_{\text{sources}}}$$

3. Plot the maximum admissible load of impurities in each sink ($M_j^{\text{sink,max}} = G_j z_j^{\text{max}}$) versus its flowrate. Therefore, each sink is represented by an arrow whose vertical distance is $M_j^{\text{sink,max}} = G_j z_j^{\text{max}}$, horizontal distance is flowrate, and slope is z_j^{max}. Start with the first sink (which has the lowest z_j^{max}). From the arrowhead of this sink, plot the second sink. Proceed to plot the rest of the sinks using superposition of the sinks arrows in ascending order. The resulting curve is referred to as the *sink composite curve* (Figure 3-18).

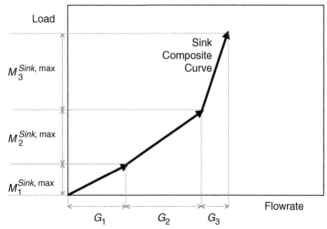

FIGURE 3-18   DEVELOPING SINK COMPOSITE DIAGRAM

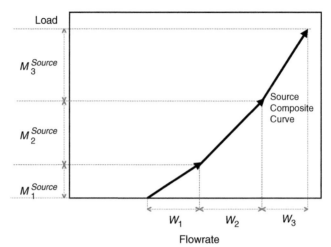

FIGURE 3-19 DEVELOPING SOURCE COMPOSITE DIAGRAM

The sink composite is a cumulative representation of all the sinks and corresponds to the upper bound on their feasibility region.

4. Represent each source as an arrow by plotting the load of each source versus its flowrate. The load of the ith source is calculated through:

$$M_i^{\text{source}} = W_i y_i \qquad (3.60)$$

Start with the source having the least composition of impurities and place its arrow tail anywhere on the horizontal axis. As will be shown later, it is irrelevant where this arrow tail is placed. Continue with the other sources and use superposition as shown in Figure 3-19 to create a *source composite curve*. The source composite curve is a cumulative representation of all process streams considered for recycle. Now, we have the two composite curves on the same diagram (Figure 3-20).

5. Move the source composite stream horizontally till it touches the sink composite stream with the source composite below the sink composite in the overlapped region. The point where they touch is the material recycle pinch point (Figure 3-21). The flowrate of sinks below which there are no sources is the target for minimum fresh usage. The flowrate in the overlapped region of process sinks and sources represents the directly recycled flowrate. Finally, the flowrate of the sources above which there are no sinks is the target for minimum waste discharge. Those targets are shown on Figure 3-22.

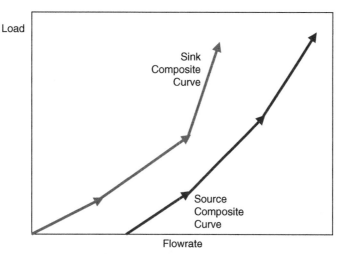

FIGURE 3-20   SINK AND SOURCE COMPOSITE DIAGRAMS

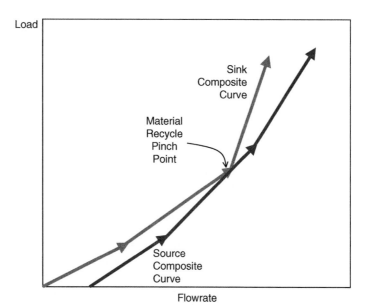

FIGURE 3-21   MATERIAL RECYCLE PINCH DIAGRAM (EL-HALWAGI ET AL., 2003)

3.5 DESIGN RULES FROM THE MATERIAL RECYCLE PINCH DIAGRAM

A key insight can be observed from the material recycle pinch diagram. The pinch point distinguishes two zones. Below that point, fresh resource is used in the sinks while above that point unused process sources are discharged. The primary characteristic for the pinch point is based on

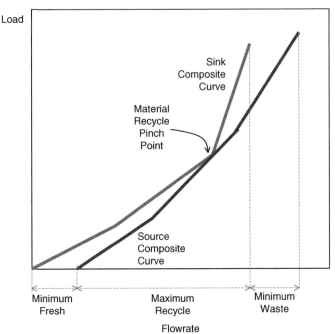

FIGURE 3-22 IDENTIFYING TARGETS FOR MINIMUM FRESH USAGE, MAXIMUM DIRECT RECYCLE, AND MINIMUM WASTE DISCHARGE (EL-HALWAGI ET AL., 2003)

the following observation: the pinch point is the point where the load of recycled/reused sources match that of the sink. Hence, it corresponds to the most constrained point in the recycle system. If the two composite curves are not touched at the pinch (e.g., by moving the source composite to the right, thereby passing a flowrate of α through the pinch), the fresh usage and waste discharge are both increased by the same magnitude of the flowrate passed through the pinch (α). Additionally, the extent of recycled flowrate is also reduced by the same magnitude (α) as shown by Figure 3-23. On the other hand, if we move the source composite to the left of the pinch, a portion of the source composite will lie above the sink composite thereby leading to the violation of constraint (3.57). This situation is shown in Figure (3-24).

The above discussion indicates that in order to achieve the minimum usage of fresh resources, maximum reuse of the process sources, and minimum discharge of waste, the following *three design rules* are needed:

- No flowrate should be passed through the pinch (i.e., the two composites must touch).
- No waste should be discharged from sources below the pinch.
- No fresh should be used in any sink above the pinch.

The targeting procedure identifies the targets for fresh, waste, and material reuse without commitment to the detailed design of the network

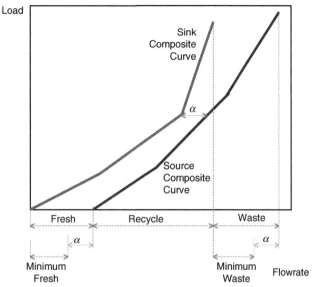

FIGURE 3-23 PASSING FLOWRATE (α) THROUGH THE PINCH POINT LEADS TO LESS INTEGRATION

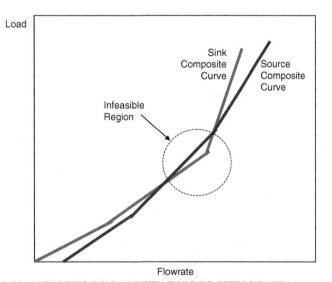

FIGURE 3-24 VIOLATING PINCH POINT LEADS TO INFEASIBILITY

matching the sources and sinks. In detailing the solution, there can be more than one solution satisfying the identified targets. Those solutions can be identified using the source–sink mapping diagram. To compare the multiple solutions having the same target of fresh usage and waste discharge, other objectives should be used (e.g., capital investment, safety, flexibility, operability, etc.).

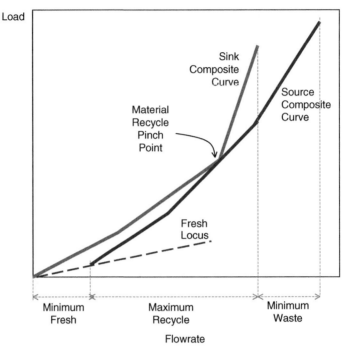

FIGURE 3-25   MATERIAL RECYCLE PINCH DIAGRAM WHEN FRESH RESOURCE IS IMPURE

3.5.1 Extension to Case of Impure Fresh

The same targeting procedure can be extended for cases when the fresh resource is impure. In the case of pure fresh, the source composite curve was slid on the horizontal axis. The reason for this is that with no impurities in the fresh, it does not contribute to the load of impurities regardless of how much flowrate of fresh is used. Consequently, the horizontal axis serves as a locus for the fresh resource. When the fresh is impure but cleaner than the rest of the sources, its locus becomes a straight line emanating from the origin and having a slope of y_{Fresh} (composition of impurities in the fresh). Hence, the source composite curve is slid on the fresh locus until it touches the sink composite while lying below it in the overlapped region. This case is shown by Figure 3-25. If the fresh resource was not the source with the least composition of impurities, the same procedure can be adopted by ranking the sources in ascending order of composition and placing the locus of the fresh at its proper rank.

3.5.2 Insights on Process Modifications

The material recycle pinch diagram and the associated design rules can be used to guide the engineer in making process modifications to enhance material reuse; for instance, the observation is that, below the pinch there is deficiency of recyclable sources, whereas above the pinch there is a surplus of sources. Therefore, sinks can be moved from below the pinch to above the pinch and sources can be moved from above the pinch to below the pinch to reduce the usage of fresh resources and the discharge of waste. Moving a sink from below the pinch to above the pinch can be achieved by increasing the upper bound on the composition constraint of the sink given by inequality (3.55). Conversely, moving a source from above the pinch to below the pinch can be accomplished by reducing its composition through changes in operating conditions or by adding an "interception" device (e.g., separator, reactor, etc.) that can lower the composition of impurities. The design of interception networks will be handled in Chapter Four, Seven, Thirteen, and Fourteen in this book. Figure 3-26a illustrates an example when a flowrate β is intercepted to reduce its content of impurities down to a composition similar to that of the fresh. Therefore, this flowrate is moved from above the pinch to below the pinch. Compared to the nominal case without interception, two benefits accrue as a result of this movement across the pinch: both the usage of fresh resource and the discharge of waste are reduced by β.

Another alternative is to reduce the load of a source below the pinch (again by altering operating conditions or adding an interception device). Consequently, the cumulative load of the source composite decreases and

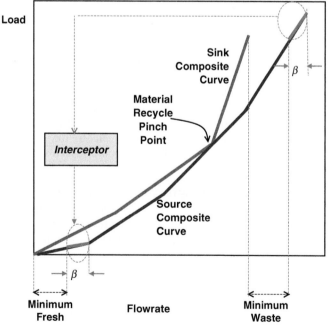

FIGURE 3-26a INSIGHTS FOR PROCESS MODIFICATION: MOVING A SOURCE ACROSS THE PINCH

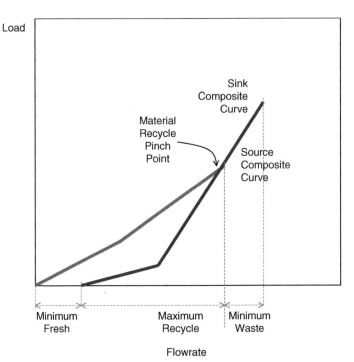

FIGURE 3-26b   EXAMPLE OF A MATERIAL RECYCLE PINCH DIAGRAM BEFORE INTERCEPTION

allows an additional recycle of process sources with the result of decreasing both the fresh consumption and waste discharge. Figure 3-26b illustrates the material recycle pinch diagram before interception. Then, the second source is intercepted to remove the load protruding above the pinch. As the intercepted load is removed, the slope of the second source decreases. The new slope is the composition of the intercepted source. Consequently, the source composite curve can be slid to the left to reduce (or in this case to eliminate) the use of fresh resource as shown in Figure 3-26c.

EXAMPLE 3-4 TARGETING FOR THE VINYL ACETATE CASE STUDY

Let us revisit the vinyl acetate case study described earlier in Example 3-2. The relevant problem data are summarized in Tables 3-1 and 3-2. The sinks and sources are ranked in ascending order of composition. Next, the sink and source composite curves are constructed as shown in Figures 3-27 and 3-28. Next, the source composite curve is slid horizontally to the right till it touches the sink composite curve. The material reuse pinch diagram is shown by Figure 3-29. The targets for minimum fresh usage and minimum waste discharge are found to be 9.6×10^3 and 4.8×10^3 kg/h, respectively. These are the same results obtained by the detailed design of the source–sink mapping diagram as shown in the results of Figure 3-17.

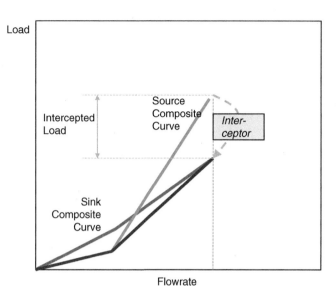

FIGURE 3-26c INSIGHTS FOR PROCESS MODIFICATION: LOWERING THE SLOPE OF SECOND SOURCE THROUGH INTERCEPTION

TABLE 3-1 SINK DATA FOR THE VINYL ACETATE EXAMPLE

Sink	Flowrate (kg/h)	Maximum inlet mass fraction	Maximum inlet load (kg/h)
Absorber I	5100	0.05	255
Acid Tower	10,200	0.10	1020

TABLE 3-2 SOURCE DATA FOR THE VINYL ACETATE EXAMPLE

Source	Flowrate (kg/h)	Inlet mass fraction	Inlet load (kg/h)
Bottoms of Absorber II	1400	0.14	196
Bottoms of Primary Tower	9100	0.25	2275

EXAMPLE 3-5 HYDROGEN RECYCLE IN A REFINERY

Hydrogen is a critical commodity in a petroleum refinery. Alves and Towler (2002) describe a case study involving the optimization of a hydrogen distribution system within a refinery, and it is comprised of four sinks and six sources. The pertinent information regarding these sinks and sources can be seen in Tables 3-3 and 3-4.

In this case study, the fresh hydrogen is available with a 5% mol/mol impurity content. Identify the target for minimum usage of fresh hydrogen and minimum discharge of gaseous waste.

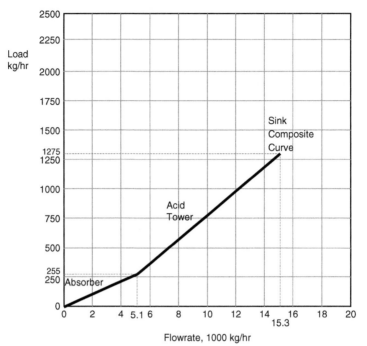

FIGURE 3-27 SINK COMPOSITE CURVE FOR THE VINYL ACETATE CASE STUDY

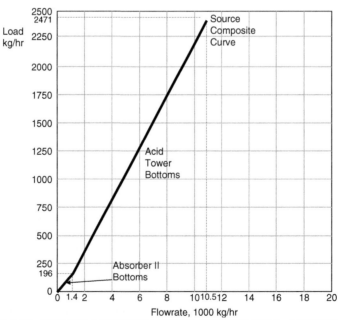

FIGURE 3-28 SOURCE COMPOSITE CURVE FOR THE VINYL ACETATE CASE STUDY

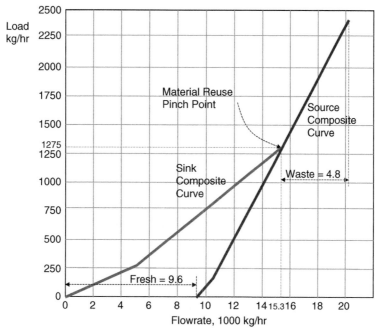

FIGURE 3-29 MATERIAL-REUSE PINCH DIAGRAM FOR THE VINYL ACETATE CASE STUDY

TABLE 3-3 SINK DATA FOR HYDROGEN PROBLEM (ALVES AND TOWLER 2002)

Sinks	Flow (mol/s)	Maximum Inlet Impurity Concentration (mol %)	Load (mol/s)
1	2495	19.39	483.8
2	180.2	21.15	38.1
3	554.4	22.43	124.4
4	720.7	24.86	179.2

TABLE 3-4 SOURCE DATA FOR THE HYDROGEN PROBLEM (ALVES AND TOWLER 2002)

Sources	Flow (mol/s)	Impurity Concentration (mol %)	Load (mol/s)
1	623.8	7	43.7
2	415.8	20	83.2
3	1801.9	25	450.5
4	138.6	25	34.7
5	346.5	27	93.6
6	457.4	30	137.2

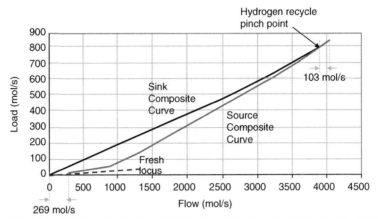

FIGURE 3-30 HYDROGEN REUSE PINCH DIAGRAM (EL-HALWAGI ET AL., 2003)

SOLUTION

Using the information in Tables 3-3 and 3-4, and by plotting a locus for fresh hydrogen (slope being 0.05), the material recycle pinch diagram can be developed and the source composite curve is slid on the fresh locus. The results are shown in Figure 3-30.

From Figure 3-30, the minimum hydrogen required and waste hydrogen to be discharged are 269 mol/s and 103 mol/s respectively.

EXAMPLE 3-6 WATER RECOVERY IN A FOOD PROCESS

Consider the food processing plant shown in the simplified flowsheet of Figure 3-31. The primary feedstocks are first pre-washed then processed throughout the facility. The gaseous waste of the process is cleaned in a water scrubber prior to discharge. Therefore, the process has two sinks that consume fresh water: the washer and the scrubber. Table 3-5 provides the data for these two sinks. The process results in two aqueous streams that are sent to biotreatment but may be considered for recycle: condensate I from the evaporator and condensate II from the stripper. The data for the two process sources are given in Table 3-6.

At present, the plant uses fresh water for the washer and the scrubber. In order to reduce the usage of fresh water and discharge of wastewater (condensate), the plant has decided to adopt direct-recycle strategies. An engineer has proposed that Condensate I be recycled to the scrubber (Figure 3-32). The result of this project is to eliminate the need for fresh water in the scrubber, reduce overall fresh water consumption to 8000 kg/h, and reduce wastewater discharge (Condensate II) to 9000 kg/h. Critique this proposed project.

SOLUTION

Before examining the details of recycle strategies, it is beneficial to establish the targets for minimum water usage and discharge. These

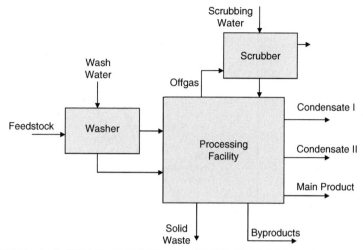

FIGURE 3-31   A SIMPLIFIED FLOWSHEET OF THE FOOD PROCESSING PLANT

TABLE 3-5 SINK DATA FOR THE FOOD PROCESSING EXAMPLE

Sink	Flowrate (kg/h)	Maximum inlet mass fraction	Maximum inlet load (kg/h)
Washer	8000	0.03	240
Scrubber	10,000	0.05	500

TABLE 3-6 SOURCE DATA FOR THE FOOD PROCESSING EXAMPLE

Source	Flowrate (kg/h)	Inlet mass fraction	Inlet load (kg/h)
Condensate I	10,000	0.02	200
Condensate II	9000	0.09	810

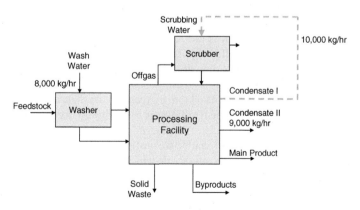

FIGURE 3-32   PROPOSED RECYCLE PROJECT

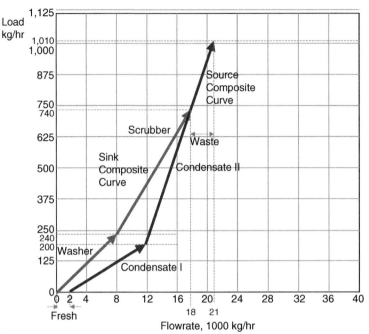

FIGURE 3-33   PINCH DIAGRAM FOR FOOD PROCESSING EXAMPLE (TARGETS FOR FRESH AND WASTE ARE 2000 AND 3000 kg/h, RESPECTIVELY)

benchmarks are obtained from the material recycle pinch diagram shown in Figure 3-33. As can be seen from the pinch diagram, the target for minimum fresh water usage is 2000 kg/h while the target for minimum wastewater discharge is 3000 kg/h[2]. Since these targets are much better than the proposed solution, we need to closely examine the suggested recycle strategy.

Next, we represent the proposed solution on the material recycle pinch diagram as shown in Figure 3-34. Since Condensate I is matched with the scrubber, it is plotted directly below the scrubber. This is a feasible solution since the flowrate is satisfied and the load of Condensate I is less than the maximum admissible load for the scrubber. However, we notice that this solution results in passing 6000 kg/h through the pinch. Therefore, we expect to see a 6000 kg/h increase in fresh usage and waste discharge. Indeed, this is the case (fresh water usage is 8000 kg/h compared to the target of 2000 kg/h and waste discharge of 9000 kg/h compared to the target of 3000 kg/h). Additionally, we notice that the proposed recycle does not follow the sink-composition rule which calls for the maximization of the inlet

---

[2]It is worth noting that in this problem, there are infinite recycle strategies that satisfy the targets for minimum fresh usage and minimum waste discharge. Can you identify some of these alternatives?

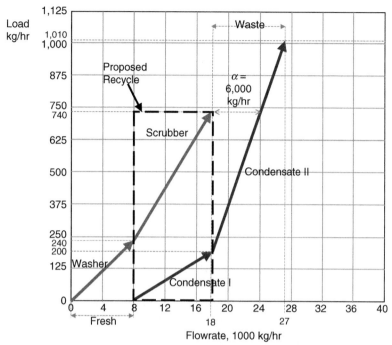

FIGURE 3-34   REPRESENTATION OF PROPOSED RECYCLE ON PINCH DIAGRAM (PROPOSED PROJECT RESULTS IN PASSING 6000 kg/h THROUGH THE PINCH)

composition to the sink (the proposed solution has the inlet to the scrubber being 2% as compared to 5%).

The above discussion indicates that the proposed project should not be implemented. Instead, an integrated solution consistent with the identified targets (Figure 3-33) should be recommended. Nonetheless, if the engineer insists on implementing the proposed recycle project, there is still an opportunity to improve upon this situation. The fact that the concentration of impurities in Condensate II is higher than the admissible concentration to the washer does not prevent us from pursuing partial recycle. Figure 3-35 is the pinch diagram for the remaining source and sink in the problem. As can be seen from the diagram, if Condensate I is used in lieu of the scrubber water, then the target for the remaining source (Condensate II) and sink (Washer) is 5300 kg/h in fresh water usage and 6300 kg/h in wastewater discharge. The foregoing discussion underscores the importance of ensuring that a short-term project be compatible with an overall integrated strategy. An individual project which is seemingly flawless may be detrimental to the overall performance of the whole process and may prevent the process from ever reaching its true target. Finally, a big-picture approach such as the material recycle pinch diagram yields insights that may not be seen by detailed engineering focusing on individual units and streams.

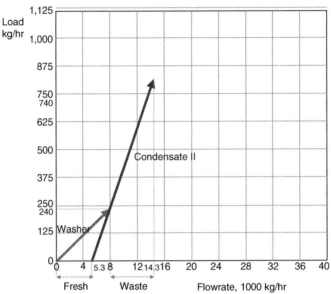

FIGURE 3-35   PINCH DIAGRAM IF CONDENSATE I AND THE SCRUBBER ARE TAKEN OUT OF THE PROBLEM (TARGETS FOR FRESH AND WASTE ARE 5300 AND 6300 kg/h, RESPECTIVELY)

3.6 MULTICOMPONENT SOURCE–SINK MAPPING DIAGRAM

The source–sink diagram can be extended to represent ternary (three-component) systems. Figure 3-36 is a ternary diagram for components A, B, and C. Each apex of the equilateral triangle represents 100% of a component. The opposite base corresponds to 0% of that component. Each line parallel to the base is a locus of a constant percentage of that component. Figure 3-36 shows the loci for various percentages of component A. The horizontal base of the triangle is the locus of any stream containing 0% of A. The other horizontal lines show the loci for compositions ranging from 10 to 90% A. Figure 3-37 illustrates the representation of a source containing 20% A, 30% B, and 50% C. Clearly, it is sufficient to represent the loci of two compositions (e.g., 20% A and 30% B) and locate the source at their intersection. The third composition is determined by passing a line parallel to the 0% C base through the source.

Composition constraints in the sinks can be represented on the ternary diagram. For instance, consider a sink with the following constraints:

$$0.30 \leq \text{mass fraction of A in stream entering the sink} \leq 0.70 \qquad (3.61a)$$

$$0.20 \leq \text{mass fraction of B in stream entering the sink} \leq 0.50 \qquad (3.61b)$$

$$0.10 \leq \text{mass fraction of C in stream entering the sink} \leq 0.60 \qquad (3.61c)$$

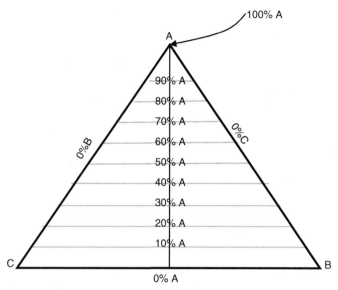

FIGURE 3-36   TERNARY COMPOSITION REPRESENTATION

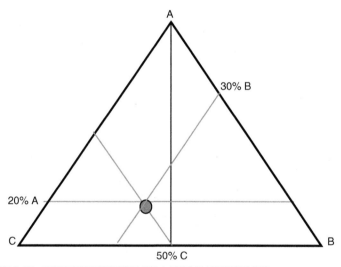

FIGURE 3-37   REPRESENTATION OF A TERNARY MIXTURE

The sink constraints are represented by Figure 3-38. Any point within the shaded region is a feasible feed to the sink. The ternary source–sink diagram has the same rules as the sink optimality and source prioritization (e.g., Parthasarathy and El-Halwagi 2000). For instance, when two process sources i and $i+1$ are considered for recycle by mixing with a fresh resource, the lever arms should be first calculated.

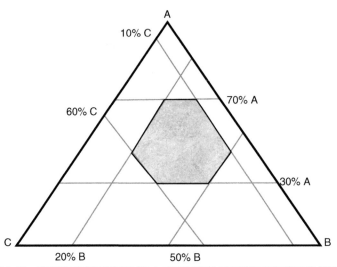

FIGURE 3-38 REPRESENTATION OF A SINK WITH TERNARY CONSTRAINTS

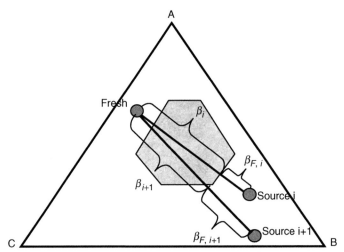

FIGURE 3-39 LEVEL-ARM RULE FOR A TERNARY SOURCE–SINK DIAGRAM

As can be seen from Figure 3-39, the lever-arm rules can be expressed as follows:

$$\frac{\text{Fresh usage when source } i \text{ is recycled}}{\text{Feed flowrate of sink}} = \frac{\beta_{F,i}}{\beta_{F,i} + \beta_i}$$

and

$$\frac{\text{Fresh usage when source } i+1 \text{ is recycled}}{\text{Feed flowrate of sink}} = \frac{\beta_{F,i+1}}{\beta_{F,i+1} + \beta_{i+1}}$$

Since source i has a shorter relative arm than that of $i+1$, it should be used first. Source $i+1$ is only used after source i has been completely recycled. This is the same concept used in source prioritization previously described in the single-component source–sink representation.

3.7 ADDITIONAL READINGS

Over the past decade, several design techniques have been developed to minimize the usage of fresh resources using network synthesis and analysis. Recent reviews can be found in literature (e.g., El-Halwagi et al., 2003; Dunn and El-Halwagi 2003). An important variation of mass exchange networks "MENs", wastewater minimization, was introduced by Wang and Smith (1994). They proposed a graphical approach to target the minimum fresh water consumption and wastewater discharged by the transfer of contaminants from process streams to water streams. Dhole et al. (1996) and El-Halwagi and Spriggs (1996) addressed the problem of material reuse and discharge through a source (supply)–sink (demand) representation. Dhole et al. (1996) created a graphical technique that represents concentration versus flowrate and creates a supply composite and a demand composite. When the two composites touch, a bottleneck (water pinch) is identified and can be eliminated by mixing of source streams. Even though the methodology had a great impact on the concept for water minimization, it has its drawbacks. The key limitation is that Dhole et al., did not provide a systematic method for elimination of pinch points by mixing. In order to overcome this limitation, Polley and Polley (2000) proposed a set of rules for sequencing, mixing, and recycle options. Additionally, Sorin and Bedard (1999) proposed an algebraic method called the evolutionary table which is based on locating the global pinch based on mixing source streams with closer concentration differences first, and then going to the stream with the next nearest concentration. Alves (1999), Alves and Towler (2002) and Hallale (2002) developed a surplus diagram with a graphical representation of purity versus flowrate. Both methods rely on extensive calculations to create the surplus diagram in order to target minimal consumption of resources (water in the case of Hallale (2002) and hydrogen in the case of Alves and Towler (2002)). Manan et al. (2004) extended the surplus diagram by developing a non-iterative targeting procedure for matter reuse. Mann and Liu (1999) provided a comprehensive coverage of the subject particularly for water reuse.

3.8 PROBLEMS

3.1 Consider a process with six sinks and five sources (Sorin and Bedard 1999). The process data are given in Tables 3-7 and 3-8. Fresh water is used in the sinks and it is desired to replace as much fresh water as possible using direct-recycle of process sources. Determine the target for minimum usage of fresh water and waste discharge after direct-recycle. How many material recycle pinch points are there? What is their significance?

TABLE 3-7 SINK INFORMATION FOR PROBLEM 3.1 (SORIN AND BEDARD 1999)

Sink	Flow (tonnes/h)	Maximum Inlet Concentration (ppm)	Load (kg/h)
1	120	0	0
2	80	50	4
3	80	50	4
4	140	140	19.6
5	80	170	13.6
6	195	240	46.8

TABLE 3-8 SOURCE INFORMATION FOR PROBLEM 3.1 (SORIN AND BEDARD 1999)

Sources	Flow (tonnes/h)	Concentration (ppm)	Load (kg/h)
1	120	100	12
2	80	140	11.2
3	140	180	25.2
4	80	230	18.4
5	195	250	48.75

TABLE 3-9 SINK DATA FOR PROBLEM 3-2 (POLLEY AND POLLEY (2000))

Sinks	Flow (tonnes/h)	Maximum inlet Concentration (ppm)	Load (kg/h)
1	50	20	1
2	100	50	5
3	80	100	8
4	70	200	14

3.2 Consider the wastewater minimization problem described by Polley and Polley (2000). The process has four sources and four sinks and information concerning them can be seen in Tables 3-9 and 3-10.

a   Find the targets for minimum fresh water and minimum wastewater discharge after direct-recycle.

b   Develop a recycle strategy that attains the identified targets.

3.3 A thermomechanical pulp and newsprint mill consisting of 54 sinks and 10 sources (Jacob et al., 2002). The source and sink data can be seen in Tables 3-11 and 3-12.

Identify a target for minimum water usage and minimum waste discharge.

TABLE 3-10 SOURCE DATA FOR PROBLEM 3.2 (POLLEY AND POLLEY (2000))

Sources	Flow (tonnes/h)	Concentration (ppm)	Load (kg/h)
1	50	50	2.5
2	100	100	10
3	70	150	10.5
4	60	250	15

TABLE 3-11 SINK DATA FOR THE PULP AND NEWSPRINT PROBLEM (JACOB ET AL., 2002)

Sink	Flow (L/min)	Max. allowable fines concentration (%)	Load (L/min)
1	200	1.000	2.0
2	400	1.000	4.0
3	355	0.020	0.1
4	150	1.000	1.5
5	13,000	1.000	130.0
6	4250	1.000	42.5
7	2800	1.000	28.0
8	4580	1.000	45.8
9	1950	1.000	19.5
10	500	1.000	5.0
11	1000	1.000	10.0
12	3000	1.000	30.0
13	435	1.000	4.4
14	310	1.000	3.1
15	60	1.000	0.6
16	1880	1.000	18.8
17	4290	1.000	42.9
18	9470	1.000	94.7
19	6500	1.000	65.0
20	620	1.000	6.2
21	55	1.000	0.6
22	70	1.000	0.7
23	320	1.000	3.2
24	1050	1.000	10.5
25	73,000	1.000	730.0
26	1765	1.000	17.7
27	235	1.000	2.4
28	95	1.000	1.0
29	20	1.000	0.2
30	180	0.000	0.0
31	160	0.018	0.0
32	30	0.018	0.0
33	20	0.018	0.0
34	315	0.000	0.0

Continued

TABLE 3-11 CONTINUED

Sink	Flow (L/min)	Max. allowable fines concentration (%)	Load (L/min)
35	315	0.000	0.0
36	930	0.018	0.2
37	460	0.018	0.1
38	30	0.018	0.0
39	30	0.018	0.0
40	315	0.000	0.0
41	315	0.000	0.0
42	110	0.018	0.0
43	110	0.018	0.0
44	190	0.000	0.0
45	190	0.000	0.0
46	100	0.000	0.0
47	20	0.000	0.0
48	15	0.000	0.0
49	60	0.018	0.0
50	30	0.018	0.0
51	100	0.000	0.0
52	20	0.000	0.0
53	100	0.000	0.0
54	20	0.000	0.0

TABLE 3-12 SOURCE DATA FOR THE PULP AND NEWSPRINT MILL (JACOB ET AL., 2002)

Source	Flow (L/min)	Fines concentration (%)	Load (L/min)
TMP clear water	25,000	0.07	17.5
TMP cloudy water	39,000	0.13	50.7
Inclined screen water	5980	0.50	29.9
Press header water	2840	0.49	13.9
Save-all clear water	6840	0.08	5.5
Save-all clear water	3720	0.1	3.7
Silo water	73,000	0.39	284.7
Machine chest whitewater	8585	0.34	29.2
Vacuum pump over-flow	2570	0.00	0.0
Residual showers	1940	0.13	2.5

Hint: To simplify the problem, notice that for the sinks, there are only four concentration levels of interest. Therefore, the sinks can be lumped into four sinks with fine concentrations (%) of 0.000, 0.018, 0.020, and 1.000.

3.4 Consider the plastic processing facility described in Problem 2.3. In addition to the information given in Problem 2.3, the following constraints are imposed on feeds to process sinks (Hamad and El-Halwagi 1998).

Sorting/Drying Unit

45,000 ≤ hot gas flowrate (kg/h) ≤ 60,000 kg/h Composition of butane in gas entering sorting unit ≤ 4000 ppmw

Furnace

380,000 ≤ Total flowrate entering furnace (kg/h) ≤ 400,000
10.0 w/w% ≤ Composition of butane in gas entering furnace ≤ 10.5 w/w%
Flowrate of air = 340,000 kg/h

On a source–sink mapping diagram, the depolymerization off-gas has the shortest arm with respect to sorter/dryer. Therefore, the depolymerization off-gas is to be mixed with the hot gas (containing no butane) and recycled to the sorter/dryer. What is the maximum flowrate (kg/h) of the depolymerization off-gas that can be recycled to the sorter/dryer?

3.9 REFERENCES

Almato, M., Sanmarti, E., Espuna, A., and Puigjaner, L. 1997, 'Rationalizing the water use in the batch process industry', *Comput. Chem. Eng.*, vol. 21, pp. S971-S976.

Alva-Argeaz, A., Vallianatos, A., and Kokossis, A. 1999, 'A multi-contaminant transhipment model for mass exchange networks and wastewater minimisation problems', *Comput. & Chem. Eng.*, vol. 23, pp. 1439-1453.

Alves, J.J. 1999, *Analysis and Design of Refinery Hydrogen Distribution Systems*, Ph.D. Thesis, UMIST, Manchester, U.K.

Alves, J.J. and Towler, G.P. 2002, 'Analysis of Refinery Hydrogen Distribution Systems', *Ind. Eng. Chem. Res.*, vol. 41, pp. 5759-5769.

Benko, N., Rev, E., and Fonyo, Z. 2000, 'The use of nonlinear programming to optimal water allocation', *Chem. Eng. Commun.*, vol. 178, pp. 67-101.

Dhole, V.R., Ramchandani, N., Tainsh, R.A., and Wasilewski, M. 1996, 'Make your process water pay for itself', *Chem. Eng.*, vol. 103, pp. 100-103.

Dunn, R.F., El-Halwagi, M.M., Lakin, J., and Serageldin, M. 1995, 'Selection of organic solvent blends for environmental compliance in the coating industries', *Proceedings of the First International Plant Operations and Design Conference*, eds. E. D. Griffith, H. Kahn and M. C. Cousins, AIChE, New York., vol. III, pp. 83-107.

Dunn, R.F., Wenzel, H., and Overcash, M.R. 2001, 'Process integration design methods for water conservation and wastewater reduction in industry', *Clean Production and Processes* vol. 3, pp. 319-329.

El-Halwagi, M.M., Gabriel, F., and Harell, D. 2003, 'Rigorous graphical targeting for resource conservation via material recycle/reuse networks', *Ind. Eng. Chem. Res.*, vol. 42, pp. 4319-4328.

El-Halwagi M.M. 1997, *Pollution Prevention Through Process Integration*, Academic Press: San Diego.

El-Halwagi, M.M. and Spriggs, H.D., 1996, 'An integrated approach to cost and energy efficient pollution prevention,' *Proceedings of the Fifth World Congress of Chemical Engineering.*, San Diego, vol. I, pp. 675-680.

Hallale, N. 2002, 'A new graphical targeting method for water minimisation', *Adv. Environ. Res.*, vol. 6, pp. 377-390.

Hamad, A.A. and El-Halwagi, M.M. 1998, 'Simultaneous synthesis of mass separation agents and interception networks, *Chem. Eng. Res. Des.*, *Trans. Inst. Chem. Eng.*, vol. 76, pp. 376-388.

Jacob, J., Kaipe, H., Couderc, F., and Paris, J. 2002, 'Water network analysis in pulp and paper processes by pinch and linear programming techniques', *Chem. Eng. Commun.*, vol. 189, pp. 184-206.

Mann, J. and Liu, Y.A., 1999, 'Industrial water reuse and wastewater minimization,' McGraw Hill, New York.

Manan, Z.A., Foo, C.Y., and Tan, Y.L. 2004. Targeting the minimum water flowrate using water cascade analysis technique, *AIChE Journal*, vol. 50, no. 12, pp. 3169-3183.

Noureldin, M.B. and El-Halwagi, M.M. 2000, 'Pollution-prevention targets through integrated design and operation', *Comp. Chem. Eng.*, vol. 24, pp. 1445-1453.

Parthasarathy, G. and El-Halwagi, M.M. 2000, 'Optimum mass integration strategies for condensation and allocation of multicomponent VOCs', *Chem. Eng. Sci.*, vol. 55, pp. 881-895.

Polley, G.T., and Polley, H.L. 2000, 'Design better water networks', *Chem. Eng. Prog.*, vol. 96, pp. 47-52.

Puigjanar, L. 1999, 'Handling the increasing complexity of detailed batch process simulation and optimisation', *Comput. Chem. Eng.*, vol. 23, pp. S929-S943.

Savelski, M.J., and Bagajewicz, M.J. 2000, 'On the optimality conditions of water utilization systems in process plants with single contaminants', *Chem. Eng. Sci.*, vol. 55, pp. 5035-5048.

Savelski, M.J., and Bagajewicz, M.J. 2001, 'Algorithmic procedure to design water utilization systems featuring a single contaminant in process plants', *Chem. Eng. Sci.*, vol. 56, pp. 1897-1911.

Sorin. M. and Bedard, S. 1999, 'The global pinch point in water reuse networks', *Trans. Inst. Chem. Eng.* vol. 77, pp. 305-308.

Wang, Y.P., and Smith, R. 1994, 'Wastewater minimisation', *Chem. Eng. Sci.*, vol. 49, pp. 981-1006.

Wang, Y.P. and Smith, R. 1995, 'Time pinch analysis', *Trans. Inst. Chem. Eng.*, vol. 73, pp. 905-914.

4

SYNTHESIS OF MASS EXCHANGE NETWORKS: A GRAPHICAL APPROACH

Mass exchange units are among the most ubiquitous separation operations used in the process industries. A mass exchanger is any direct contact mass transfer unit that employs a mass separating agent "MSA" (or a lean-stream) to selectively remove certain components (e.g., impurities, pollutants, byproducts, products) from a rich stream. The designation of a rich- or a lean-stream is not tied to the composition level of the components to be exchanged. Instead, the definition is task related. The stream from which the targeted components are removed is designated as the rich stream while the stream to which the targeted components are transferred is referred to as the lean-stream (or MSA). The MSA should be partially or completely immiscible in the rich phase. Examples of mass exchange operations include absorption, adsorption, stripping, ion exchange, solvent extraction, and leaching.

Multiple mass exchange units are typically used in a processing facility. Therefore, their collective selection, design, and operation must be coordinated and integrated. This chapter presents a systematic approach to the synthesis of networks involving multiple units of mass exchangers. First, the basics of mass exchange equilibrium and design of individual units is

FIGURE 4-1   A GENERIC MASS EXCHANGER

addressed. Next, a graphical approach will be described to illustrate how a mass exchange network can be synthesized and how the multiple MSAs and mass exchange technologies can be screened.

4.1 DESIGN OF INDIVIDUAL MASS EXCHANGERS

Consider the mass exchanger shown in Figure 4-1. A certain component is transferred from the rich stream, i, to the lean-stream, j. The rich stream has a flowrate, G_i, an inlet composition, y_i^{in}, and an outlet composition, y_i^{out}. The lean-stream has a flowrate, L_j, an inlet composition, x_j^{in}, and an outlet composition, x_j^{out}. Two important aspects govern the performance of a mass exchanger: *equilibrium function and material balance*.

Equilibrium refers to the state at which there is no net interphase transfer of the targeted species (solute). This situation corresponds to the state at which both phases have the same value of chemical potential for the solute. Mathematically, the composition of the solute in the rich phase, y_i, can be related to its composition in the lean phase, x_j, via an equilibrium distribution function, f_j^*. Hence, for a given rich-stream composition, y_i, the maximum achievable composition of the solute in the lean phase, x_j^*, is given by

$$y_i = f_j^*(x_j^*) \tag{4.1}$$

Figure 4-2 is a schematic representation of an equilibrium function. In many cases, the equilibrium function can be linearized over a specific range of operation. As shown by Figure 4-2, the linearized form has a slope of m_j and an intercept of b_j, i.e.,

$$y_i = m_j x_j^* + b_j \tag{4.2}$$

There are several important special cases of equation (4.2) when the intercept, b_j, is zero. These include Henry's law, Raoult's law, and extraction equilibrium with distribution coefficients.

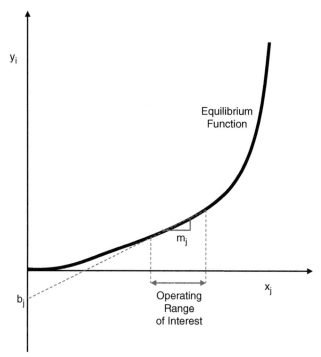

FIGURE 4-2 LINEARIZED SEGMENT OF EQUILIBRIUM FUNCTION

The material balance on the transferable solute accounts for the fact that the mass of solute lost from the rich stream is equal to the mass of solute gained by the lean-stream, i.e.

$$G_i(y_i^{in} - y_i^{out}) = L_j(x_j^{out} - x_j^{in}) \qquad (4.3)$$

The material balance equation provides the mathematical description of the *operating line*. The operating line can be graphically represented on a y–x (McCabe-Thiele) diagram. The operating line extends between the two terminal points (y_i^{in}, x_j^{out}) and (y_i^{out}, x_j^{in}) and has a slope of L_j/G_i, as shown in Figure 4-3.

Mass exchangers may be broadly classified into two categories: *stagewise units* and *differential (continuous) contactors*. Stagewise units are characterized by discrete solute transfer where mass exchange takes place in a stage followed by disengagement between the rich and lean phases then mass exchange and so on. Examples of stagewise units include tray columns and multistage mixer-settler arrangements. An important concept in stagewise operations is the notion of an *equilibrium stage or a theoretical plate*. With sufficient mixing time, the two phases leaving the theoretical stage are essentially in equilibrium; hence the name equilibrium stage. Each theoretical stage can be represented by a step between the operating line and the

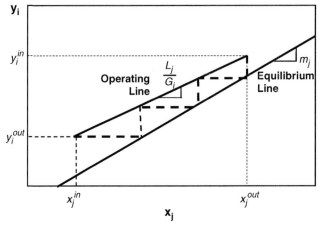

FIGURE 4-3   THE McCABE-THIELE DIAGRAM FOR A MASS EXCHANGER

equilibrium line. Hence, the number of theoretical plates NTP can be determined by "stepping off" stages between the two ends of the exchanger, as shown by Figure 4-3.

For the case of isothermal, dilute mass exchange with linear equilibrium, NTP can be determined through the Kremser (1930) equation:

$$\text{NTP} = \frac{\ln[(1 - (m_j G_i / L_j))(y_i^{\text{in}} - m_j x_j^{\text{in}} - b_j / y_i^{\text{out}} - m_j x_j^{\text{in}} - b_j) + (m_j G_i / L_j)]}{\ln(L_j / m_j G_i)}$$

(4.4)

Other forms of the Kremser equation include

$$\text{NTP} = \frac{\ln[(1 - (L_j / m_j G_i))(x_i^{\text{in}} - x_j^{\text{out},*} / x_j^{\text{out}} - x_j^{\text{out},*}) + (L_i / m_j G_i)]}{\ln(m_j G_i / L_j)}$$

(4.5)

where

$$x_j^{\text{out},*} = \frac{y_i^{\text{in}} - b_j}{m_j}$$

(4.6)

Also,

$$\frac{y_i^{\text{in}} - m_j x_j^{\text{out}} - b_j}{y_i^{\text{out}} - m_j x_j^{\text{in}} - b_j} = \left(\frac{L_j}{m_j G_i}\right)^{\text{NTP}}$$

(4.7)

Other expressions for the number of theoretical plates avoiding logarithmic terms (for mathematical programming purposes) can also be found in literature (Fraser and Shenoy 2004; Szitkai et al., 2002).

Equilibrium requires long-enough (infinite) contact time between the two phases. Therefore, to relate actual performance to equilibrium behavior, it is necessary to calculate the number of actual plates "NAP" by incorporating contacting efficiency. A common method to account for efficiency is to use an overall exchanger efficiency, η_o, which can be used to relate NAP and NTP as follows

$$NAP = NTP/\eta_o \qquad (4.8)$$

Once the number of plates is determined, the height of the mass exchanger can be determined by allowing a plate spacing distance between each two consecutive plates. The column diameter is normally determined by selecting a superficial velocity for one (or both) of the phases. This velocity is intended to ensure proper mixing while avoiding hydrodynamic problems such as flooding, weeping, or entrainment. Once a superficial velocity is determined, the cross-sectional area of the column is obtained by dividing the volumetric flowrate by the velocity.

Continuous (differential) mass exchangers include packed units, spray exchangers, and bubble columns. The height of a differential contactor, H, may be estimated using

$$H = HTU_y NTU_y \qquad (4.9a)$$

$$= HTU_x NTU_x \qquad (4.9b)$$

where HTU_y and HTU_x are the overall height of transfer units based on the rich and the lean phases, respectively, while NTU_y and NTU_x are the overall number of transfer units based on the rich and the lean phases, respectively.

The overall height of a transfer unit may be provided by the packing (or unit) manufacturer or estimated using empirical correlations (typically by dividing superficial velocity of one phase by its overall mass transfer coefficient). On the other hand, the number of transfer units can be theoretically estimated for the case of isothermal and dilute mass exchangers with linear equilibrium as follows:

$$NTU_y = \frac{y_i^{in} - y_i^{out}}{(y_i - y_i^*)_{\text{log mean}}} \qquad (4.10a)$$

where

$$(y_i - y_i^*)_{\text{log mean}} = \frac{(y_i^{in} - m_j x_j^{out} - b_j) - (y_i^{out} - m_j x_j^{in} - b_j)}{\ln[(y_i^{in} - m_j x_j^{out} - b_j)/(y_i^{out} - m_j x_j^{in} - b_j)]} \qquad (4.10b)$$

and

$$\text{NTU}_x = \frac{x_j^{\text{in}} - x_j^{\text{out}}}{(x_j - x_j^*)_{\text{log mean}}} \tag{4.11a}$$

where

$$(x_j - x_j^*)_{\text{log mean}} = \frac{[x_j^{\text{out}} - (y_i^{\text{in}} - b_j/m_j)] - [x_j^{\text{in}} - (y_i^{\text{out}} - b_j/m_j)]}{\ln\{[x_j^{\text{out}} - (y_i^{\text{in}} - b_j/m_j)]/[x_j^{\text{in}} - (y_i^{\text{out}} - b_j/m_j)]\}} \tag{4.11b}$$

If the terminal compositions or L_j/G_i are unknown, it is convenient to use the following form:

$$\text{NTU}_y = \frac{\ln[(1 - (m_j\,G_i/L_j))(y_i^{\text{in}} - m_jx_j^{\text{in}} - b_j)/(y_i^{\text{out}} - m_jx_j^{\text{in}} - b_j) + (m_jG_i/L_j)]}{1 - (m_jG_i/L_j)} \tag{4.12}$$

4.2 COST OPTIMIZATION OF MASS EXCHANGERS

In assessing the economics of a mass exchanger, two types of cost must be considered: fixed and operating. The *fixed cost (investment)* refers to the cost of the mass exchanger (e.g., shell, trays, etc.), ancillary devices (e.g., pump, compressor), installation, insulation, instrumentation, electric work, piping, engineering work, and construction. Fixed capital investments are characterized by the fact that they have to be replaced after a number of years commonly referred to as service life or useful life period because of wear and tear or by virtue of becoming obsolete or inefficient. Therefore, it is useful to evaluate an annual cost associated with the capital investment of the mass exchanger, referred to as the *annualized fixed cost* "AFC". A simplified method for evaluating AFC is to consider the initial fixed cost of the equipment (FC_o) and its salvage value (FC_s) after n years of useful life period. Using an annual depreciation scheme, we get

$$\text{AFC} = \frac{FC_o - FC_s}{n} \tag{4.13}$$

In addition to the fixed capital investment needed to purchase and install the mass exchange system and auxiliaries, there is a continuous expenditure referred to as operating cost which is needed to operate the mass exchanger. The operating cost includes mass separating agents (makeup, regeneration, etc.) and utilities (heating, cooling, etc.).

By combining the fixed and operating costs, we get the total annualized cost (TAC) of a mass exchange system:

$$\text{Total annualized cost} = \text{Annualized fixed cost} + \text{Annual operating cost}$$
(4.14)

In order to minimize TAC, it is necessary to trade off the fixed cost versus the operating cost. Such trade offs can be established by identifying the role of the mass exchange driving force between the actual operation and the equilibrium limits. In order to reach equilibrium compositions, an infinitely large mass exchanger is required. Therefore, the operating line must have a positive driving force with respect to the equilibrium line. The minimum driving force between the operating line and the equilibrium line is referred to as the *minimum allowable composition difference* and is designated by ε_j as shown by Figure 4-4.

The minimum allowable composition difference can be used to trade off capital versus operating costs. In order to demonstrate this concept, let us consider the mass exchanger represented on the y-x diagram of Figure 4-4. For the rich stream, the inlet and outlet compositions as well as the flowrate are all given. For the lean-stream, the inlet composition is given while the flowrate and the outlet compositions are unknown. Hence, the operating line described by equation (4.3) has two unknowns: x_j^{out} and L_j. The maximum theoretically attainable outlet composition in the lean phase ($x_j^{\text{out},*}$) is the equilibrium value corresponding to the inlet composition of the rich stream (Figure 4-4). As mentioned earlier, achieving this equilibrium value requires an infinitely large mass exchanger. Once the minimum allowable

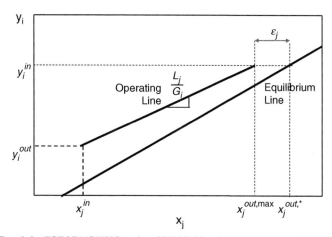

FIGURE 4-4 ESTABLISHING A MINIMUM ALLOWABLE COMPOSITION DIFFERENCE

composition difference is selected, the maximum practically feasible outlet composition in the lean-stream ($x_j^{out,max}$) can be determined as:

$$x_j^{out,max} = x_j^{out,*} - \varepsilon_j \tag{4.15}$$

but

$$y_i^{in} = m_j x_j^{out,*} + b_j \tag{4.16}$$

Combining equations (4.15) and (4.16), one obtains

$$x_j^{out,max} = \frac{y_i^{in} - b_j}{m_j} - \varepsilon_j \tag{4.17}$$

To examine the effect of ε_j on the cost of the mass exchanger, let us assess the effect of selecting two values of the minimum allowable composition difference: ε_1 and ε_2 (Figure 4-5). As ε_j increases, the slope of the operating line increases and the flowrate of the MSA increases leading to an increase in the operating cost of the mass exchanger. Meanwhile, as ε_j increases, the number of theoretical plates decreases thereby leading to a reduction in the fixed cost. By varying ε_j and evaluating the corresponding annualized fixed cost, annual operating cost, and total annualized cost, we can determine the optimum value of minimum allowable composition difference, $\varepsilon_j^{Optimum}$, which corresponds to the minimum total annualized cost (Figure 4-6). It is also worth noting that when ε_j is set to zero, the annualized fixed cost is infinity while the annual operating cost is at its minimum value.

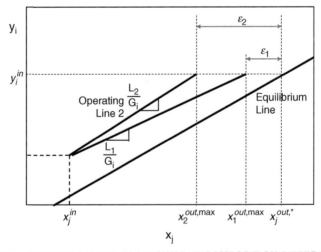

FIGURE 4-5   CHANGING MINIMUM ALLOWABLE COMPOSITION DIFFERENCE

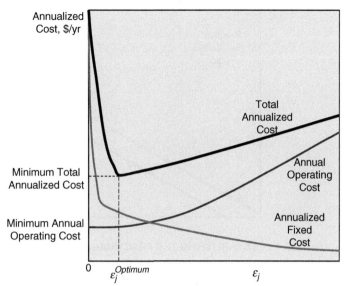

FIGURE 4-6 SELECTION OF OPTIMUM MINIMUM ALLOWABLE COMPOSITION DIFFERENCE

The foregoing analysis can be used to incorporate thermodynamic and economic issues into practical-feasibility constraints. To demonstrate this concept, let us consider a mass exchanger for which the equilibrium relation governing the transfer of the solute from the rich stream, i, to the MSA, j, is given by the linear expression

$$y_i = m_j x_j^* + b_j \tag{4.18}$$

which indicates that for a rich stream composition of y_i, the maximum theoretically attainable composition of the MSA is x_j^*. By employing a minimum allowable composition difference of ε_j, one can draw a "practical-feasibility line" that is parallel to the equilibrium line but offset to its left by a distance ε_j (Figure 4-7). Therefore, the lean-stream composition on the practical-feasibility line can be mathematically represented as:

$$x_j^{\mathrm{max}} = x_j^* - \varepsilon_j \tag{4.19}$$

i.e.,

$$x_j^{\mathrm{max}} = \frac{y_i - b_j}{m_j} - \varepsilon_j \tag{4.20a}$$

or

$$y_i = m_j(x_j^{\mathrm{max}} + \varepsilon_j) + b_j \tag{4.20b}$$

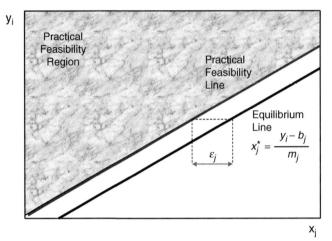

FIGURE 4-7   ESTABLISHING THE PRACTICAL-FEASIBILITY REGION

In order for an operating line to be practically feasible, it must lie in the region to the left of the practical-feasibility line. Hence, for any point lying on the practical-feasibility line, two statements can be made. For a given $\bar{y}_i$, the value x_j^{max} corresponds to the maximum composition of the solute that is practically achievable in the MSA. Therefore, the solute can be practically transferred from $\bar{y}_i$ to x_j^{max} or any lean-stream composition lower than x_j^{max} (lies to its left on the graph). Conversely, for a given x_j^{max}, the value $\bar{y}_i$ corresponds to the minimum composition of the solute in the rich stream that is needed to practically transfer the solute from the rich stream to the MSA. Therefore, the solute can be transferred to x_j^{max} from $\bar{y}_i$ or any higher rich-stream composition. These two aspects are shown by Figure 4-8.

EXAMPLE 4-1 BENZENE RECOVERY FROM A GASEOUS EMISSION (EL-HALWAGI 1997)

Benzene is to be removed from a gaseous emission by contacting it with an absorbent (wash oil, molecular weight 300). The gas flowrate is 0.2 kg-mole/s (about 7700 ft^3/min) and it contains 0.1% mol/mol (1000 ppm) of benzene. The molecular weight of the gas is 29, its temperature is 300 K, and it has a pressure of 141 kPa (approximately 1.4 atm). It is desired to reduce the benzene content in the gas to 0.01% mol/mol using the system shown in Figure 4-9. Benzene is first absorbed into oil. The oil is then fed to a regeneration system in which oil is heated and passed to a flash column that recovers benzene as a top product. The bottom product is the regenerated oil, which contains 0.08 % mol/mol benzene. The regenerated oil is cooled and pumped back to the absorber.

What is the optimal flowrate of recirculating oil that minimizes the TAC of the system?

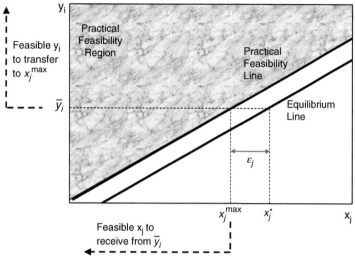

FIGURE 4-8   IDENTIFYING PRACTICALLY FEASIBLE COMPOSITION REGIONS

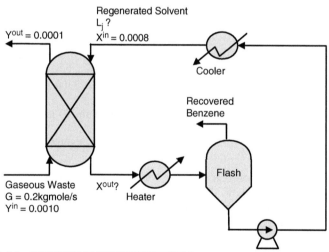

FIGURE 4-9   BENZENE RECOVERY PROCESS

The following data may be used:

Equilibrium Data

- The absorption operation is assumed to be isothermal (at 300 K) and to follow Raoult's law:

$$y = \left(\frac{p^0}{P^{\text{Total}}}\right)x \qquad (4.21)$$

where y is the mole fraction of benzene in air and x is the mole fraction of benzene in oil.

- The vapor pressure of benzene, p^0, is given by:

$$\ln p^0 = 20.8 - \frac{2789}{T - 52} \tag{4.22}$$

where p^0 is in Pascal and T is in Kelvin.

Absorber Sizing Criteria

- The overall-gas height of transfer unit for the packing is 0.6 m.
- The superficial velocity of the gas in the absorber is taken as 1.5 m/s to avoid flooding.
- The mass velocity of oil in the absorber should be kept above $2.7 \, kg/m^2 \cdot s$ to insure proper wetting.

Cost Information

- The operating cost (including pumping, oil makeup, heating, and cooling) is \$0.05/kg-mole of recirculating oil.
- The system is to be operated for 8000 h/annum.
- The installed cost (\$) of the absorption column (including auxiliaries, but excluding packing) is given by

$$\text{Installed cost of column} = 2300 \, H^{0.85} D^{0.95} \tag{4.23}$$

where H is the packing height (m) and D is the column diameter (m).

$$\text{The packing cost is } \$800/m^3 \tag{4.24}$$

- The oil-regeneration system is to be salvaged from a closing unit in the plant. Hence, its fixed cost will not be accounted for in the optimization calculations.
- The absorber and packing are assumed to depreciate linearly over five years with negligible salvage values.

SOLUTION

At 300 K, the vapor pressure of benzene can be calculated from equation (4.22):

$$\ln p^0 = 20.8 - \frac{2789}{300 - 52} \tag{4.25a}$$

i.e.,

$$p^0 = 14{,}101 \, \text{Pa} \tag{4.25b}$$

Since the system is assumed to follow Raoult's law, then equation (4.21) can be used to give

$$m = \frac{14,101}{141,000}$$

$$\approx 0.1 \frac{\text{mole fraction of benzene in air}}{\text{mole fraction of benzene in oil}} \tag{4.26}$$

As has been previously mentioned, the minimum TAC can be identified by iteratively varying ε. Since the inlet and outlet compositions of the rich stream as well as the inlet composition of the MSA are fixed, one can vary ε at the rich end of the exchanger (and consequently the outlet composition of the lean-stream) to minimize the TAC of the system. In order to demonstrate this optimization procedure, let us first select a value of ε at the rich end of the exchanger equal to 1.5×10^{-3} and evaluate the system size and cost for this value.

Outlet Composition of Benzene in Oil

Let us set the outlet mole fraction of benzene in oil equal to its maximum practically feasible value given by equations (4.15) and (4.16). Since the intercept (b) in this case is zero, we get:

$$x^{\text{out}} = \frac{y^{\text{in}}}{m} - \varepsilon \tag{4.27a}$$

$$x^{\text{out}} = \frac{10^{-3}}{0.1} - 1.5 \times 10^{-3}$$

$$= 8.5 \times 10^{-3} \tag{4.27b}$$

Flowrate of Oil

A component material balance on benzene gives

$$L(8.5 \times 10^{-3} - 8 \times 10^{-4}) = 0.2(10^{-3} - 10^{-4}) \tag{4.28a}$$

or

$$L = 0.0234 \text{ kg-mol/s} \tag{4.28b}$$

Operating Cost

$$\begin{aligned}\text{Annual operating cost} = {} & .05 \text{ \$/kg-mole oil} \times 0.0234 \text{ kg-mole oil/s} \\ & \times 3600 \times 8000 \text{ s/year} = \$33,700/\text{year}\end{aligned} \tag{4.29}$$

Column Height
   According to equation (4.10b)

$$(y - y^*)_{\text{log mean}} = \frac{(10^{-3} - 0.1 \times 8.5 \times 10^{-3}) - (10^{-4} - 0.1 \times 8.0 \times 10^{-4})}{\ln\,[(10^{-3} - 0.1 \times 8.5 \times 10^{-3})/(10^{-4} - 0.1 \times 8.0 \times 10^{-4})]}$$

$$= 6.45 \times 10^{-5}$$

(4.30)

Therefore, NTU_y can be calculated using equation (4.10a) as

$$\text{NTU}_y = \frac{10^{-3} - 10^{-4}}{6.45 \times 10^{-5}}$$

$$= 13.95$$

(4.31)

and the height is obtained from equation (4.9a)

$$H = 0.6 \times 13.95$$
$$= 8.37 \text{ m}$$

(4.32)

Column Diameter

$$D = \sqrt{\frac{4 \times \text{Volumetric Flowrate of Gas}}{\pi\ \text{Gas Superficial Velocity}}}$$

(4.33)

But,

$$\text{Molar density of gas} = \frac{P}{RT}$$

$$= \frac{141}{8.3143 \times 300}$$

$$= 0.057 \text{ kg-mole/m}^3$$

(4.34)

Therefore,

$$\text{Volumetric flowrate of gas} = \frac{0.2(\text{kg-mole/s})}{0.057(\text{kg-mole/m}^3)}$$

$$= 3.51 \text{ m}^3/\text{s}$$

(4.35)

and

$$D = \sqrt{\frac{4 \times 3.51}{3.14 \times 1.5}}$$

$$= 1.73\text{m}$$

(4.36)

It is worth pointing out that the mass velocity of oil is

$$\frac{0.0234(\text{kg-mole/s}) \times (300 \text{ kg/kg-mole})}{(\pi/4) (1.73)^2} \approx 3\text{kg/s} \tag{4.37}$$

which is acceptable since it is greater than the minimum wetting velocity (2.7 kg/m^{2}s).

Fixed Cost

Fixed cost of installed shell and auxiliaries $= 2300(8.37)^{0.85}(1.73)^{0.95}$

$$= \$23,600 \tag{4.38}$$

$$\text{Cost of packing} = (800)\frac{\pi}{4}(1.73)^2(8.37)$$

$$= \$15,700 \tag{4.39}$$

Total Annualized Cost

$$\text{TAC} = \text{Annual operating cost} + \text{annualized fixed cost}$$

$$= 33,700 + \frac{(23,600 + 15,700)}{5}$$

$$= \$41,560/\text{year} \tag{4.40}$$

This procedure is carried out for various values of ε until the minimum TAC is identified. The results shown in Figure 4-10 indicate that the value of $\varepsilon = 1.5 \times 10^{-3}$ used in the preceding calculations is the optimum one ■ ■ ■ leading to a minimum TAC of \$41,560/year.

4.3 PROBLEM STATEMENT FOR SYNTHESIS OF MASS EXCHANGE NETWORKS

In many processing facilities, mass exchangers are used to separate targeted species from a number of rich streams. More than one mass exchange technology and more than one MSA may be considered. In such situations, it is necessary to integrate the decisions and design of the multiple mass exchangers. This requires a holistic approach to consider all separation tasks from all rich streams, simultaneously screen all candidate mass exchange operations and MSAs, and identify the optimum network of mass exchangers. El-Halwagi and Manousiouthakis (1989a) introduced the problem of synthesizing *mass exchange network* "MENs" and developed systematic techniques for their optimal design. The problem of synthesizing MENs can be stated as follows: Given a number N_R of rich streams

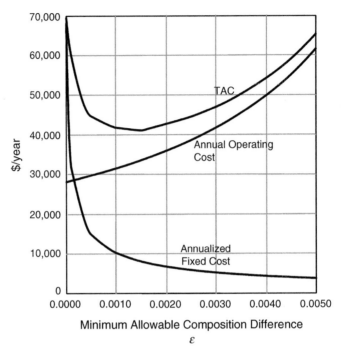

FIGURE 4-10 COST MINIMIZATION FOR THE BENZENE RECOVERY MASS EXCHANGE PROCESS

(sources) and a number N_S of MSAs (lean-streams), it is desired to synthesize a cost-effective network of mass exchangers that can preferentially transfer certain species from the rich streams to the MSAs. Given also are the flowrate of each rich stream, G_i, its supply (inlet) composition y_i^s, and its target (outlet) composition y_i^t, where $i = 1, 2, \ldots, N_R$. In addition, the supply and target compositions, x_j^s and x_j^t are given for each MSA, where $j = 1, 2, \ldots, N_S$. The flowrate of each MSA is unknown and is to be determined so as to minimize the network cost. Figure 4-11 is a schematic representation of the MEN problem statement.

The candidate lean-streams can be classified into N_{SP} process MSAs and N_{SE} external MSAs (where $N_{SP} + N_{SE} = N_S$). The process MSAs already exist on plant site and can be used for the removal of the undesirable species at a very low cost (virtually free). The flowrate of each process MSA that can be used for mass exchange is bounded by its availability in the plant, i.e.,

$$L_j \leq L_j^c \quad j = 1, 2, \ldots, N_{SP} \tag{4.41}$$

where L_j^c is the flowrate of the jth MSA that is available in the plant. On the other hand, the external MSAs can be purchased from the market. Their flowrates are to be determined according to the overall economic considerations of the MEN.

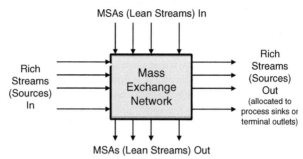

FIGURE 4-11 SCHEMATIC REPRESENTATION OF THE MEN SYNTHESIS PROBLEM (EL-HALWAGI AND MANOUSIOUTHAKIS 1989a)

Typically, rich streams leaving the MEN are either allocated to process sinks (equipment) or assigned to be terminal streams (e.g., products, wastes). When the outlet rich streams are allocated to process sinks, the target composition of the rich stream are selected so as to satisfy the constraints on the feed to these sinks. In case of final discharge, the target composition of the undesirable species in each rich stream corresponds to the environmental regulations. Finally, if the outlet rich stream corresponds to a terminal product, the target composition is set to satisfy quality requirements for the product.

The target composition of each MSA is an upper bound on the actual outlet composition of the MSA. The value of the target composition is selected based on a number of factors whose nature may be:

- *Physical* (e.g., saturation compositions, solubility limits, precipitation conditions)
- *Operational*: If the outlet MSA is used in a subsequent unit, its content of certain species must conform to the constraints on the feed to the subsequent unit
- *Safety* (e.g., to stay away from flammability/explosion limits)
- *Health* (e.g., to avoid reaching toxic compositions)
- *Environmental* (e.g., to satisfy emission regulations)
- *Economic* (e.g., to minimize the cost of the mass exchange and regeneration systems such as Example 4-1)
- *Technical feasibility* (e.g., to satisfy thermodynamic constraints and minimum driving force such as equation (4.17)

The MEN synthesis task entails answering several design questions and challenges:

- Which mass exchange technologies should be utilized (e.g., adsorption, solvent extraction ion exchange, etc.)?
- Which MSAs should be selected (e.g., which solvents, adsorbents)?
- What is the optimal flowrate of each MSA?
- How should these MSAs be matched with the rich streams?

- What is the optimal system configuration (e.g., how should these mass exchangers be arranged? Is there any stream splitting and mixing?)?

In responding to these questions, one must consider the numerous (infinite) number of alternative solutions. Instead of attempting an exhaustive enumeration technique (which would be hopelessly complicated), it is necessary to extract the optimum solution from among the numerous alternatives without enumeration or trial and error. The next section presents an integrated method to the synthesis of MENs referred to as the mass exchange pinch analysis.

4.4 MASS EXCHANGE PINCH DIAGRAM

The mass exchange pinch analysis (El-Halwagi and Manousiouthakis 1989a) provides a holistic and systematic approach to synthesizing MENs. It also enables the identification of rigorous targets such as minimum cost of MSAs. The first step in the analysis is to develop an integrated view of all the separation tasks for the rich streams. This can be achieved by developing a composite representation of mass exchanged from all the rich streams. Mass of targeted species removed from the ith rich stream is given by:

$$\text{MR}_i = G_i(y_i^s - y_i^t) \quad i = 1, 2, \ \ldots, N_R \quad\quad (4.42)$$

By plotting mass exchanged versus composition, each rich stream is represented as an arrow whose tail corresponds to its supply composition and its head to its target composition. The slope of each arrow is equal to the stream flowrate. The vertical distance between the tail and the head of each arrow represents the mass of targeted species that is lost by that rich stream. In this representation, the vertical scale is only relative. Any stream can be moved up or down while preserving the same vertical distance between the arrow head and tail and maintaining the same supply and target compositions. A stream cannot be moved left or right, otherwise stream composition will be altered. A convenient way of vertically placing each arrow is to rank the rich streams in ascending order of their targeted composition then we stack the rich streams on top of one another, starting with the rich stream having the lowest target composition. Once the first rich stream is represented, we draw a horizontal line passing through the arrow tail of the stream. Next, the second rich stream is represented as an arrow extending between its supply and target compositions and having a vertical distance equal to the mass of the targeted species to be removed from this stream. The arrow head of the second rich stream is placed on the horizontal line passing through the arrowtail of the first rich stream. The procedure is continued for all the rich streams (Figure 4-12).

After all the rich streams have been represented, it is necessary to develop a combined representation of the rich streams that allows us to observe the separation tasks of all rich streams as a function of composition.

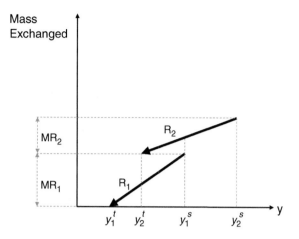

FIGURE 4-12 REPRESENTATION OF MASS EXCHANGED BY TWO RICH STREAMS

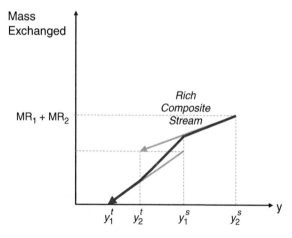

FIGURE 4-13 CONSTRUCTING A RICH COMPOSITE STREAM USING SUPERPOSITION

A rich composite stream can be constructed using "diagonal rule" for superposition to add up mass in the overlapped regions of streams (Figure 4-13). In the region between y_1^t and y_2^t, there is only R_1. Therefore, the composite representation is exactly the same as R_1. Similarly, in the region between y_1^s and y_2^s there is only R_2 and, hence, the composite representation is exactly the same as R_2. In the overlapping region of the two rich streams (between y_2^t and y_1^s), the composite representation of the two streams is the diagonal (hence the name diagonal rule). By connecting these three linear segments, we now have a rich composite stream which represents the cumulative mass of the targeted species removed from all the rich streams. It captures the relevant characteristics of the rich streams

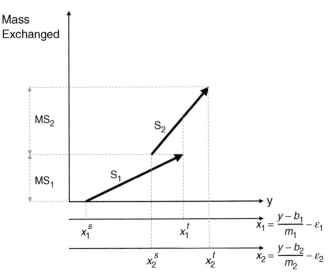

FIGURE 4-14 REPRESENTATION OF MASS EXCHANGED BY TWO PROCESS MSAs

and enables the simultaneous consideration of all rich streams and developing an integrated mass exchange strategy for all of them.

Next, attention is turned to the lean-streams. Since the process MSAs are available on-site and may be used virtually for no or little operating cost, we will first consider maximizing their use. The remaining load will then be removed using external MSAs. Therefore, we first establish N_{SP} lean composition scales (one for each process MSA) that are in one-to-one correspondence with the rich scale according to equation (4.20). Next, the mass of targeted species that can be gained by each process MSA is plotted versus the composition scale of that MSA. Hence, each process MSA is represented as an arrow extending between supply and target compositions (see Figure 4-14 for a two-MSA example). The vertical distance between the arrow head and tail is given by

Mass of solute that can be gained by the jth process MSA

$$MS_j = L_j^c(x_j^t - x_j^s) \quad j = 1, 2, \ldots, N_{SP} \tag{4.43}$$

Once again, the vertical scale is only relative and any stream can be moved up or down on the diagram. A convenient way of vertically placing each arrow is to stack the process MSAs on top of one another starting with the MSA having the lowest supply composition (Figure 4-15). Hence, a lean composite stream representing the cumulative mass of the targeted species gained by all the MSAs is obtained by using the diagonal rule for superposition.

Next, both composite streams are plotted on the same diagram (Figure 4-16). On this diagram, thermodynamic feasibility of mass exchange

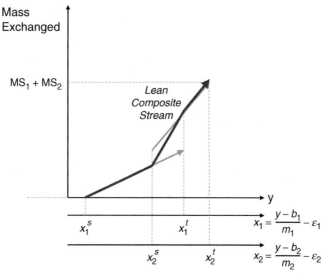

FIGURE 4-15 CONSTRUCTION OF THE LEAN COMPOSITE STREAM USING SUPERPOSITION

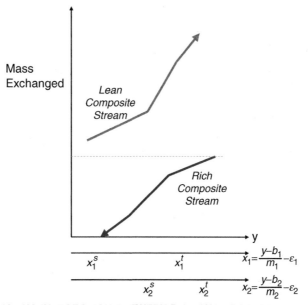

FIGURE 4-16 NO INTEGRATION BETWEEN RICH AND PROCESS MSAs

is guaranteed when at any mass exchange level (which corresponds to a horizontal line), the composition of the lean composite stream is located to the left of the rich composite stream. Because of the use of equation (4.20) in mapping the composition scales, when mass is transferred to the left, constraints on thermodynamics as well as the minimum driving force are satisfied. For a given set of corresponding composition scales $\{y,\ x_1,\ x_2 \ldots x_j \ldots x_{NS}\}$, it is thermodynamically and practically feasible to transfer the targeted species from any rich stream to any MSA. In addition, it is also feasible to transfer the targeted species from any rich stream of a composition y_i to any MSA which has a composition less than the x_j obtained from equation (4.20a).

The lean composite stream can be moved up and down which implies different mass exchange decisions. For instance, if we move the lean composite stream upwards in a way that leaves no horizontal overlap with the rich composite stream, then there is no integrated mass exchange between the rich composite stream and the process MSAs as seen in Figure 4-16. When the lean composite stream is moved downwards so as to provide some horizontal overlap (Figure 4-17), some integrated mass exchange can be achieved. The remaining load of the rich composite stream has to be removed by the external MSAs. However, if the lean composite stream is moved downwards such that a portion of the lean is placed to the right of the rich composite stream, thereby creating infeasibility (Figure 4-18). Therefore, the optimal situation is constructed when the lean composite stream is slid vertically until it touches the rich composite stream while lying completely to the left of the rich composite stream at any horizontal level. The point where the two composite streams touch is called the "mass exchange pinch point"; hence the name "pinch diagram" (Figure 4-19).

On the pinch diagram, the vertical overlap between the two composite streams represents the maximum amount of the targeted species that can be transferred from the rich streams to the process MSAs. It is referred to as the "integrated mass exchange". The vertical distance of the lean composite stream which lies above the upper end of the rich composite stream is

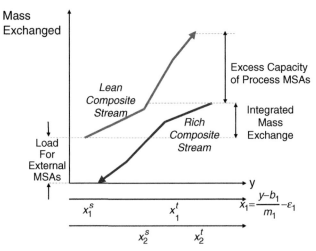

FIGURE 4-17   PARTIAL INTEGRATION OF RICH AND LEAN-STREAMS (PASSING MASS THROUGH THE PINCH)

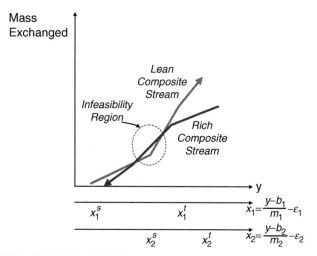

FIGURE 4-18 CAUSING INFEASIBILITY BY PLACING LEAN TO THE RIGHT OF RICH

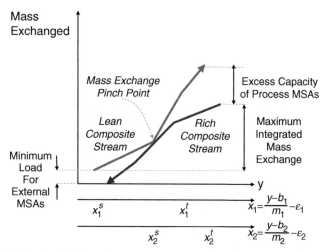

FIGURE 4-19 THE MASS EXCHANGE PINCH DIAGRAM

referred to as "excess process MSAs". It corresponds to that capacity of the process MSAs to remove the targeted species that cannot be used because of thermodynamic infeasibility. According to the designer's preference or to the specific circumstances of the process, such excess can be eliminated from service by lowering the flowrate and/or the outlet composition of one or more of the process MSAs. Finally, the vertical distance of the waste composite stream which lies below the lower end of the lean composite stream corresponds to the mass of the targeted species to be removed by external MSAs. These targets have been identified rigorously using little data and without detailing the design of the MEN.

The above discussion indicates that in order to achieve the targets for maximum integration of mass exchange from rich stream to process MSAs and minimum load to be removed by the external MSAs, the following *three design rules* are needed:

- No mass should be passed through the pinch (i.e., the two composites must touch).
- No excess capacity should be removed from MSAs below the pinch.
- No external MSAs should be used above the pinch.

Above the pinch, exchange between the rich and the lean process streams takes place. External MSAs are not required. Using an external MSA above the pinch will incur a penalty eliminating an equivalent amount of process lean-streams from service. On the other hand, below the pinch, both the process and the external lean-streams should be used. Furthermore, Figure 4-17 indicates that if any mass is transferred across the pinch, the composite lean-stream will move upward and, consequently, external MSAs in excess of the minimum requirement will be used. Therefore, to minimize the cost of external MSAs, mass should not be transferred across the pinch. It is worth pointing out that these observations are valid only for the class of MEN problems covered in this chapter. When the assumptions employed in this chapter are relaxed, more general conclusions can be made. For instance, it is shown later that the pinch analysis can still be undertaken even when there are no process MSAs in the plant.

EXAMPLE 4-2 SYNTHESIZING A NETWORK OF MASS EXCHANGERS FOR BENZENE RECOVERY (EI-HALWAGI 1997)

Figure 4-20 shows a simplified flowsheet of a copolymerization plant. The copolymer is produced via a two-stage reaction. The monomers are first dissolved in a benzene-based solvent. The mixed-monomer mixture is fed to the first stage of reaction where a catalytic solution is added. Several additives (extending oil, inhibitors, and special additives) are mixed in a mechanically stirred column. The resulting solution is fed to the second-stage reactor, where the copolymer properties are adjusted. The stream leaving the second-stage reactor is passed to a separation system which produces four fractions: copolymer, unreacted monomers, benzene, and gaseous waste. The copolymer is fed to a coagulation and finishing section. The unreacted monomers are recycled to the first-stage reactor, and the recovered benzene is returned to the monomer-mixing tank. The gaseous waste, R_1, contains benzene as the primary pollutant that should be recovered. The stream data for R_1 are given in Table 4-1.

Two process MSAs and one external MSA are considered for recovering benzene from the gaseous waste. The two process MSAs are the additives, S_1, and the liquid catalytic solution, S_2. They can be used for benzene recovery at virtually no operating cost. In addition to its positive environmental impact, the recovery of benzene by these two MSAs offers an economic incentive since it reduces the benzene makeup needed to

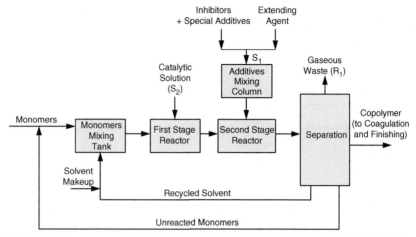

FIGURE 4-20   A SIMPLIFIED FLOWSHEET OF A COPOLYMERIZATION PROCESS

TABLE 4-1 DATA OF WASTE STREAM FOR THE BENZENE REMOVAL EXAMPLE

Stream	Description	Flowrate G_i (kg-mol/s)	Supply composition (mole fraction) y_i^s	Target composition (mole fraction) y_i^t
R_1	Off-gas from product separation	0.2	0.0020	0.0001

TABLE 4-2 DATA OF PROCESS LEAN-STREAMS FOR THE BENZENE REMOVAL EXAMPLE

Stream	Description	Upper bound on flowrate (kg-mol/s) L_j^C	Supply composition of benzene (mole fraction) x_j^s	Target composition of benzene (mole fraction) x_j^t
S_1	Additives	0.08	0.003	0.006
S_2	Catalytic solution	0.05	0.002	0.004

compensate for the processing losses. Furthermore, the additives mixing column can be used as an absorption column by bubbling the gaseous waste into the additives. The mixing pattern and speed of the mechanical stirrer can be adjusted to achieve a wide variety of mass transfer tasks. The stream data for S_1 and S_2 are given in Table 4-2. The equilibrium data for benzene in the two process MSAs are given by:

$$y_1 = 0.25x_1 \tag{4.44}$$

TABLE 4-3 DATA FOR THE EXTERNAL MSA FOR THE BENZENE REMOVAL EXAMPLE

Stream	Description	Upper bound on flowrate (kg-mole/s) L_j^c	Supply composition of benzene (mole fraction) x_j^s	Target composition of benzene (mole fraction) x_j^t
S_3	Organic oil	?	0.0008	0.0100

and

$$y_1 = 0.50x_2 \tag{4.45}$$

where y_1, x_1, and x_2 are the mole fractions of benzene in the gaseous waste, S_1 and S_2, respectively. For control purposes, the minimum allowable composition difference for S_1 and S_2 should not be less than 0.001.

The external MSA, S_3, is an organic oil that can be regenerated using flash separation. The operating cost of the oil (including pumping, makeup, and regeneration) is \$0.05/kg mol of recirculating oil. The equilibrium relation for transferring benzene from the gaseous waste to the oil is given by

$$y_1 = 0.10x_3 \tag{4.46}$$

The data for S_3 are given in Table 4-3. The absorber sizing equations and fixed cost were given in Example 4-2. Using the graphical pinch approach, synthesize a cost-effective MEN that can be used to remove benzene from the gaseous waste (Figure 4-21a).

SOLUTION
Constructing the Pinch Diagram

As has been described earlier, the rich composite stream is first plotted as shown in Figure 4-21b. Next, the lean composite stream is constructed for the two process MSAs. Equation (4.20) is employed to generate the correspondence among the composition scales y, x_1, and x_2. The least permissible values of the minimum allowable composition differences are used ($\varepsilon_1 = \varepsilon_2 = 0.001$). Later, it will be shown that these values are optimum for a minimum operating cost "MOC" solution. Next, the mass exchangeable by each of the two process lean-streams is represented as an arrow versus its respective composition scale (Figure 4-22a). As demonstrated by Figure 4-22b, the lean composite stream is obtained by applying superposition to the two lean arrows. Finally, the pinch diagram is constructed by combining Figures 4-21b and 4-22b. The lean composite stream is slid vertically until it is completely above the rich composite stream.

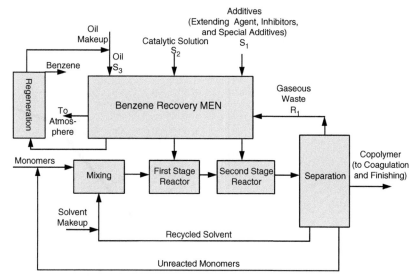

FIGURE 4-21a THE COPOLYMERIZATION PROCESS WITH A BENZENE RECOVERY MEN

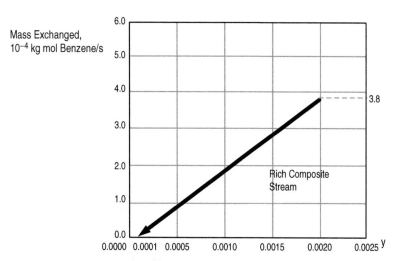

FIGURE 4-21b RICH COMPOSITE STREAM FOR THE BENZENE RECOVERY EXAMPLE

Interpreting Results of the Pinch Diagram

As can be seen from Figure 4-23, the pinch is located at the corresponding mole fractions $(y, x_1, x_2) = (0.0010, 0.0030, 0.0010)$. The excess capacity of the process MSAs is 1.4×10^{-4} kg-mol benzene/s and cannot be used because of thermodynamic and practical-feasibility limitations. This excess can be eliminated by reducing the outlet compositions and/or

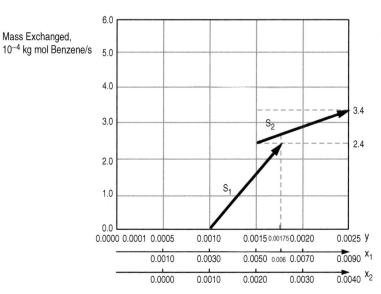

FIGURE 4-22a REPRESENTATION OF THE TWO PROCESS MSAs FOR THE BENZENE RECOVERY EXAMPLE

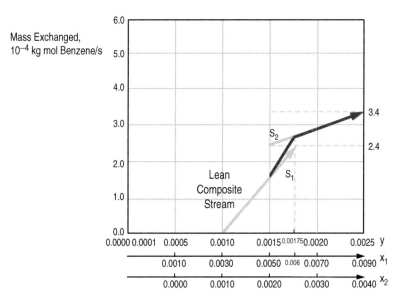

FIGURE 4-22b CONSTRUCTION OF THE LEAN COMPOSITE STREAM FOR THE TWO PROCESS MSAs OF THE BENZENE RECOVERY EXAMPLE

flowrates of the process MSAs. Since the inlet composition of S_2 corresponds to a mole fraction of 0.0015 on the y scale, the waste load immediately above the pinch (from $y = 0.0010$ to $y = 0.0015$) cannot be removed by S_2. Therefore, S_1 must be included in an MOC solution. Indeed, S_1 alone can be used

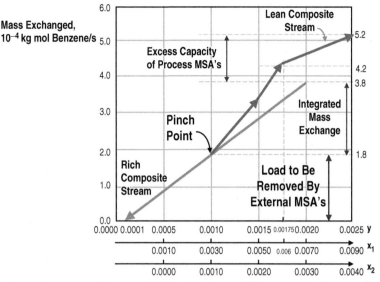

FIGURE 4-23 THE PINCH DIAGRAM FOR THE BENZENE RECOVERY EXAMPLE ($\varepsilon_1 = \varepsilon_2 = 0.001$)

to remove all the waste load above the pinch (2×10^{-4} kg-mol benzene/s). To reduce the fixed cost by minimizing the number of mass exchangers, it is preferable to use a single solvent above the pinch rather than two solvents. This is particularly attractive if given the availability of the mechanically stirred additives-mixing column for absorption. Hence, the excess capacity of the process MSAs is eliminated by avoiding the use of S_2 and reducing the flowrate and/or outlet composition of S_1. There are infinite combinations of L_1 and x_1^{out} that can be used to remove the excess capacity of S_1 according to the following material balance:

Benzene load above the pinch to be removed by $S_1 = L_1(x_1^{out} - x_1^s)$

(4.47a)

i.e.,

$$2 \times 10^{-4} = L_1(x_1^{out} - 0.003)$$
(4.47b)

Nonetheless, since the additives-mixing column will be used for absorption, the whole flowrate of S_1 (0.08 kg-mol/s) should be fed to the column. Hence, according to equation (4.47b), the outlet composition of S_1 is 0.0055. The same result can be obtained graphically as shown in Figure 4-24. It is worth recalling that the target composition of an MSA is only an upper bound on the actual value of the outlet composition. As shown in this example, the outlet composition of an MSA is typically selected so as to optimize the cost of the system.

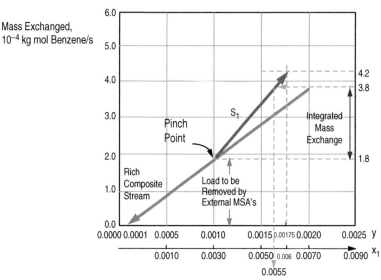

FIGURE 4-24　GRAPHICAL IDENTIFICATION OF x_1^{out}

Selection of the Optimal Value of ε_1

Since S_1 is a process MSA with almost no operating cost and since it is to be used in process equipment (the mechanically stirred column) that does not require additional capital investment for utilization as an absorption column, S_1 should be utilized to its maximum practically feasible capacity for absorbing benzene. The remaining benzene load (below the pinch) is to be removed using the external MSA. The higher the benzene load below the pinch, the higher the operating and fixed costs. Therefore, in this example, it is desired to maximize the integrated mass exchanged above the pinch. As can be seen on the pinch diagram when ε_1 increases, the x_1 axis moves to the right relative to the y axis. Consequently, the extent of integrated mass exchange decreases leading to a higher cost of external MSAs. For instance, Figure 4-25 demonstrates the pinch diagram when ε_1 is increased to 0.002. The increase of ε_1 to 0.002 results in a load of 2.3×10^{-4} kg-mol benzene/s to be removed by external MSAs (compared to 1.8×10^{-4} kg-mol benzene/s for $\varepsilon_1 = 0.001$), an integrated mass exchange of 1.5×10^{-4} kg-mol benzene/s (compared to 2.0×10^{-4} kg-mol benzene/s for $\varepsilon_1 = 0.001$) and an excess capacity of process MSAs of 1.9×10^{-4} kg-mol benzene/s (compared to 1.4×10^{-4} kg-mol benzene/s for $\varepsilon_1 = 0.001$). Thus, the optimum ε_1 in this example is the smallest permissible value given in the problem statement to be 0.001.

It is worth noting that there is no need to optimize over ε_2. As previously shown, when ε_2 was set equal to its lowest permissible value (0.001), S_1 was selected as the optimal process MSA above the pinch. On the pinch diagram, as ε_2 increases, S_2 moves to the right, and the same arguments for selecting S_1 over S_2 remain valid.

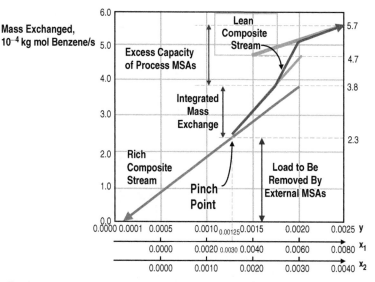

FIGURE 4-25   THE PINCH DIAGRAM WHEN ε_1 IS INCREASED TO 0.002

Optimizing the Use of the External MSA

The pinch diagram (Figure 4-24) demonstrates that below the pinch, the load of the waste stream has to be removed by the external MSA, S_3. This renders the remainder of this example identical to Example 4-1. Therefore, the optimal flowrate of S_3 is 0.0234 kg-mol/s and the optimal outlet composition of S_3 is 0.0085. Furthermore, the minimum total annualized cost of the benzene recovery system is $41,560/year (see Figure 4-10).

Constructing the Synthesized Network

The previous analysis shows that the MEN comprises two units: one above the pinch in which R_1 is matched with S_1, and one below the pinch in which the remainder load of R_1 is removed using S_3. Figure 4-26 illustrates the network configuration.

4.5 SCREENING OF MULTIPLE EXTERNAL MSAs AND CONSTRUCTING THE PINCH DIAGRAM WITHOUT PROCESS MSAs

So far, construction of the mass exchange pinch diagram started by maximizing the use of process MSAs with the consequence of minimizing the external MSAs. Then, MSAs must be screened so as to determine which one(s) should be used and the task of each MSA. Two questions arise:

- If the process has no process MSAs, how can the mass exchange pinch analysis be carried out? Since the flowrate of each external MSA is unknown, how can the lean composite stream be constructed?
- If there are multiple external MSAs, how can they be screened?

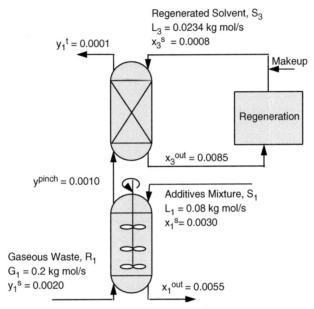

Regenerated Solvent, S_3
$L_3 = 0.0234$ kg mol/s
$y_1^t = 0.0001$ $x_3^s = 0.0008$

Makeup

Regeneration

$x_3^{out} = 0.0085$

$y^{pinch} = 0.0010$

Additives Mixture, S_1
$L_1 = 0.08$ kg mol/s
$x_1^s = 0.0030$

Gaseous Waste, R_1
$G_1 = 0.2$ kg mol/s
$y_1^s = 0.0020$ $x_1^{out} = 0.0055$

FIGURE 4-26   OPTIMAL MEN FOR THE BENZENE RECOVERY EXAMPLE

TABLE 4-4 DATA FOR TWO ADSORBENTS

MSA	x^s kg H_2/kg MSA	x^t kg H_2/kg MSA
Sand	0.0	10^{-9}
Activated carbon	0.0	0.1

In order to answer these questions, let us first consider the following motivating example. Suppose that it is desired to recover 10 kg/h of hydrogen from an industrial gas stream. Two candidate adsorbents are being screened: sand, whose cost is $0.01/kg of sand and activated carbon, whose cost is $1.00/kg of carbon. So, which adsorbent is cheaper? Is the cost per kilogram of the MSA an appropriate criterion to screen the candidate MSAs? The cost of the two adsorbents should be linked with the separation task. The following Table 4-4 summarizes the supply and target compositions for the two adsorbents.

A material balance for the MSA around a mass exchanger can be written as follows:

$$\text{Mass removed by one kilogram of the MSA} = L_j(x_j^{out} - x_j^s) \qquad (4.48)$$

Therefore,

$$\text{Flowrate of sand} = 10/(10^{-9} - 0.0) = 10^{10} \text{ kg sand/h} \qquad (4.49)$$

Similarly,

$$\text{Flowrate of activated carbon} = 10/(0.1 - 0.0) = 100 \text{ kg carbon/h} \quad (4.50)$$

Therefore,

$$\text{Cost of sand} = 10^{10} \text{ kg sand/h} \times \$0.01/\text{kg sand} = \$100 \text{ million per hour!} \quad (4.51)$$

On the other hand,

$$\text{Cost of activated carbon} = 100 \text{ kg carbon/hr} \times \$1.0/\text{kg carbon} = \$100/\text{h} \quad (4.52)$$

The foregoing discussion illustrates the cost of the MSA expressed as $/kg MSA is inappropriate in screening MSAs. Instead, the cost must be tied to the separation task. This can be achieved by identifying cost per kilogram removed of the targeted species. Both costs are related through material balance on the MSA. In a mass exchanger, mass of the targeted species removed by unit mass of the MSA (e.g., 1.0 kg of MSA) is given by:

$$\text{Mass of the targeted species removed by unit mass of the MSA} = x_j^t - x_j^s \quad (4.53)$$

Therefore,

$$c_j^r = \frac{c_j(\$/\text{kg MSA})}{x_j^{out} - x_j^s(\text{kg species/kg MSA})} \quad \text{where } j = 1, 2, 3 \quad (4.54)$$

where c_j^r is the removal cost of unit mass of the targeted species using the jth MSA. Now that the appropriate cost criterion for the MSA has been determined, let us proceed to the screening of the MSAs. First, the rich composite line is plotted. Then, equation (4.20) is employed to generate the correspondence among the rich composition scale, y, and the lean composition scales for all external MSAs. Each external MSA is then represented versus its composition scale as a horizontal arrow extending between its supply and target compositions (Figure 4-27). Several useful insights can be gained from this diagram. Let us consider three MSAs; S_1, S_2, and S_3 whose costs ($/kg of recirculating MSA) are c_1, c_2, and c_3, respectively. These costs can be converted into $/kg of removed solute, c_j^r, through equation (4.54). If arrow S_2 lies completely to the left of arrow S_1 and c_2^r is less than c_1^r, one can eliminate S_1 from the problem since it is thermodynamically and economically inferior to S_2. On the other hand, if arrow S_3 lies completely

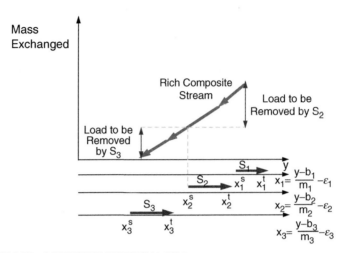

FIGURE 4-27   SCREENING EXTERNAL MSAs

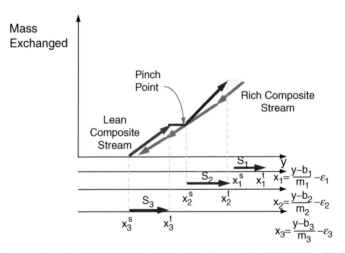

FIGURE 4-28   CONSTRUCTING THE PINCH DIAGRAM FOR EXTERNAL MSAs

to the left of arrow S_2 but c_3^r is greater than c_2^r, one should retain both MSAs. In order to minimize the operating cost of the network, separation should be staged to use the cheapest MSA where it is feasible. Hence, S_2 should be used to remove all the rich load to its left while the remaining rich load is removed by S_3 (Figure 4-28). The flowrates of S_2 and S_3 are calculated by simply dividing the rich load removed by the composition difference for the MSA. Now that the MSAs have been screened and their optimal

TABLE 4-5 DATA FOR MSAs OF TOLUENE REMOVAL EXAMPLE

Stream	Upper bound on flowrate (kg/s) L_j^C	Supply composition (ppmw) x_j^s	Target composition (ppmw) x_j^t	m_j	ε_j (ppmw)	C_j ($/kg MSA)
S_1	∞	0	9000	0.015	5000	3.1×10^{-3}
S_2	∞	70	8000	0.001	15,000	0.08
S_3	∞	50	2100	0.002	10,000	0.06

flowrates have been determined, one can construct the pinch diagram as shown in Figure 4-28.

4.6 EXAMPLE – WASTEWATER TREATMENT

An organic pollutant is to be removed from a wastewater stream. The flowrate of the waste stream is 20 kg/s and its inlet composition of toluene is 330 ppmw. It is desired to reduce the toluene composition in water to 30 ppmw. Three external MSAs are considered: air (S_1) for stripping, activated carbon (S_2) for adsorption, and a solvent extractant (S_3). The data for the candidate MSAs are given in Table 4-5. The equilibrium data for the transfer of the pollutant from the waste stream to the jth MSA is given by

$$y_1 = m_j x_j \tag{4.55}$$

where y_1 and x_j are the mass fractions of the toluene in the wastewater and the jth MSA, respectively.

Use the pinch diagram to determine the minimum operating cost of the MEN.

SOLUTION

Based on the given data and equation (4.54), we can calculate c_j^r to be 0.344, 10.088, and 29.268 $/kg of pollutant removed for air, activated carbon, and extractant, respectively. Since air is the least expensive, it will be used to remove all the load to its right (0.0051 kg pollutant/s, as can be seen from Figure 4-29a). Therefore, the flowrate of air can be calculated as

$$\text{Flowrate of air} = \frac{0.0051}{9000 \times 10^{-6} - 0} = 0.567 \text{ kg/s} \tag{4.56}$$

Since both the activated carbon and the extractant are feasible MSAs (lying to the left of the remaining load of the rich stream), while the adsorbent has a lower c_j^r, it will be used to remove the remaining load (0.0009 kg pollutant/s, as shown by Figure 4-29b). The flowrate of activated carbon is 0.114 kg/s. For 8000 operating hours per year, the annual operating cost of the system is $312,000/year.

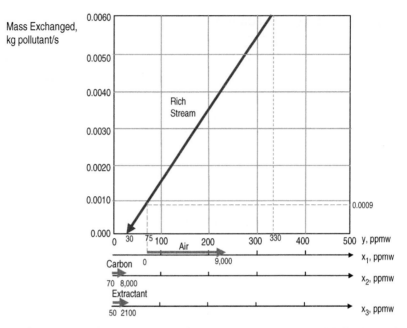

FIGURE 4-29a SCREENING EXTERNAL MSAs FOR THE WASTEWATER TREATMENT EXAMPLE

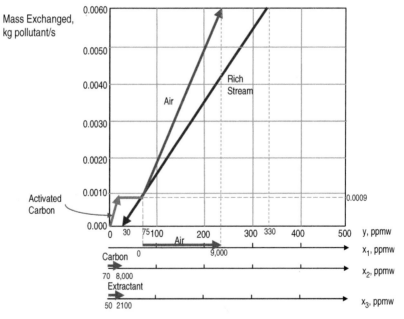

FIGURE 4-29b THE PINCH DIAGRAM FOR THE WASTEWATER REMOVAL EXAMPLE

4.7 ADDITIONAL READINGS

The original MEN work was introduced by El-Halwagi and Manousiouthakis (1989a). Since then, numerous papers have been published on the subject. Multicomponent MENs can also be systematically synthesized (e.g., El-Halwagi and Manousiouthakis 1998b); Alva-Argaez et al. (1999). El-Halwagi and Manousiouthakis (1990a,b) developed mathematical programming techniques to synthesize MENs as well as the regeneration systems. Genetic algorithms techniques were employed to synthesize MENs (Garrard and Fraga 1998; Xue et al., 2000). Srinivas and El-Halwagi introduced the problem of reactive mass exchange networks (El-Halwagi and Srinivas 1992; Srinivas and El-Halwagi 1994a). The design of MEN-hybrid systems was addressed by El-Halwagi et al. (1992) and Hamad et al. (1996). The simultaneous design of mass- and heat exchange networks was addressed by Srinivas and El-Halwagi (1994b) and Sebastian et al. (1996). Kiperstok and Sharratt (1995) solved the problem of synthesizing MENs with fixed-load removal. Mass exchange networks with variable supply and target compositions were tackled by (Garrison et al., 1995). Fixed-cost targeting techniques were developed by Hallale and Fraser (1997, 2000). Other classes of MENs include those providing flexible performance (Zhu and El-Halwagi 1995). Papalexandri and Pistikopoulos 1994), and controllable MENs (Huang and Edgar 1995; Huang and Fan 1995). Batch MENs have been synthesized by Foo et al. (2004). Furthermore, mass-pinch diagrams have been developed for a single lean-stream for resource conservation such as minimizing water use (Wang and Smith 1994; Dhole et al., 1996; Mann and Liu 1999) and managing process hydrogen (Alves and Towler 2002).

Many industrial applications of species interception have also been published including petroleum refining (El-Halwagi and El-Halwagi 1992; El-Halwagi et al., 1992), pulp and paper (Hamad et al., 1995, 1998; Dunn and El-Halwagi 1993), synthetic fuels (Warren et al., 1995), petrochemicals (Stanley and El-Halwagi 1995), and metal finishing (El-Halwagi and Manousiouthakis 1990a). In addition, many examples illustrating the detailed application of MEN's and mass integration to pollution prevention are published in the recent (El-Halwagi 1997; El-Halwagi and Spriggs 1998; Dunn and El-Halwagi 2003).

4.8 PROBLEMS

4.1 A processing facility converts scrap tires into fuel via pyrolysis (El-Halwagi 1997). Figure 4-30 is a simplified block flow diagram of the process. The discarded tires are fed to a high-temperature reactor where heat breaks down the hydrocarbon content of the tires into oils and gaseous fuels. The oils are further processed and separated to yield transportation fuels. The reactor off-gases are cooled to condense light oils. The condensate is decanted into two layers: organic and aqueous. The organic layer is mixed with the liquid products of the reactor. The aqueous layer is a wastewater

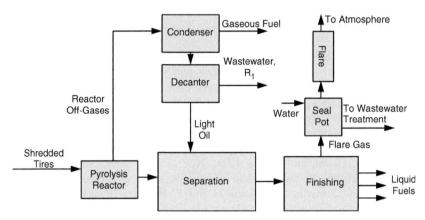

FIGURE 4-30 A SIMPLIFIED BLOCK FLOW DIAGRAM OF A TIRE-TO-FUEL PROCESS (EL-HALWAGI 1997)

TABLE 4-6 DATA FOR THE WASTEWATER STREAM OF TIRE PYROLYSIS PLANT

Stream	Description	Flowrate (kg/s) G_i	Supply composition (ppmw) y_i^s	Target composition (ppmw) y_j^t
R_1	Aqueous layer from decanter	0.2	500	50

stream whose organic content must be reduced prior to discharge. The primary pollutant in the wastewater is a heavy hydrocarbon. The data for the wastewater stream are given in Table 4-6.

A process lean-stream and three external MSAs are considered for removing the pollutant. The process lean-stream is a flare gas (a gaseous stream fed to the flare) which can be used as a process stripping agent. To prevent the back-propagation of fire from the flare, a seal pot is used. An aqueous stream is passed through the seal pot to form a buffer zone between the fire and the source of the flare gas. Therefore, the seal pot can be used as a stripping column in which the flare gas strips the organic pollutant off the wastewater while the wastewater stream constitutes a buffer solution for preventing back-propagation of fire.

Three external MSAs are considered: a solvent extractant (S_2), an adsorbent (S_3), and a stripping agent (S_4). The data for the candidate MSAs are given in Table 4-7. The equilibrium data for the transfer of the pollutant from the waste stream to the jth MSA is given by

$$y_1 = m_j x_j \qquad (4.57)$$

where y_1 and x_j are the mass fractions of the organic pollutant in the wastewater and the jth MSA, respectively.

TABLE 4-7 DATA FOR THE MSAs OF THE TIRE PYROLYSIS PROBLEM

Stream	Upper bound on flowrate L_j^C (kg/s)	Supply composition (ppmw) x_j^s	Target composition (ppmw) x_j^t	m_j	ε_j (ppmw)	C_j ($/kg MSA)
S_1	0.15	200	900	0.5	200	–
S_2	?	300	1000	1.0	100	0.001
S_3	?	10	200	0.8	50	0.020
S_4	?	20	600	0.2	50	0.040

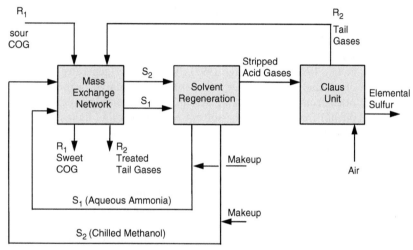

FIGURE 4-31 SWEETENING OF COG (FROM EL-HALWAGI AND MANOUSIOUTHAKIS 1989a)

For the given data, use the pinch diagram to determine the minimum operating cost of the MEN.

4.2 If the fixed cost is disregarded in the previous problem, what is the lowest target for operating cost of the MEN? *Hint*: Set all the ε_j's equal to zero.

4.3 Consider the coke-oven gas "COG" sweetening process shown in Figure 4-31 (El-Halwagi and Manousiouthakis 1989a; El-Halwagi 1997). The basic objective of COG sweetening is the removal of acidic impurities, primarily hydrogen sulfide from COG (a mixture of H_2, CH_4, CO, N_2, NH_3, CO_2, and H_2S). Hydrogen sulfide is an undesirable impurity, because it is corrosive and contributes to SO_2 emission when the COG is burnt. The existence of ammonia in COG and the selectivity of aqueous ammonia in absorbing H_2S suggests that aqueous ammonia is a candidate solvent (process lean-stream, S_1). It is desirable that the ammonia recovered from the sour gas compensates for a large portion of the ammonia losses throughout the system and, thus, reduces the need for ammonia makeup.

Besides ammonia, an external MSA (chilled methanol, S_2) is also available for service to supplement the aqueous ammonia solution as needed.

The purification of the COG involves washing the sour COG, R_1, with sufficient aqueous ammonia and/or chilled methanol to absorb the required amounts of hydrogen sulfide. The acid gases are subsequently stripped from the solvents and the regenerated MSAs are recirculated. The stripped acid gases are fed to a "Claus unit" where elemental sulfur is recovered from hydrogen sulfide. In view of air pollution control regulations, the tail gases leaving the Claus unit, R_2, should be treated for partial removal of the unconverted hydrogen sulfide. Table 4-8 summarizes the stream data.

Using the pinch diagram with $\varepsilon_1 = \varepsilon_2 = 0.0001$, find the minimum cost of MSAs required to handle the desulfurization of R_1 and R_2. Where is the pinch located?

4.4 Figure 4-32 is a simplified flow diagram of an oil refinery (El-Halwagi et al., 1992; El-Halwagi 1997). The process generates two major sources of phenolic wastewater; one from the catalytic cracking unit and the other from the visbreaking system. Two technologies can be used to remove phenol from R_1 and R_2: solvent extraction using light gas oil S_1 (a process MSA) and adsorption using activated carbon S_2 (an external MSA). Table 4-9 provides data for the streams. A minimum allowable composition difference of 0.01 can be used for the two MSAs.

By constructing a pinch diagram for the problem, find the minimum cost of MSAs needed to remove phenol from R_1 and R_2. How do you characterize the point at which both composite streams touch? Is it a true pinch point?

4.5 A processing facility has one rich stream, R_1, which contains a valuable byproduct and two process lean-streams (S_1 and S_2), that can recover the byproduct. Three external MSAs (S_3, S_4, and S_5) are also considered for recovering the byproduct. The data for the rich stream are given in Table 4-10. The data for the candidate MSAs are given in Table 4-11. The equilibrium data for the transfer of the pollutant from the waste stream to the jth MSA is given by

$$y = m_j x_j$$

where y and x_j are the mass fractions of the byproduct in the rich stream and the jth MSA, respectively.

TABLE 4-8 STREAM DATA FOR THE COG-SWEETENING PROBLEM

	Rich stream				MSAs					
Stream	G_i (kg/s)	y_i^s	y_i^t	Stream	L_j^C (kg/s)	x_j^s	x_j^t	m_j	b_j	c_j ($/kg)
R_1	0.90	0.0700	0.0003	S_1	2.3	0.0006	0.0310	1.45	0.000	0.00
R_2	0.10	0.0510	0.0001	S_2	?	0.0002	0.0035	0.26	0.000	0.10

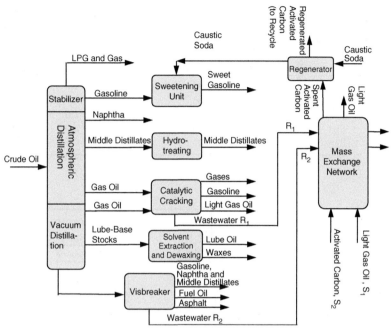

FIGURE 4-32   DEPHENOLIZATION OF REFINERY WASTES (FROM EL-HALWAGI ET AL., 1992)

TABLE 4-9 STREAM DATA FOR REFINERY PROBLEM

	Rich stream				MSAs					
Stream	G_i (kg/s)	y_i^s	y_i^t	Stream	L_i^c (kg/s)	x_j^s	x_j^t	m_j	b_j	c_j ($/kg)
R_1	8.00	0.10	0.01	S_1	10.00	0.01	0.02	2.00	0.00	0.00
R_2	6.00	0.08	0.01	S_2	?	0.00	0.11	0.02	0.00	0.08

TABLE 4-10 DATA FOR THE RICH STREAM

Stream	Description	Flowrate G_i (kg/s)	Supply composition y_i^s (ppmw)	Target composition y_i^t (ppmw)
R_1	Mixture containing byproduct	0.1	1200	100

What is the minimum operating cost of the system ($/year)? Assume there are 8760 operating hours per year.

4.6 Figure 4-33 shows the process flowsheet for an ethylene/ethylbenzene plant (Stanley and El-Halwagi 1994; El-Halwagi 1997). Gas oil is cracked with steam in a pyrolysis furnace to form ethylene, low BTU gases, hexane, heptane, and heavier hydrocarbons. The ethylene is then

TABLE 4-11 DATA FOR THE MSAs

Stream	Upper bound on flowrate (kg/s) L_i^c	Supply composition (ppmw) x_j^s	Target composition (ppmw) x_i^t	m_j	ε_j (ppmw)	C_j ($/kg MSA)
S_1	0.1	500	700	1.0	300	-
S_2	0.3	200	350	2.0	200	-
S_3	∞	50	800	1.0	50	0.010
S_4	∞	400	1200	3.0	100	0.002
S_5	∞	50	2950	0.5	50	0.030

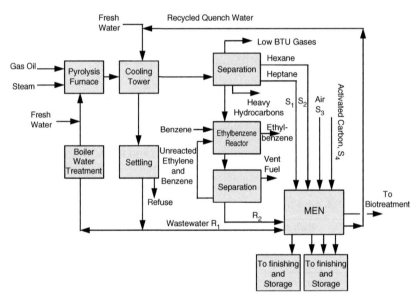

FIGURE 4-33 PROCESS FLOWSHEET FOR AN ETHYLENE/ETHYLBENZENE PLANT (STANLEY AND EL-HALWAGI 1994)

reacted with benzene to form ethylbenzene. Two wastewater streams are formed: R_1 which is the quench water recycle for the cooling tower and R_2 which is the wastewater from the ethylbenzene portion of the plant. The primary pollutant present in the two wastewater streams is benzene. Benzene must be removed from Stream R_1 down to a concentration of 200 ppm before R_1 can be recycled back to the cooling tower. Benzene must also be removed from stream R_2 down to a concentration of 360 ppm before R_2 can be sent to biotreatment. The data for streams R_1 and R_2 are shown in Table 4-12.

There are two process MSAs available to remove benzene from the wastewater streams. These process MSAs are hexane (S_1) and heptane (S_2). Hexane is available at a flowrate of 0.8 kg/s and supply composition

TABLE 4-12 DATA FOR THE WASTE STREAMS OF THE ETHYLBENZENE PLANT (STANLEY AND EL-HALWAGI 1994)

Stream	Description	Flowrate (kg/s) G_i	Supply composition (ppmw) y_i^s	Target composition (ppmw) y_i^t
R_1	Wastewater from settling	100	1000	200
R_2	Wastewater from ethylbenzene separation	50	1800	360

of 10 ppmw, while heptane is available at a flowrate of 0.3 kg/s and supply composition of 15 ppmw. The target compositions for hexane and heptane are unknown and should be determined by the engineer designing the MEN. The mass transfer driving forces, ε_1 and ε_2, should be at least 30,000 and 20,000 ppmw, respectively. The equilibrium data for benzene transfer from wastewater to hexane and heptane are

$$y = 0.011x_1 \tag{4.58}$$

and

$$y = 0.008x_2 \tag{4.59}$$

where y, x_1, and x_2 are given in mass fractions.

Two external MSAs are considered for removing benzene; air (S_3) and activated carbon (S_4). Air is compressed to 3 atm before stripping. Following stripping, benzene is separated from air using condensation. Henry's law can be used to predict equilibrium for the stripping process. Activated carbon is continuously regenerated using steam in the ratio of 1.5 kg steam:1 kg of benzene adsorbed on activated carbon. Makeup at the rate of 1% of recirculating activated carbon is needed to compensate for losses due to regeneration and deactivation. Over the operating range, the equilibrium relation for the transfer of benzene from wastewater onto activated carbon can be described by:

$$y = 7.0 \times 10^{-4}x_4 \tag{4.60}$$

a. Using the pinch diagram determine the pinch location, minimum load of benzene to be removed by external MSAs and excess capacity of process MSAs. How do you remove this excess capacity?
b. Considering the four candidate MSAs, what is the MOC needed to remove benzene?

4.7 Consider the magnetic-tape manufacturing process (Dunn et al., 1995; El-Halwagi 1997) shown in Figure 4-34. First, coating ingredients are dissolved in 0.09 kg/s of organic solvent and mixed to form a slurry. The slurry is suspended with resin binders and special additives. Next, the coating slurry is deposited on a base film. Nitrogen gas is used to induce evaporation rate of solvent that is proper for deposition. In the coating chamber, 0.011 kg/s of solvent are decomposed into other organic species. The decomposed organics are separated from the exhaust gas in a membrane unit. The retentate stream leaving the membrane unit has a flowrate of 3.0 kg/s and is primarily composed of nitrogen that is laden with 1.9 wt/wt% of the organic solvent. The coated film is passed to a dryer where nitrogen gas is employed to evaporate the remaining solvent. The exhaust gas leaving the dryer has a flowrate of 5.5 kg/s and contains 0.4 wt/wt% solvent. The two exhaust gases are mixed and disposed off.

Due to environmental regulations, it is required to reduce the total solvent emission to 0.06 kg/s (by removing 25% of current emission). Three MSAs can be used to remove the solvent from the gaseous emission. The equilibrium data for the transfer of the organic solvent to the jth lean-stream is given by $y = m_j x_j$, where the values of m_j are given in Table 4-13. Throughout this problem, a minimum allowable composition difference of 0.001(kg organic solvent)/(kg MSA) is to be used.

a. Using the pinch diagram, determine which solvent(s) should be employed to remove the solvent? What is the MOC for the solvent removal task? *Hint*: Consider segregating the two waste streams and removing solvent from one of them. The annualized fixed cost

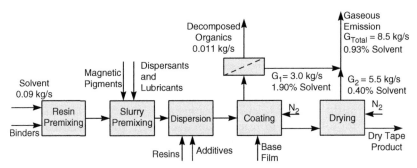

FIGURE   4-34 SCHEMATIC   REPRESENTATION   OF   A   MAGNETIC-TAPE MANUFACTURING PROCESS (EL-HALWAGI 1997)

TABLE 4-13 DATA FOR THE MSAs

Stream	Upper bound on flowrate (kg/s) L_i^C	Supply composition (mass fraction) x_j^s	Target composition (mass fraction) x_j^t	m_j	Mass fraction ε_j	C_j ($/kg MSA)
S_1	∞	0.014	0.040	0.4	0.001	0.002
S_2	∞	0.020	0.080	1.5	0.001	0.001
S_3	∞	0.001	0.010	0.1	0.001	0.002

of a mass exchanger, $/year, may be approximated by 18,000 (Gas Flowrate, kg/s)$^{0.65}$.

b.   The value of the recovered solvent is $0.80/kg of organic solvent. What is the annual gross revenue (annual value of recovered solvent – total annualized cost of solvent recovery system)?

4.8   Consider the oil-recycling plant shown in Figure 4-35 (El-Halwagi 1997). In this plant, two types of waste oil are handled: gas oil and lube oil. The two streams are first deashed and demetallized. Next, atmospheric distillation is used to obtain light gases, gas oil, and a heavy product. The heavy product is distilled under vacuum to yield lube oil. Both the gas oil and the lube oil should be further processed to attain desired properties. The gas oil is steam stripped to remove light and sulfur impurities, then hydrotreated. The lube oil is dewaxed/deasphalted using solvent extraction followed by steam stripping.

The process has two main sources of wastewater. These are the condensate streams from the steam strippers. The principal pollutant in both wastewater streams is phenol. Phenol is of concern primarily because of its toxicity, oxygen depletion, and turbidity. In addition, phenol can cause objectionable taste and odor in fish flesh and potable water.

Several techniques can be used to separate phenol. Solvent extraction using gas oil or lube oil (process MSAs: S_1 and S_2, respectively) is a potential option. Besides the purification of wastewater, the transfer of phenol to gas oil and lube oil is a useful process for the oils. Phenol tends to act as an oxidation inhibitor and serves to improve color stability and reduce sediment formation. The data for the waste streams and the process MSAs are given in Tables 4-14 and 4-15, respectively.

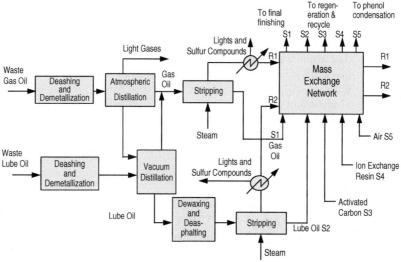

FIGURE 4-35   SCHEMATIC REPRESENTATION OF AN OIL RECYCLING PLANT (EL-HALWAGI 1997)

TABLE 4-14 DATA OF WASTE STREAMS FOR THE DEPHENOLIZATION EXAMPLE (EL-HALWAGI 1997)

Stream	Description	Flowrate G_i (kg/s)	Supply composition y_i^s	Target composition y_i^t
R_1	Condensate from first stripper	2	0.050	0.010
R_2	Condensate from second stripper	1	0.030	0.006

TABLE 4-15 DATA OF PROCESS MSAs FOR THE DEPHENOLIZATION EXAMPLE (EL-HALWAGI 1997)

Stream	Description	Upper Bound on Flowrate (kg/s) L_j^C	Supply composition, x_j^s	Target composition, x_j^t
S_1	Gas oil	5	0.005	0.015
S_2	Lube oil	3	0.010	0.030

TABLE 4-16 DATA FOR THE EXTERNAL MSAs (EL-HALWAGI 1997)

Stream	Upper bound on flowrate (kg/s) L_j^C	Supply composition (mass fraction) x_j^s	Target composition (mass fraction) x_j^t	C_j (\$/kg MSA)
S_3	∞	0.000	0.110	0.081
S_4	∞	0.000	0.186	0.214
S_5	∞	0.000	0.029	0.060

Three external technologies are also considered for the removal of phenol. These processes include adsorption using activated carbon, S_3, ion exchange using a polymeric resin, S_4, and stripping using air, S_5. The equilibrium data for the transfer of phenol to the jth lean-stream is given by $y = m_j x_j$, where the values of m_j are 2.00, 1.53, 0.02, 0.09, and 0.04 for S_1, S_2, S_3, S_4, and S_5, respectively. Throughout this example, a minimum allowable composition difference, ε_j, of 0.001(kg phenol)/ (kg MSA) will be used. The data for the three external MSAs are given in Table 4-16.

What is the minimum operating cost of the MEN?

4.9 SYMBOLS

b_j	intercept of equilibrium line for the jth MSA
C_j	unit cost of the jth MSA including regeneration and makeup, \$/unit flowrate of recirculating MSA
c_j^r	unit cost of the jth MSA required to remove a unit mass/mole of the key pollutant
G_i	flowrate of the ith waste stream

i	index for waste streams
j	index for MSAs
L_j	flowrate of the jth MSA
L_j^C	upper bound on the available flowrate of the jth MSA
m_j	slope of equilibrium line for the jth MSA
MR_i	mass/moles of pollutant lost from the ith waste stream
MS_j	mass/moles of pollutant gained by the jth MSA as defined
N_i	number of independent synthesis problems
N_S	number of MSAs
N_{SE}	number of external MSAs
N_{SP}	number of process MSAs
N_R	number of rich (waste) streams
R_i	the ith waste stream
S_j	the jth MSA
x_j	composition of key component in the jth MSA
x_j^{in}	inlet composition of key component in the jth MSA
$x_j^{in.max}$	maximum practically feasible inlet composition of key component in the jth MSA
$x_j^{in,*}$	maximum thermodynamically feasible inlet composition of key component in the jth MSA
x_j^{out}	outlet composition of key component in the jth MSA
x_j^s	supply composition of the key component in the jth MSA
x_j^t	target composition of the key component in the jth MSA
x_j^*	composition of key component in the jth MSA which is in equilibrium with y_i
y	composition scale for the key component in any waste stream
y_i	composition of key component in the ith waste stream
y_i^s	supply composition of key component in the ith waste stream
y_i^t	target composition of key component in the ith waste stream

Greek

ε_j	minimum allowable composition difference for the jth MSA

4.10 REFERENCES

Alva-Argaez, A., Vallianatos, A., and Kokossis, A. 1999, 'A multi-contaminant transhipment model for mass exchange networks and wastewater minimisation problems', *Comp. Chem. Eng.*, vol. 23, no. 10, pp. 1439-1453.

Alves, J.J. and Towler, G.P. 2002, 'Analysis of Refinery Hydrogen Distribution Systems', *Ind. Eng. Chem. Res.*, no. 41, 5759-5769.

Dhole, V.R., Ramchandani, N., Tainsh, R.A., Wasilewski, M. 1996, 'Make your process water pay for itself', *Chem. Eng.*, no. 103, 100-103.

Dunn, R.F., El-Halwagi, M.M., Lakin, J., and Serageldin, M. 1995, 'Selection of Organic Solvent Blends for Environmental Compliance in the Coating Industries', *Proceedings of the First International Plant Operations and Design Conference*, eds. E.D. Griffith, H. Kahn and M.C. Cousins, vol. III, pp. 83-107

Dunn, R.F. and El-Halwagi, M.M. 1993, 'Optimal recycle/reuse policies for minimizing the wastes of pulp and paper plants', *Environ. Sci. Health* vol. A28, no. 1, pp. 217-234.

Dunn, R.F., and El-Halwagi, M.M. 2003, 'Process integration technology review, background and applications in the chemical process industry', *J. Chem. Tech. And Biotech*, vol. 78, pp.1011-1021.

El-Halwagi, A.M., and El-Halwagi, M.M. 1992, 'Waste minimization via computer aided chemical process synthesis-A new design philosophy', *TESCE J.* vol. 18, no. 2, pp. 155-187.

El-Halwagi, M.M. 1997, *Pollution Prevention through Process Integration: Systematic Design Tools*, Academic Press, San Diego.

El-Halwagi, M.M., Hamad, A.A., and Garrison, G.W. 1996, 'Synthesis of Waste Interception and Allocation Networks', *AIChE J.*, vol. 42, no. 11, pp. 3087-3101.

El-Halwagi, M.M., El-Halwagi, A.M., and Manousiouthakis, V. 1992, 'Optimal design of dephenolization networks for petroleum-refinery wastes', *Trans. Inst. Chem. Eng.* vol. 70, Part B, pp. 131-139.

El-Halwagi, M.M., and Manousiouthakis, V. 1989a, 'Synthesis of mass exchange networks', *AIChE J.* vol. 35, no. 8, pp. 1233-1244.

El-Halwagi, M.M., and Manousiouthakis, V. 1989b, *Design and Analysis of Mass-Exchange Networks with Multicomponent Targets*, The AIChE Annual Meeting, San Francisco, Nov. 5-10.

El-Halwagi, M.M., and Manousiouthakis, V. 1990a, 'Automatic synthesis of mass exchange networks with single-component targets', *Chem. Eng. Sci.* vol. 45, no. 9, pp. 2813-2831.

El-Halwagi, M.M., and Manousiouthakis, V. 1990b, 'Simultaneous synthesis of mass exchange and regeneration networks', *AIChE J.* vol. 36, no. 8, pp. 1209-1219.

El-Halwagi, M.M., and Spriggs, H.D. 1998, 'Solve design Puzzles with Mass Integration', *Chem. Eng. Prog.*, vol. 94, August, pp. 25-44.

El-Halwagi, M.M., and Srinivas, B.K. 1992, 'Synthesis of reactive mass-exchange networks', *Chem. Eng. Sci.* vol. 47. no. 8, pp. 2113-2119.

Foo, C.Y., Manan, A.A., Yunus, R.M., and Aziz, R.A. 2004. 'Synthesis of mass exchange network for batch processes - Part 1: Utility targeting', *Chem. Eng. Sci.*, vol. 59, no. 5, pp.1009-1026.

Fraser, D.M., and Shenoy, U.V. 2004, 'A new method for sizing mass-exchange units without the singularity of the kremser equation', *Comp. Chem. Eng.*, vol. 15, pp. 2331-2335.

Garrard, A., and Fraga, E.S. 1998, 'Mass exchange network synthesis using genetic algorithms', *Comp. Chem. Eng.*, vol. 22, no. 12, pp. 1837-1850.

Garrison, G.W., Cooley, B.L., and El-Halwagi, M. M. 1995, 'Synthesis of mass exchange networks with multiple target mass separating agents', *Dev. Chem. Eng. Miner. Proc.* vol. 3, no. 1, 31-49.

Hallale, N., and Fraser, D.M. 1997, 'Synthesis of cost optimum gas treating process using pinch analysis', *Proceedings, Top. Conf. on Sep. Sci. and Tech.*, eds. W.S. Ho, and R.G. Luo, *AIChE*, New York, Part II, pp. 1,708–1,713.

Hallale, N. and Fraser, D. 2000, 'Supertargeting for mass exchange networks Part I: targeting and design techniques', *Chem. Eng. Res. Des.*, vol. 78, no. A2, pp. 202-207.

Hamad, A.A., Varma, V., Krishnagopalan, G., and El-Halwagi, M.M. 1998, 'Mass integration anaylsis: a technique for reducing methanol and effluent discharge in pulp mills', *TAPPI J*, no. 81, pp. 170-179.

Hamad, A.A., Varma, V. El-Halwagi, M.M., and Krishnagopalan, G. 'Systematic integration of source reduction and recycle/reuse for the cost-effective compliance with the cluster rules', *AIChE Annual Meeting*, Miami, November 1995.

Hamad, A.A., Garrison, G.W., Crabtree, E.W., and El-Halwagi, M.M. 1996, 'Optimal design of hybrid separation systems for waste reduction', *Proceedings of the Fifth World Congress of Chemical Engineering*, San Diego, Vol. III, pp. 453-458.

Huang, Y.L. and Fan, L.T. 1995, 'Intelligent process design and control for in-plant waste minimization' in *Waste Minimization Through Process Design*, eds. A.P. Rossiter, McGraw Hill, New York, pp. 165-180.

Huang, Y.L. and Edgar, T.F. 1995, 'Knowledge based design approach for the simultaneous minimization of waste generation and energy consumption in a petroleum refinery' in *Waste Minimization Through Process Design*, eds. A.P. Rossiter, McGraw Hill, New York, pp. 181-196.

Kiperstok, A. and Sharratt, P.N. 1995, 'On the optimization of mass exchange networks for removal of pollutants', *Trans. Inst. Chem. Eng.* vol. 73, Part B, pp. 271-277.

Mann, J. and Liu, Y.A. 1999, 'Industrial Water Reuse and Wastewater Minimization,' McGraw Hill.

Papalexandri, K.P. and Pistikopoulos, E.N. 1994, 'A multiperiod MINLP model for the synthesis of heat and mass exchange networks', *Comput. Chem. Eng.* vol. 18, no. 12, pp. 1125-1139.

Sebastian P., Nadeau, J.P., and Puiggali, J.R. 1996, 'Designing dryers using heat and mass exchange networks: An application to conveyor belt dryers', *Chem. Eng. Res. Des.*, vol. 74, no. A8: pp. 934-943.

Srinivas, B.K. and El-Halwagi, M.M. 1994a, 'Synthesis of reactive mass-exchange networks with general nonlinear equilibrium functions' *AIChE J.* vol. 40, no. 3, pp. 463-472.

Srinivas, B.K. and El-Halwagi, M.M. 1994b, 'Synthesis of combined heat reactive mass-exchange networks', *Chem. Eng. Sci.* vol. 49, no. 13, pp. 2059-2074.

Stanley, C. and El-Halwagi, M.M. 1995, 'Synthesis of mass exchange networks using linear programming techniques' in *Waste Minimization Through Process Design*, eds. A.P. Rossiter, and McGraw Hill, New York, pp. 209-224..

Szitkai Z., Lelkes Z., Rev., Z.E., and Fonyo, Z. 2002, 'Handling of removable discontinuities in MINLP models for process synthesis problems, formulations of the Kremser equation', *Comp. Chem. Eng.*, vol. 26, no. 11, pp. 1501-1516.

Wang, Y.P., and Smith, R. 1994, Wastewater minimization. *Chem. Eng. Sci.* vol. 49, no. 7, pp. 981-1006.

Warren, A., Srinivas, B.K., and El-Halwagi, M.M. 1995, 'Design of cost-effective waste-reduction systems for synthetic fuel plants'. *J. Environ. Eng.* vol. 121, no. 10, pp.742-747.

Xue, D.F., S.J. Li, Yuan Y., and Yao, P.J. 2000, 'Synthesis of waste interception and allocation networks using genetic-alopex algorithm', *Comp. Chem. Eng.*, vol. 24, no. 2-7, pp.1455-1460

Zhu, M. and El-Halwagi, M.M. 1995, 'Synthesis of flexible mass exchange networks', *Chem. Eng. Commun.* vol. 138, pp. 193-211.

5

VISUALIZATION TECHNIQUES FOR THE DEVELOPMENT OF DETAILED MASS-INTEGRATION STRATEGIES

Chapter Two provided holistic techniques for identifying overall targets for benchmarking mass consumption and discharge. These targets are determined ahead of detailed design. Once a target is determined, it is necessary to develop cost-effective strategies to reach the target. For a given target, there are numerous design decisions that must be judiciously made. These include addressing the following challenging questions:

- What are the optimum stream-rerouting strategies?
- Should any streams be segregated, mixed, or rerouted? Which ones?
- Which streams should be recycled/reused? To which units?
- Should design variable and/or operating conditions of existing units be altered? Which ones? To what extent?

- Is there a need to add/replace units? Which ones? Where to add/replace?
- Should interception (e.g., separation) devices be added? Which streams should be intercepted? To remove what? To what extent?
- Which separating agents should be selected for interception?
- What is the optimal flowrate of each separating agent?
- How should these separating agents be matched with the rich streams (i.e., stream pairings)?

Because of the prohibitively large number of alternatives involved in answering these questions, a systematic procedure is needed to extract the optimum solution(s) without enumerating them. This is the role provided by mass integration. Mass integration is a holistic and systematic methodology that provides a fundamental understanding of the global flow of mass within the process and employs this understanding in identifying performance targets and optimizing the allocation, separation, and generation of streams and species. Mass integration is based on fundamental principles of chemical engineering combined with system analysis using graphical and optimization-based tools. In order to develop detailed mass-integration strategies, let us represent the process flowsheet from a species viewpoint (e.g., El-Halwagi and Spriggs 1998; El-Halwagi et al., 1996) as shown in Figure 5-1. For each targeted species, there are sources (streams that carry the species) and process sinks (units that can accept the species). Process sinks include reactors, separators, heaters/coolers, biotreatment facilities, and discharge media. Streams leaving the sinks become, in turn, sources. Therefore, sinks are also generators of the targeted species. Each sink/generator may be manipulated via design and/or operating changes to affect the flowrate and composition of what each sink/generator accepts and discharges.

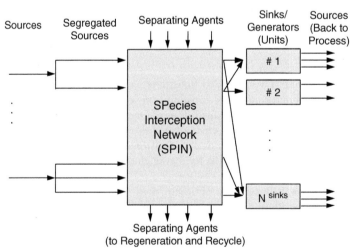

FIGURE 5-1   PROCESS FROM A SPECIES PERSPECTIVE (EL-HALWAGI ET AL., 1996; GARRISON ET AL., 1996)

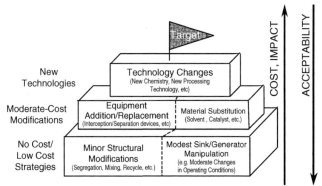

FIGURE 5-2   HIERARCHY OF MASS-INTEGRATION STRATEGIES (EL-HALWAGI 1999)

Properties of sources (e.g., flowrate, composition, pressure, temperature, etc.) can be modified by adding new units that intercept the streams prior to being fed to the process sinks and condition their properties to the desired values. This is performed by a species-interception network (SPIN) that may use mass and energy separating agents.

There are three main factors used in screening candidate mass-integration strategies: economics, impact, and acceptability. Economics can be assessed by a variety of criteria such as cost, return on investment, payback period, etc. Impact is a measure of the extent of the solution strategy to (partially) attain the desired target. Acceptability is a measure of how likely the proposed strategies to be accepted and implemented by the plant. This factor is plant based as it differs from one company to the other depending on their operating policies, corporate environment, and level of comfort within the plant to adopt process modifications. Based on these factors, one can classify candidate strategies via mass integration into three categories (Figure 5-2):

- No/low cost changes
- Moderate cost modifications, and
- New technologies

These strategies are typically in ascending order of cost and impact and in descending order of acceptability. The following sections provide more details on these strategies.

5.1 LOW/NO COST STRATEGIES

These strategies can be broadly classified into two categories: structural and parametric modifications. The structure-based changes pertain to low/no cost in process configuration such as stream rerouting (e.g., segregation and recycle) which involves piping and pumping primarily. Parametric changes include moderate adjustments in process variables and operating conditions (e.g., temperature, pressure, etc.) which require no or modest capital expenditure. Let us further define these strategies:

Segregation simply refers to avoiding the mixing of streams. In many cases, segregating waste streams at the source renders several streams environmentally acceptable and hence reduces the pollution prevention cost. Furthermore, segregating streams with different compositions avoid unnecessary dilution of streams. This may reduce the cost of removing the targeted species from the segregated streams. It may also provide composition levels that allow the streams to be recycled directly to process units.

Recycle refers to the utilization of a process stream (a source) in a process unit (a sink). As described in Chapter Three, each sink has a number of constraints on the characteristics (e.g., flowrate and composition) of feed that it can process. If a source satisfies these constraints, it may be directly recycled to or reused in the sink. However, if the source violates these constraints, segregation, mixing, and/or interception may be used to prepare the stream for recycle. Source–sink mapping diagram and material recycle pinch diagram can be used to determine targets and detailed options for fresh usage, recycle/reuse, and waste discharge.

5.2 MODEST CHANGES IN PROCESS VARIABLES AND OPERATING CONDITIONS

In order to relate process performance with changes in operating conditions, it is necessary to use an analytical tool that simulates the input–output relations of selected units and can track components in the process and determine what design and operating condition changes are required to meet process targets.

5.3 MEDIUM-COST STRATEGIES AND MAIN TECHNOLOGY CHANGES

In addition to the aforementioned low-cost strategies, it is to attain the target by replacing or adding new equipment and using alternative species. Substantial technology changes can also be implemented. These include major changes in processing scheme including the application of new chemical pathways and core changes in process chemistry. Alternative reactions and molecular design techniques will be discussed later. Here, we focus on equipment addition along with their associated material utilities (e.g., separating agents). *Interception* denotes the utilization of new unit operations to adjust the composition, flowrate, and other properties of certain process streams to make them acceptable for existing process sinks. A particularly important class of interception devices is separation systems. These separations may be induced by the use of mass separating agents (MSAs) and/or energy separating agents (ESAs). A systematic technique is needed to screen the multitude of separating agents and separation technologies to find the optimal separation system. The synthesis of MSA-induced physical-separation systems has been covered in Chapter Four through the synthesis of mass exchange networks (MENs). Other interception systems are covered

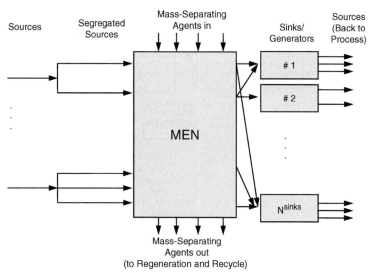

Sources   Segregated   Mass-Separating   Sinks/   Sources
           Sources   Agents in   Generators (Back to
                                                       Process)

MEN

#1

#2

N^{sinks}

Mass-Separating
Agents out
(to Regeneration and Recycle)

FIGURE 5-3   A PROCESS FROM A SPECIES VIEWPOINT WHEN MSAs ARE USED FOR INTERCEPTION (EL-HALWAGI ET AL., 1996)

throughout the book. The MEN can be used to prepare sources for recycle. When MSAs are used for interception, Figure 5-1 is revised as shown in Figure 5-3 by placing a MEN instead of the SPIN.

EXAMPLE 5-1 APPLICATION OF MASS INTEGRATION TO DEBOTTLENECK AN ACRYLONITRILE PROCESS AND REDUCE WATER USAGE AND DISCHARGE (EL-HALWAGI 1997)

Acrylonitrile (AN, C_3H_3N) is manufactured via the vapor-phase ammoxidation of propylene:

$$C_3H_6 + NH_3 + 1.5O_2 \overset{catalyst}{\rightarrow} C_3H_3N + 3H_2O$$

The reaction takes place in a fluidized-bed reactor in which propylene, ammonia, and oxygen are catalytically reacted at 450°C and 2 atm. The reaction is a single pass with almost complete conversion of propylene. The reaction products are cooled using an indirect-contact heat exchanger which condenses a fraction of the reactor off-gas. The remaining off-gas is scrubbed with water, then decanted into an aqueous layer and an organic layer. The organic layer is fractionated in a distillation column under slight vacuum which is induced by a steam-jet ejector. Figure 5-4 shows the process flowsheet along with pertinent material balance data.

The wastewater stream of the plant is composed of the off-gas condensate, the aqueous layer of the decanter, the bottom product of the distillation column, and the condensate from the steam-jet ejector. This wastewater stream is fed to the biotreatment facility. Since the biotreatment facility is currently operating at full hydraulic capacity, it constitutes

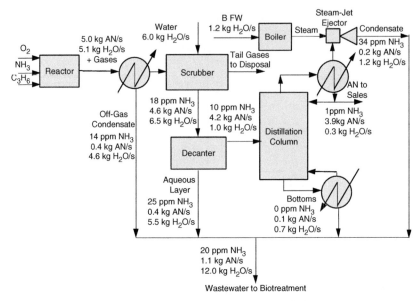

■■■■■■   **FIGURE 5-4   FLOWSHEET OF AN PRODUCTION (EL-HALWAGI 1997)**

a bottleneck for the plant. Plans for expanding production of AN are contingent upon debottlenecking of the biotreatment facility by reducing its influent or installing an additional treatment unit. The new biotreatment facility will cost about $4 million in capital investment and $360,000/year in annual operating cost, leading to a TAC of $760,000/year with a 10-year linear depreciation. The objective of this case study is to use mass integration techniques to devise cost-effective strategies to debottleneck the biotreatment facility.

The following technical constraints should be observed in any proposed solution:

Scrubber

$$5.8 \leq \text{flowrate of wash feed (kg/s)} \leq 6.2 \qquad (5.1)$$

$$0.0 \leq \text{ammonia content of wash feed (ppm NH}_3) \leq 10.0 \qquad (5.2)$$

Boiler Feed Water (BFW)

$$\text{Ammonia content of BFW (ppm NH}_3) = 0.0 \qquad (5.3)$$

$$\text{AN content of BFW (ppm AN)} = 0.0 \qquad (5.4)$$

Decanter

$$10.6 \leq \text{flowrate of feed (kg/s)} \leq 11.1 \qquad (5.5)$$

Distillation Column

$$5.2 \leq \text{flowrate of feed (kg/s)} \leq 5.7 \qquad (5.6)$$

$$0.0 \leq \text{ammonia content of feed (ppm NH}_3) \leq 30.0 \qquad (5.7)$$

$$80.0 \leq \text{AN content of feed (wt\% AN)} \leq 100.0 \qquad (5.8)$$

Furthermore, for quality and operability objectives, the plant does not wish to recycle the AN product stream (top of distillation column), the feed to the distillation column and, the feed to the decanter.

Three external MSAs are considered for removing ammonia from water; air (S_1), activated carbon (S_2), and an adsorbing resin (S_3). The data for the candidate MSAs are given in Table 5-1. The equilibrium funtion for the transfer of the pollutant from the waste stream to the jth MSA is given by,

$$y_1 = m_j x_j \qquad (5.9)$$

where y_1 and x_j are weight-based parts per million of ammonia in the wastewater and the jth MSA, respectively.

SOLUTION

The first step in the analysis is to identify the target for debottlenecking the biotreatment facility. An overall water balance for the plant (Figure 5-5a) can be written as follows:

$$\text{Water in} + \text{Water generated by chemical reaction}$$
$$= \text{Wastewater out} + \text{Water losses}$$

Since the wastewater discharge is larger than fresh water flowrate, it is possible, in principle, to bring wastewater to a quality that can substitute fresh water using segregation, mixing, recycle, and interception. Furthermore, sink/generator manipulation can be employed to reduce the flowrate of fresh water. Hence, fresh water usage in this example can in principle be completely eliminated, and for the same reaction conditions

TABLE 5-1 DATA FOR MSAs OF THE AN PROBLEM

Stream	Upper bound on flowrate L_j^C	Supply composition (ppmw) x_j^s	Target composition (ppmw) x_j^t	m_j	ε_j (ppmw)	C_j ($/kg MSA)	C_j'($/kg NH$_3$) removed
S_1	∞	0	6	1.4	2	0.004	667
S_2	∞	10	400	0.02	5	0.070	180
S_3	∞	3	1100	0.01	5	0.100	91

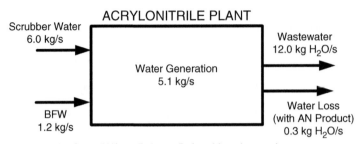

(a) Overall Water Balance Before Mass Integration

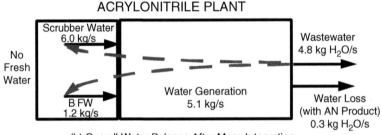

(b) Overall Water Balance After Mass Integration

FIGURE 5-5 ESTABLISHING TARGETS FOR BIOTREATMENT INFLUENT: OVERALL WATER BALANCE (a) BEFORE AND (b) AFTER MASS INTEGRATION (EL-HALWAGI 1997)

and water losses, the target for wastewater discharge can be calculated from the overall water balance as follows (Figure 5-5b):

$$
\text{Target of minimum discharge to biotreatment} = 5.1 - 0.3
$$
$$
= 4.8 \text{ kg } H_2O/s \tag{5.10}
$$

As can be seen from Figure 5-5b, the corresponding target for minimum usage of fresh water is potentially zero.

Figure 5-6 illustrates the gap between the current process performance and the benchmarked mass targets. Having identified this target, let us now determine how to best attain the target. It is also necessary to sequence and integrate the solution strategies.

First, we develop no/low cost strategies starting with minor process modifications. An operating parameter that can be altered to reduce fresh water usage (and consequently wastewater discharge) is the flowrate of the feed to the scrubber. As given by constraint, equation (5.1), the flowrate of fresh water fed to the first scrubber may be reduced to 5.8 kg/s. This is a minor process modification which involves setting the flow control valve. Compared to the current usage of 6.0 kg/s, the net result is a reduction of 0.2 kg/s in water usage (and discharge). These results are shown in Figure 5-7. To track water, the wastewater discharge is described

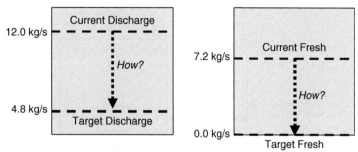

FIGURE 5-6   BENCHMARKING WATER USAGE AND DISCHARGE

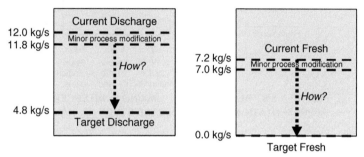

FIGURE 5-7   REDUCTION IN WATER USAGE AND DISCHARGE (EXPRESSED AS Kg H₂O/s) AFTER MINOR PROCESS MODIFICATION

in terms of kg H_2O/s (not as total flowrate of wastewater including other species).

Next, we consider other no/low cost strategies including segregation, mixing, and direct-recycle opportunities. First, we identify the relevant sources and sinks. Once the streams composing the terminal wastewater are segregated, we get four sources that can be potentially recycled. Fresh water used in the scrubber and the boiler provides two more sources. In order to reduce wastewater discharge to biotreatment, fresh water must be reduced. Hence, we should focus our attention on recycling opportunities to sinks that employ fresh water, namely, the scrubber and the boiler. Figure 5-8 illustrates the sources and sinks involved in the analysis.

Direct-recycle strategies are based on segregating, rerouting, and mixing of sources without the use of new equipment. Hence, there is no SPIN involved in direct recycle. Therefore, for direct recycle, Figure 5-8 is revised by eliminating the SPIN as shown in Figure 5-9. Because of the stringent limitation on the BFW (no ammonia or AN), no recycled stream can be used in lieu of fresh water (segregation, mixing, recycle, and interception can reduce, but not eliminate, ammonia/AN content). Consequently, in Figure 5-9 none of the process sources are allocated to the boiler.

Hence, the boiler should not be considered as a sink for recycle (with or without interception). Instead, it should be handled at the stage of

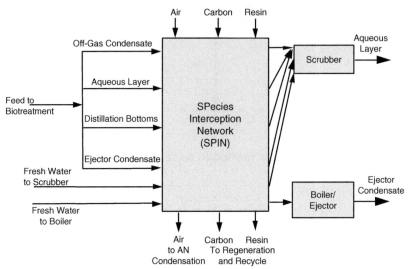

FIGURE 5-8 SEGREGATION, MIXING, INTERCEPTION, AND RECYCLE REPRESENTATION FOR THE AN CASE STUDY (EL-HALWAGI 1997)

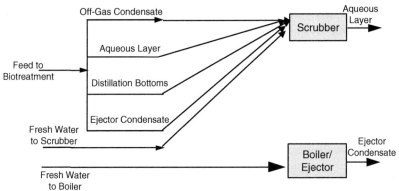

FIGURE 5-9 SCHEMATIC REPRESENTATION OF SEGREGATION, DIRECT RECYCLE, AND MIXING FOR THE AN EXAMPLE (EL-HALWAGI 1997)

sink/generator manipulation. This leaves us with the five segregated sources and two sinks (boiler and scrubber). The data for the sources and sinks are summarized in Tables 5-2 and 5-3, respectively. Using these data, the material recycle pinch diagram is constructed by first developing the sink composite curve (Figure 5-10), then the source composite curve (Figure 5-11), and the material recycle pinch diagram (Figure 5-12). As can be seen from Figure 5-13, when direct recycle is used, the following targets can be obtained:

$$\text{Minimum fresh water} = 2.1 \text{ kg/s} \qquad (5.11a)$$

TABLE 5-2 SINK DATA FOR THE AN EXAMPLE

Sink	Flowrate (kg/s)	Maximum inlet composition of NH_3 (ppm)	Maximum inlet load (10^{-6} kg NH_3/s)
BFW	1.2	0.0	0.0
Scrubber	5.8	10.0	58.0

TABLE 5-3 SOURCE DATA FOR THE AN EXAMPLE

Source	Flowrate (kg/s)	Inlet composition of NH_3(ppm)	Inlet load (10^{-6} kg NH_3/s)
Distillation bottoms	0.8	0.0	0.0
Off-gas condensate	5.0	14	70.0
Aqueous layer	5.7[1]	25	142.5
Jet-ejector condensate	1.4	34	47.6

$$\text{Maximum direct recycle} = 4.9 \text{ kg/s} \qquad (5.11b)$$

$$\text{Minimum waste discharge} = 8.0 \text{ kg/s} \qquad (5.11c)$$

As mentioned earlier, to track water, it is useful to express the flowrate of waste discharge as kg H_2O/s (not as total flow including other species). This can be readily calculated by noting that the reduction in fresh water usage is equal to the reduction in wastewater discharge (expressed as kg H_2O/s). Since,

$$\text{Reduction in fresh water usage} = 7.2 - 2.1 = 5.1 \text{ kg/s} \qquad (5.11d)$$

We have

$$\text{Minimum waste discharge (expressed as kg } H_2O)/s) = 12.0 - 5.1$$
$$= 6.9 \text{ kg } H_2O/s^2 \qquad (5.11e)$$

---

[1]Since the flowrate of the feed to the scrubber has been reduced by 0.2 kg/s as a result of minor process modification, the flowrate of the aqueous layer is assumed to decrease from 5.9 to 5.7 kg/s.

[2]The difference between the wastewater discharge expressed as total flowrate (8.0 kg/s) versus that expressed in terms of H_2O (6.9 kg/s) accounts for the 1.1 kg/s of collective AN losses in the wastewater.

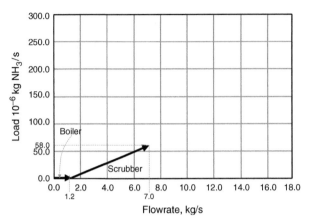

FIGURE 5-10   SINK COMPOSITE DIAGRAM FOR THE AN EXAMPLE

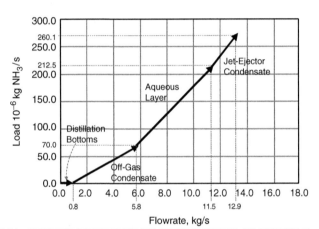

FIGURE 5-11   SOURCE COMPOSITE DIAGRAM FOR THE AN EXAMPLE

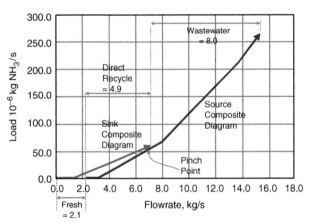

FIGURE 5-12   MATERIAL RECYCLE PINCH DIAGRAM FOR THE AN EXAMPLE

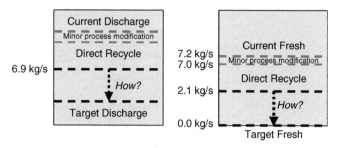

FIGURE 5-13   REDUCTION IN WATER USAGE AND DISCHARGE (EXPRESSED AS kg H_2O/s) AFTER MINOR PROCESS MODIFICATION AND DIRECT RECYCLE

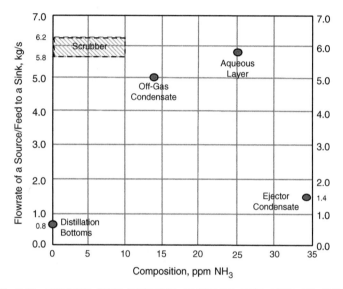

FIGURE 5-14   SOURCE–SINK MAPPING DIAGRAM FOR THE AN EXAMPLE (EL-HALWAGI 1997)

It is beneficial to detail the direct-recycle strategies to attain the identified targets. Towards that end, the source–sink mapping diagram can be used to identify which sources should be recycled to which sinks. Figure 5-14 is a source–sink mapping representation of the problem. As mentioned earlier, because of the stringent requirements on the feed to the boiler sink (0.0 composition of any pollutant), it is not possible to replace any of the feed to the boiler through recycle (even with interception). Consequently, we will not represent the boiler sink on the source–sink mapping diagram. Following the lever-arm rules including sink-feed conditions and source prioritization rules described in Chapter Three, the following statements can be made:

- The composition of the feed to the scrubber should be set to its maximum value (10 ppm).

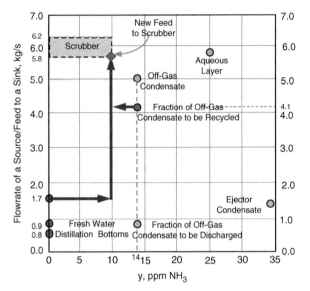

FIGURE 5-15 DIRECT-RECYCLE STRATEGIES FOR THE AN EXAMPLE (EL-HALWAGI 1997)

- The use of the sources should be prioritized as follows: distillation bottoms, off-gas condensate, aqueous layer, then jet-ejector condensate.

Hence, we start by considering the use of the first two sources (shortest fresh arms): distillation bottoms and off-gas condensate. The flowrate resulting from combining these two sources (5.8 kg/s) is sufficient to run the scrubber. However, its ammonia composition[3] as determined by the lever-arm principle is 12.1 ppm, which lies outside the zone of permissible recycle to the scrubber. As shown in Figure 5-15, the maximum flowrate of the off-gas condensate to be recycled to the scrubber[4] is determined as follows:

$$\frac{\text{Arm of gas condensate}}{\text{Total arm}} = \frac{\text{Flowrate of recycled gas condensate}}{\text{Flowrate of scrubber feed}} \qquad (5.12)$$

i.e.,

$$\frac{10-0}{14-0} = \frac{\text{Flowrate of recycled gas condensate}}{5.8}$$

---

[3]Algebraically, this composition can be calculated as follows: ((5.0 kg/s)(14 ppm NH)+0)/(5.8 kg/s) = 12.1 ppm NH$_3$.

[4]Again, algebraically this flowrate can be calculated as follows: (Flowrate of recycled off-gas condensate x 14 ppm   NH$_3$ + 0 + 0)/(5.8 kg/s) = 10 ppm NH$_3$. Where the numerator represents the ammonia in recycled off-gas condensate, distillation bottoms (none), and fresh water (none). Hence, flowrate of recycled off-gas condensate = 4.14 kg/s.

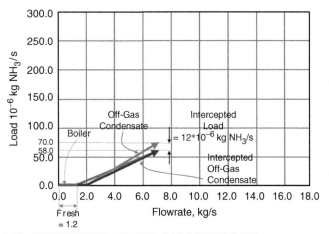

FIGURE 5-16   INTERCEPTING THE OFF-GAS CONDENSATE

Hence, the flowrate of recycled gas condensate $= 4.1\,\text{kg/s}$ and the flowrate of fresh water is $0.9\,\text{kg/s}$ $(5.8 - 0.8 - 4.1)$. Therefore, direct recycle can reduce the fresh water consumption (and consequently the influent to biotreatment) by $5.1\,\text{kg/s}$. This is exactly the same target identified from the material recycle pinch analysis as described by equation (5.11d).

With the details of the direct-recycle strategies determined, it is possible to determine the implementation cost. The primary cost of direct recycling is pumping and piping. Assuming that the TAC for pumping and piping is $80/\text{m} \cdot \text{year}$ and assuming that the total length of piping is $600\,\text{m}$, the TAC for pumping and piping is \$48,000/year.

Since not all the off-gas condensate has been recycled, there is no need to consider recycle from any other source (since they have longer fresh arms). Therefore, there are no more direct-recycle opportunities and we have exhausted the no/low cost strategies. Next, we move to adding new units and we consider interception.

Before screening interception devices, it is necessary to determine the interception task. For all of the off-gas condensate to be recycled to the boiler, its ammonia content has to be reduced. As can be seen from Figure 5-16, in order to fully recycle the off-gas condensate, the ammonia load to be removed is $12 \times 10^{-6}\,\text{kg/s}$. The composition of the intercepted off-gas condensate is the slope of the intercepted stream which is 11.6 ppm. Therefore, the interception task is to reduce the composition of ammonia in the off-gas condensate from 14.0 ppm to 11.6 ppm. The same result can be obtained algebraically as follows:

Load removed from off-gas condensate = flowrate of off-gas condensate

$\times$ (supply composition − target composition)

$$(5.13a)$$

i.e.

$$12 \times 10^{-6} = 5.0 \times (14.0 - \text{target composition})$$

Therefore,

Target (intercepted) composition of ammonia in off-gas condensate

$$= 11.6 \text{ ppm}$$

(5.13b)

The same result may also be obtained through the source–sink mapping diagram as shown in Figure 5-17. Alternatively, it may be calculated as follows:

$$\frac{(5.0 \text{ kg/s})y^t \text{ ppm NH}_3 + 0}{5.8 \text{ kg/s}} = 10.0 \text{ ppm NH}_3$$

(5.13c)

i.e. $y^t = 11.6$ ppm

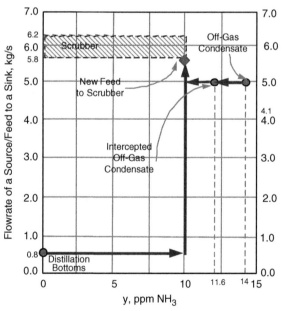

FIGURE 5-17   DETERMINATION OF INTERCEPTION TASK FOR THE OFF-GAS CONDENSATE (EL-HALWAGI 1997)

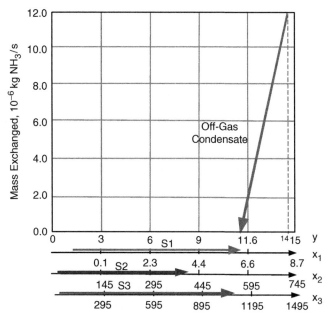

FIGURE 5-18   THE PINCH DIAGRAM FOR THE AN CASE STUDY (EL-HALWAGI 1997)

In order to synthesize an optimal MEN for intercepting the off-gas condensate, we construct the pinch diagram as shown in Figure 5-18. Since the three MSAs lie completely to the left of the rich stream, they are all thermodynamically feasible. Hence, we choose the one with the least cost ($/kg NH_3 removed); namely the resin. The annual operating cost for removing ammonia using the resin is:

$$5 \, \text{kg Liquid/s} \times (14.0 \times 10^{-6} - 11.6 \times 10^{-6}) \, \text{kg NH}_3/\text{kg Liquid}$$
$$\times 91\$/\text{kg NH}_3 \times 3600 \times 8760 \, \text{s/year} = \$34,437/\text{year} \tag{5.14}$$

The annualized fixed cost of the adsorption column along with its ancillary equipment (e.g., regeneration, materials handling, etc.) is estimated to be about $85,000/year. Therefore, the TAC for the interception system is $119,437/year.

As a result of minor process modification, segregation, interception, and recycle, we have eliminated the use of fresh water in the scrubber, leading to a reduction in fresh water consumption (and influent to biotreatment) by 6.0 kg/s. Therefore, the target for segregation, interception, and recycle has been realized (Figure 5-19).

Next, we focus our attention on sink/generator manipulation to remove fresh water consumption in the steam-jet ejector. The challenge here is to alter the design and/or operation of the boiler, the ejector, or the distillation

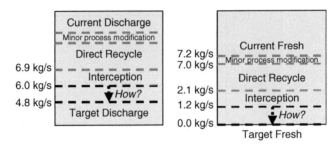

FIGURE 5-19   REDUCTION IN WATER USAGE AND DISCHARGE (EXPRESSED AS kg H$_2$O/s) AFTER MINOR PROCESS MODIFICATION, DIRECT RECYCLE, AND INTERCEPTION

column to reduce or eliminate the use of steam. Several solutions may be proposed including:

- Replacing of the steam-jet ejector with a vacuum pump. The distillation operation will not be affected. The operating cost of the ejector and the vacuum pump are comparable. However, a capital investment of $75,000 is needed to purchase the pump. For a five-year linear depreciation with negligible salvage value, the annualized fixed cost of the pump is $15,000/year.
- Operating the column under atmospheric pressure, thereby eliminating the need for the vacuum pump. Here a simulation study is needed to examine the effect of pressure change.
- Relaxing the requirement on BFW quality to a few parts per million of ammonia and AN. In this case, recycle and interception techniques can be used to significantly reduce the fresh water feed to the boiler and, consequently, the net wastewater generated.

Figure 5-20 illustrates the revised flowsheet with segregation, interception, recycle, and sink/generator manipulation. As can be seen from the figure, the flowrate of the terminal wastewater stream has been reduced to 4.8 kg H$_2$O/s. This is exactly the same *target* predicted in Figure 5-5b. In order to refine the material balance throughout the plant, a simulation study is needed, as discussed in Chapter One.

Figures 5-21 and 5-22 are impact diagrams (sometime referred to as Pareto charts) for the reduction in wastewater and the associated TAC. These figures illustrate the cumulative impact of the identified strategies on biotreatment influent and cost.

We are now in a position to discuss the merits of the identified solutions. As can be inferred from Figure 5-20, the following benefits can be achieved:

- Acrylonitrile production has increased from 3.9 kg/s to 4.6 kg/s, which corresponds to an 18% yield enhancement for the plant.

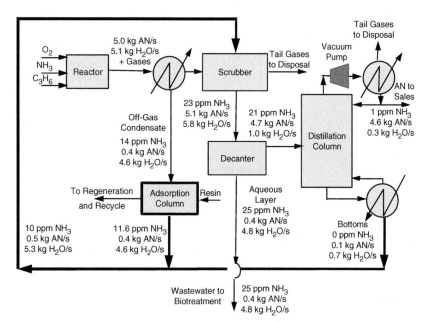

FIGURE 5-20   OPTIMAL SOLUTION TO THE AN CASE STUDY (EL-HALWAGI 1997)

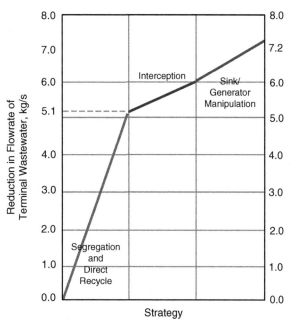

FIGURE 5-21   IMPACT DIAGRAM FOR REDUCING BIOTREATMENT INFLUENT (EL-HALWAGI 1997)

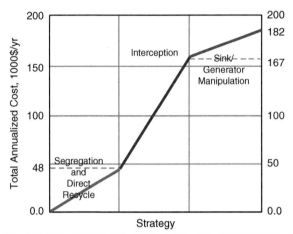

FIGURE 5-22 IMPACT DIAGRAM FOR TAC OF IDENTIFIED SOLUTIONS (EL-HALWAGI 1997)

This production increase is a result of better allocation of process streams; the essence of mass integration. For a selling value of $0.6/kg of AN, the additional production of 0.7 kg AN/s can provide an annual revenue of $13.3 million/year!

• Fresh water usage and influent to biotreatment facility are decreased by 7.2 kg/s. The value of fresh water and the avoidance of treatment cost are additional benefits.

• Influent to biotreatment is reduced to 40% of current level. Therefore, the plant production can be expanded 2.5 times the current capacity before the biotreatment facility is debottlenecked again.

Clearly, this is a superior solution to the installation of an additional biotreatment facility.

It is instructive to draw some conclusions from this case study and emphasize the design philosophy of mass integration. First, the target for debottlenecking the biotreatment facility was determined ahead of design. Then, systematic tools were used to generate optimal solutions that realized the target. Next, an analysis to study the performance was needed to refine the results. This is an efficient approach to understanding the global insights of the process, setting performance targets, realizing these targets, and saving time and effort by focusing on the big picture first and dealing with the details later. This is a fundamentally different approach than using the designer's subjective decisions to alter the process and check the consequences using detailed analysis. It is also different from using simple end-of-pipe treatment solutions. Instead, the various species are optimally allocated throughout the process. Therefore, objectives such as yield enhancement, pollution prevention, and cost savings can be simultaneously addressed. Indeed, pollution prevention (when undertaken with the proper techniques) can be a source of profit for the company, not an economic burden.

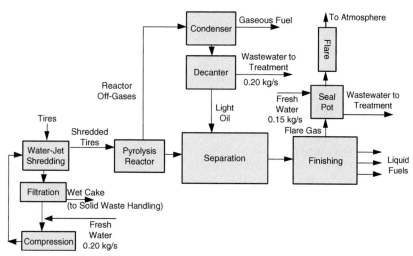

FIGURE 5-23 SCHEMATIC FLOWSHEET OF TIRE-TO-FUEL PROCESS (EL-HALWAGI 1997)

5.4 PROBLEMS

5.1 Let us revisit the tire-to-fuel process described in Problem 4.1. Figure 5-23 is a more detailed flowsheet. Tire shredding is achieved by using high-pressure water jets. The shredded tires are fed to the process while the spent water is filtered. The wet cake collected from the filtration system is forwarded to solid waste handling. The filtrate is mixed with 0.20 kg/s of fresh water makeup to compensate for water losses with the wet cake (0.08 kg H_2O/s) and the shredded tires (0.12 kg H_2O/s). The mixture of filtrate and water makeup is fed to a high-pressure compression station for recycle to the shredding unit. Due to the pyrolisis reactions, 0.08 kg H_2O/s are generated.

The plant has two primary sources for wastewater; the decanter (0.20 kg H_2O/s) and the seal pot (0.15 kg/s). The plant has been shipping the wastewater for off-site treatment. The cost of wastewater transportation and treatment is $0.01/kg leading to a wastewater treatment cost of approximately $110,000/year. The plant wishes to stop off-site treatment of wastewater to avoid cost of off-site treatment ($110,000/year) and alleviate legal-liability concerns in case of transportation accidents or inadequate treatment of the wastewater. The objective of this problem is to eliminate or reduce to the extent feasible off-site wastewater treatment. For capital budget authorization, the plant has the following economic criterion:

$$\text{Payback period} = \frac{\text{Fixed capital investment}}{\text{Annual savings}} \leq 3 \text{ years} \qquad (5.15)$$

where

$$\text{Annual Savings} = \text{Annual avoided cost of off-site treatment} -$$

$$\text{Annual operating cost of on-site system}$$

In addition to the information provided by Problem 4.1, the following data are available:

Economic Data

- Fixed cost of extraction system associated with S_2, $\$ = 120{,}000$ (Flowrate of wastewater, kg/s)$^{0.58}$
- Fixed cost of adsorption system associated with S_3, $\$ = 790{,}000$ (Flowrate of wastewater, kg/s)$^{0.70}$
- Fixed cost of stripping system associated with S_4, $\$ = 270{,}000$ (Flowrate of wastewater, kg/s)$^{0.65}$
- A biotreatment facility that can handle 0.35 kg/s wastewater has a fixed cost of \$240,000 and an annual operating cost of \$60,000/year.

Technical Data

Water may be recycled to two sinks; the seal pot and the water-jet compression station. The following constraints on flowrate and composition of the pollutant (heavy organic) should be satisfied:

Seal Pot

$$0.10 \leq \text{Flowrate of feed water (kg/s)} \leq 0.20$$

$$0 \leq \text{Pollutant content of feed water (ppmw)} \leq 500$$

Makeup to Water-Jet Compression Station

$$0.18 \leq \text{Flowrate of makeup water (kg/s)} \leq 0.20$$

$$0 \leq \text{Pollutant content of makeup water (ppmw)} \leq 50$$

5.2 Consider the magnetic-tape manufacturing process previously described by Problem 4.7. In this process (Dunn et al., 1995), several binders used in coating are dissolved in 0.09 kg/s of organic solvent. The mixture is additionally mixed and suspended with magnetic pigments, dispersants, and lubricants. A base file is coated with the suspension. Nitrogen gas is used for blanketing and to aid in evaporation of the solvent. As a result of thermal decomposition in the coating chamber, 0.011 kg/s of solvent are decomposed into other organic species. The decomposed organics are separated from the exhaust gas in a membrane unit. The reject stream leaving the membrane unit has a flowrate of 3.0 kg/s and is primarily composed of nitrogen which containts 1.9 wt/wt% of the organic solvent. The coated

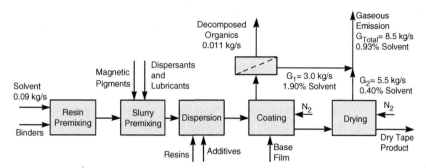

FIGURE 5-24 SCHEMATIC REPRESENTATION OF A MAGNETIC TAPE MANUFACTURING PROCESS (DUNN ET AL., 1995; EL-HALWAGI 1997)

TABLE 5-4 DATA FOR THE MSAs (EL-HALWAGI 1997)

Stream	Upper bound on flowrate L_j^C (kg/s)	Supply composition (mass fraction) x_j^s	Target composition (mass fraction) x_j^t	m_j	Mass fraction ε_j	C_j ($/kg MSA)
S_1	∞	0.014	0.040	0.4	0.001	0.002
S_2	∞	0.020	0.080	1.5	0.001	0.001
S_3	∞	0.001	0.010	0.1	0.001	0.002

film is passed to a dryer where nitrogen gas is employed to evaporate the remaining solvent. The exhaust gas leaving the dryer has a flowrate of 5.5 kg/s and contains 0.4 wt/wt% solvent. The two exhaust gases are mixed and disposed off.

In order to recover the solvent from the exhaust gases, there are three candidate MSAs. The equilibrium data for the transfer of the organic solvent to the jth lean stream is given by $y = m_j x_j$ where the values of m_j are given in Table 5-4. A minimum allowable composition difference, of 0.001(kg organic solvent)/(kg MSA) is to be used.

The annualized fixed cost of a mass exchanger, $/year, may be approximated by 18,000 (Gas flowrate, kg/s)$^{0.65}$. The value of the recovered solvent is $0.80/kg of organic solvent.

In addition to the environmental problem, there is also an economic incentive to recover the solvent from the exhaust gases. The value of 0.09 kg/s of solvent is approximately $2.3 million/year.

The objective of this problem is to develop a minimum-cost solution which minimizes the usage of fresh solvent in the process. The solution strategies may include segregation, mixing, recycle, and interception. To assess the economic potential of the solution, evaluate the paypack period for your solution.

5.5 REFERENCES

Dunn, R.F., El-Halwagi, M.M., Lakin, J., and Serageldin, M. 1995, 'Selection of organic solvent blends for environmental compliance in the coating industries', *Proceedings of the First International Plant Operations and Design Conference*, eds. E.D. Griffith, H. Kahn, and M.C. Cousins, *AIChE*, New York, vol. III, pp. 83-107.

El-Halwagi, M. M. 1997, *Pollution Prevention Through Process Integration*, Academic Press, San Diego.

El-Halwagi, M.M. 1999, 'Sustainable Pollution Prevention through Mass Integration', in *Tools and Methods for Pollution Prevention*, eds. S. Sikdar, and U. Diwekar, Kluwer, pp. 233-275.

El-Halwagi, M.M., Hamad, A.A., and Garrison, G.W. 1996, 'Synthesis of Waste Interception and Allocation Networks', *AIChE J.*, vol. 42, no. 11, pp. 3087-3101.

El-Halwagi, M.M. and Spriggs, H.D., 1998, 'Solve design puzzles with mass integration', *Chem. Eng. Prog.*, vol. 94, August, pp. 25-44.

Garrison, G.W., Spriggs, H.D., and El-Halwagi, M.M. 1996, 'A global approach to integrating environmental, energy, economic and technological objectives', *Proceedings of Fifth World Congr. of Chem. Eng.*, San Diego, vol. I, pp. 675-680.

6

ALGEBRAIC APPROACH TO TARGETING DIRECT RECYCLE

Chapter Three presented the material recycle pinch diagram as a graphical tool to identify targets for direct recycle problems. In spite of the usefulness and insights of the graphical methods, it is beneficial to develop an algebraic procedure which is particularly useful in the following cases:

- Numerous sources and sinks: As the number of sources and sinks increase, it becomes more convenient to use spreadsheets or algebraic calculations to handle the targeting.
- Scaling problems: If there is a significant difference in values of flowrates and/or loads for some of the sources and/or sinks, the graphical representation becomes inaccurate since the larger flows/ loads will skew the scale for the other streams.
- If the targeting is tied with a broader design task that is handled through algebraic computations, it is desirable to use consistent algebraic tools for all the tasks.

This chapter presents the algebraic analogue to the material recycle pinch diagram.

6.1 PROBLEM STATEMENT

The problem can be expressed as follows:

Given a process with a number ($N_{sources}$) of process sources (e.g., process streams, wastes) that may be considered for possible recycle and replacement of the fresh material and/or reduction of waste discharge. Each source, i, has a given flowrate, W_i, and a given composition of a targeted species, y_i. Available for service is a fresh (external) resource that can be purchased to supplement the use of process sources in sinks. The sinks are N_{sinks} process units that employ a fresh resource. Each sink, j, requires a feed whose flowrate, G_j^{in}, and an inlet composition of a targeted species, z_j^{in}, must satisfy the following bounds:

$$0 \leq z_j^{in} \leq z_j^{max} \quad \text{where } j = 1, 2, \ldots, N_{sinks} \tag{6.1}$$

Fresh (external) resource may be purchased to supplement the use of process sources in sinks. The objective is to develop a non-iterative algebraic procedure aimed at minimizing the purchase of fresh resource, maximizing the usage of process sources, and minimizing waste discharge.

6.2 ALGEBRAIC TARGETING APPROACH

In this section, the algebraic procedure developed by Almutlaq and El-Halwagi (2006) and Almutlaq et al. (2005) is presented. First, it is necessary to recall the two direct-recycle optimality conditions derived in Chapter Three.

Sink Composition Rule: When a fresh resource is mixed with process source(s), the composition of the mixture entering the sink should be set to a value that minimizes the fresh arm. For instance, when the fresh resource is a pure substance that can be mixed with pollutant-laden process sources, the composition of the mixture should be set to the maximum admissible value.

Source Prioritization Rule: In order to minimize the usage of the fresh resource, recycle of the process sources should be prioritized in order of their fresh arms starting with the source having the shortest fresh arm.

These rules constitute the basis for the material recycle pinch diagram as shown in Figure 6-1. As described in Chapter Three, the sink composite is a cumulative representation of all the sinks and corresponds to the upper bound on their feasibility region, whereas the source composite curve is a cumulative representation of all process streams considered for recycle. The source composite stream may be represented anywhere and is then slid horizontally (in the case of pure fresh) on the flowrate axis until it touches the sink composite stream with the source composite below the sink composite in the overlapped region.

Now, suppose that we start plotting both composite streams from the origin point (Figure 6-2). If the source composite is completely below

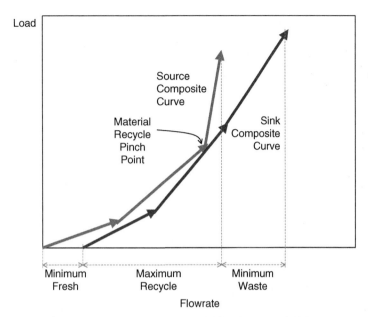

FIGURE 6-1   MATERIAL RECYCLE PINCH DIAGRAM (EL-HALWAGI ET AL., 2003)

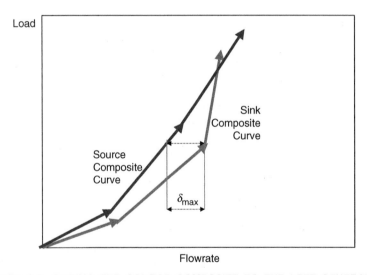

FIGURE 6-2   SLIDING THE SOURCE COMPOSITE TO THE LEFT GENERATES INFEASIBILITY (ALMUTLAQ AND EL-HALWAGI 2006)

the sink composite, then the process does not require a fresh resource. Nonetheless, if there is any portion of the source composite lying above the sink composite, then infeasibilities exist and must be removed by using a fresh resource. Such infeasibility may be described in a couple of ways

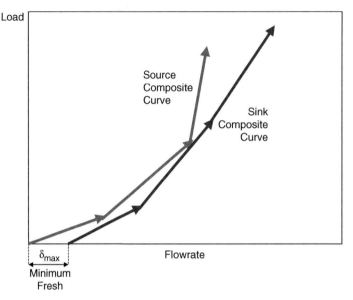

FIGURE 6-3   MINIMUM FRESH TARGET CORRESPONDS TO MAXIMUM FLOW SHORTAGE (ALMUTLAQ AND EL-HALWAGI 2006)

by looking vertically and horizontally. By looking vertically at a given flowrate, if the source composite lies above the sink composite, then the source composite violates the maximum load admissible to the sink. Alternatively, by looking horizontally at a given load, if the source composite lies to the left of the sink composite, then there is a shortage of the flowrate necessary for the sink. The maximum horizontal infeasibility corresponds to the maximum shortage of flowrate which is designated as δ_{max}. Indeed, all infeasibilities are eliminated by sliding the source composite curve to the right a distance equal to δ_{max} (Figure 6-3). Consequently, the target for minimum fresh usage is equal to the maximum shortage, i.e.,

$$\text{Target for minimum fresh consumption} = \delta_{max} \qquad (6.2)$$

The targeting question involves the algebraic identification of this maximum shortage without the need to resort to the graphical representation.

Let us revisit Figure 6-2 and draw horizontal lines at corner points (kinks) of the source and sink composite curves (Figure 6-4). Let us use an index k to designate those horizontal lines starting with $k = 0$ at the zero load level and going up at each horizontal level. The load at each horizontal level, k, is referred to as M_k. The vertical distance between each two horizontal lines is referred to as a *load interval* and is given the index k as well. The load within interval k is calculated as follows:

$$\Delta M_k = M_k - M_{k-1} \qquad (6.3)$$

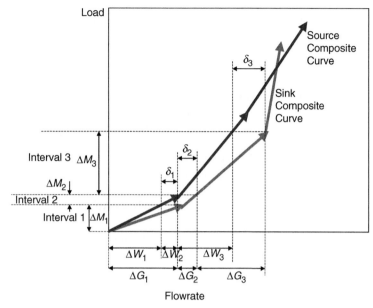

FIGURE 6-4   LOAD INTERVALS, FLOWS, AND RESIDUALS (ALMUTLAQ AND EL-HALWAGI 2006)

The next step is to calculate the flowrates of the source and the sink within each load interval. These flowrates correspond to the horizontal distances on the source and sink composite curves contained within the interval. Therefore, the following expressions may be used to calculate the source and sink flowrates (respectively) within the kth interval:

$$\Delta W_k = \frac{\Delta M_k}{y_{\text{source in interval } k}} \tag{6.4}$$

and

$$\Delta G_k = \frac{\Delta M_k}{z^{\max}_{\text{sink in interval } k}} \tag{6.5}$$

Figure 6-4 illustrates the concepts of a load interval and flowrates of sources and sinks within an interval. Additionally, Figure 6-4 illustrates that at any horizontal level $(\bar{k})$, the horizontal distance between the source and sink composite curves is given by:

$$\delta_{\bar{k}} = \sum_{k=1}^{\bar{k}} W_k - \sum_{k=1}^{\bar{k}} G_k \tag{6.6}$$

Equation (6.6) indicates that at any value of load, the horizontal distance between the source and sink composite curves is the difference in cumulative flowrates. As mentioned earlier, a negative value of δ implies that the source composite lies to the left of the sink composite which is infeasible.

To illustrate equation (6.6), let us apply it to the first interval:

$$\delta_1 = \Delta W_1 - \Delta G_1 \qquad (6.7)$$

This result can be verified by Figure 6-4. Similarly, applying equation (6.6) to the second interval, we have:

$$\delta_2 = \Delta W_1 + \Delta W_2 - \Delta G_1 - \Delta G_2 \qquad (6.8)$$

Substituting equation (6.7) into equation (6.8), we obtain

$$\delta_2 = \delta_1 + \Delta W_2 - \Delta G_2 \qquad (6.9)$$

and, for the kth interval:

$$\delta_k = \delta_{k-1} + \Delta W_k - \Delta G_k \qquad (6.10)$$

with $\delta_0 = 0$. Equation (6.10) is represented by Figure 6-5. The flow balances can be carried out for all intervals resulting in the cascade diagram shown in Figure 6-6. As mentioned earlier, the most negative value of δ on the cascade diagram (δ_{max}) represents the target for minimum fresh consumption as indicated by equation (6.2). In order to remove the infeasibilities, a flowrate

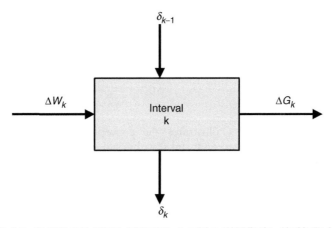

FIGURE 6-5   FLOW BALANCE AROUND A LOAD INTERVAL (ALMUTLAQ AND EL-HALWAGI 2006)

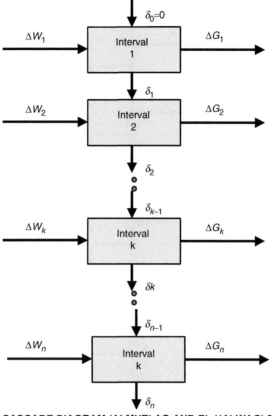

FIGURE 6-6   CASCADE DIAGRAM (ALMUTLAQ AND EL-HALWAGI 2006)

of the fresh resource equal to δ_{max} is added to the top of the cascade (i.e., $\delta_0 = \delta_{max}$). The residuals are accordingly increased by δ_{max}, i.e.,

$$\delta'_k = \delta_k + \delta_{max} \tag{6.11}$$

where δ'_k is the revised residual flow leaving the kth interval. Consequently, the most negative residual becomes zero thereby designating the pinch location. Additionally, the revised residual leaving the last interval is the target for minimum wastewater discharge since it represents the unrecycled/ unreused flowrates of the sources. Figure 6-7 is an illustration of the revised cascade diagram. As can be seen from this illustration, the following residuals determine the targets:

$$\delta_{max} = \text{Target for minimum fresh usage} \tag{6.12}$$

$$\delta'_n = \text{Target for minimum waste discharge} \tag{6.13}$$

where n is the last load interval.

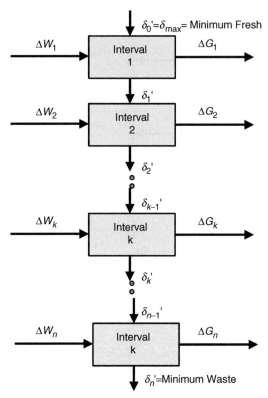

FIGURE 6-7   REVISED CASCADE DIAGRAM (ALMUTLAQ AND EL-HALWAGI 2006)

6.3 ALGEBRAIC TARGETING PROCEDURE

Based on the foregoing analysis, the algebraic procedure can be summarized as follows:

1. Rank the sinks in ascending order of maximum admissible composition, $z_1^{max} \leq z_2^{max} \leq \ldots z_j^{max} \ldots \leq z_{N_{Sinks}}^{max}$
2. Rank sources in ascending order of pollutant composition, i.e., $y_1 < y_2 < \ldots y_i \ldots < y_{N_{sources}}$
3. Calculate the load of each sink ($M_j^{sink, max} = G_j z_j^{max}$) and source ($M_i^{source} = W_i y_i$).
4. Compute the cumulative loads for the sinks and for the sources (by summing up their individual loads).
5. Rank the cumulative loads in ascending order.
6. Develop the load-interval diagram (LID) shown in Figure 6-8. First, the loads are represented in ascending order starting with zero load. The scale is irrelevant. Next, each source (and each

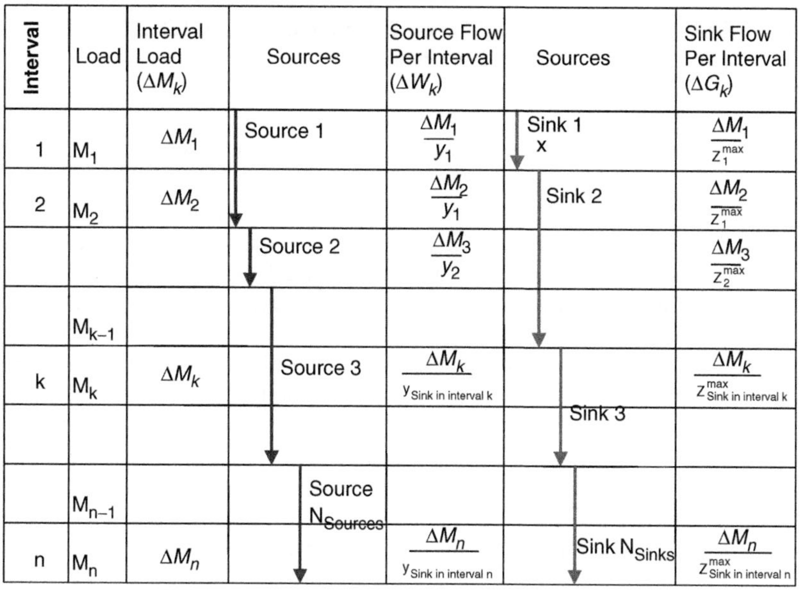

Interval	Load	Interval Load (ΔM_k)	Sources	Source Flow Per Interval (ΔW_k)	Sources	Sink Flow Per Interval (ΔG_k)
1	M_1	ΔM_1	Source 1	$\dfrac{\Delta M_1}{y_1}$	Sink 1 ✕	$\dfrac{\Delta M_1}{z_1^{max}}$
2	M_2	ΔM_2		$\dfrac{\Delta M_2}{y_1}$	Sink 2	$\dfrac{\Delta M_2}{z_1^{max}}$
			Source 2	$\dfrac{\Delta M_3}{y_2}$		$\dfrac{\Delta M_3}{z_2^{max}}$
	M_{k-1}					
k	M_k	ΔM_k	Source 3	$\dfrac{\Delta M_k}{y_{\text{Sink in interval } k}}$		$\dfrac{\Delta M_k}{z_{\text{Sink in interval } k}^{max}}$
					Sink 3	
	M_{n-1}		Source N_{Sources}			
n	M_n	ΔM_n		$\dfrac{\Delta M_n}{y_{\text{Sink in interval } n}}$	Sink N_{Sinks}	$\dfrac{\Delta M_n}{z_{\text{Sink in interval } n}^{max}}$

FIGURE 6-8   LOAD INTERVAL DIAGRAM (ALMUTLAQ AND EL-HALWAGI 2006)

sink) is represented as an arrow whose tail corresponds to its starting load and head corresponds to its ending load. Equations (6.3)–(6.5) are used to calculate the intervals load, source flowrate and sink flowrate.

7. Based on the interval source and sink flowrates, develop the cascade diagram and carry out flow balances around the intervals to calculate the values of the flow residuals (δ_k's). The most negative δ_k is the target for minimum fresh consumption, i.e.,

$$\delta_{\max} = \text{Target for minimum fresh usage}$$

8. Revise the cascade diagram by adding the maximum δ_k to the first interval and calculate the revised residuals. The interval with the first zero residual is the material recycle/reuse global pinch point. The residual flow leaving the last interval is the target for minimum waste discharge, i.e.,

$$\delta_n' = \text{Target for minimum waste discharge}$$

It is worth noting that when the fresh source is impure, the foregoing procedure can be modified to account for the concentration of contaminants. This procedure is described in detail by Almutlaq et al. (2005).

6.4 CASE STUDY: TARGETING FOR ACETIC ACID USAGE IN A VINYL ACETATE PLANT

Here, we revisit Example 3-4 on the recovery of AA from a VAM facility. The data for the problem are shown in Tables 6-1 and 6-2. The last column is calculated as the cumulative load.

The LID is illustrated in Figure 6-9. The cascade diagram is given by Figure 6-10(a). As can be seen, the most negative residual is −9584 kg/h. Therefore, the target for minimum fresh AA is 9584 kg/h. When this value is added to the first interval, we can carry out the revised cascade calculations leading to a target of minimum waste discharge (residual leaving last interval) of 4784 kg/h. The zero residual designates the pinch location. Hence, the material recycle pinch point is located at the horizontal lines separating intervals 3 and 4. As can be seen from the LID, this location corresponds to a cumulative load of 1275 kg/h and a source mass fraction of water being 0.25 (between intervals 3 and 4 which corresponds to a mass fraction of 0.25 on the source side). These results are consistent with the ones determined graphically in Chapter Three.

TABLE 6-1 SOURCE DATA FOR THE VINYL ACETATE EXAMPLE

Source	Flowrate (kg/h)	Inlet mass fraction	Inlet load (kg/h)	Cumulative load (kg/h)
Bottoms of absorber II	1400	0.14	196	196
Bottoms of primary tower	9100	0.25	2275	2471

TABLE 6-2 SINK DATA FOR THE VINYL ACETATE EXAMPLE

Sink	Flowrate (kg/h)	Maximum inlet mass fraction	Maximum inlet load (kg/h)	Cumulative maximum load (kg/h)
Absorber I	5100	0.05	255	255
Acid tower	10,200	0.10	1020	1275

Interval	Load, kg/hr	Interval Load (ΔM_k) kg/hr	Sources	Source Flow per Interval (ΔW_k),ton/hr	Sources	Sink Flow per Interval (ΔG_k),ton/hr
	0					
1	196	196	Source 1 $y = 0.14$	1,400	Sink 1 $z^{max}=0.05$	3,920
2	255	59		236		1,180
3	1275	1,020	Source 2 $y = 0.25$	4,080	Sink 2 $z^{max}=0.10$	10,200
4	2471	1,196		4,784		0

FIGURE 6-9 LID FOR THE VAM CASE STUDY

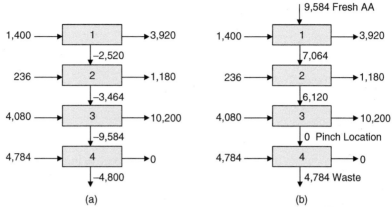

6.5 PROBLEMS

6.1 Solve Problem 3.1 using the algebraic method.

6.2 Solve Problem 3.2 using the algebraic method.

6.3 Solve Problem 3.3 using the algebraic method.

6.4 Solve Problem 3.4 using the algebraic method.

6.5 Solve the food processing case study (Example 3-6) using the algebraic method.

6.6 Describe the algebraic procedure for direct-recycle target when the fresh resource is impure.

6.6 SYMBOLS

G	sink (unit) flowrate, mass/time
M	load of contaminant, mass/time
$M^{\text{sink, max}}$	maximum admissible load to the sink, mass/time
$M^{\text{source, max}}$	contaminant load in source, mass/time
N_{sources}	number of process streams (or sources)
N_{sinks}	number of process units (sinks)
W	sink (unit) flow, mass or volume/time
y	contaminant composition of process streams (or sources)
z	contaminant composition of process streams (or sources)
$\overline{k}$	total number of intervals

Superscripts

min	lower bound of allowable contaminant concentration to the sink
max	upper bound of allowable contaminant concentration to the sink

Subscripts

i	index for sources
j	index for sinks
k	index for load intervals

Greek

δ	interval residual
Δ	difference between two consecutive intervals

6.7 REFERENCES

Almutlaq, A.M. and El-Halwagi, M.M. 'An Algebraic Targeting Approach to Resource Conservation via Material Recycle/Reuse', *Int. J. Environ. & Pollution* (in press, 2006).

Almutlaq, A.M., Kazantzi, V., and El-Halwagi, M.M. 2005, 'An Algebraic Approach to Targeting Waste Discharge and Impure-Fresh Usage via Material Recycle/Reuse Networks', *J. Clean Tech. & Environ. Policies*, vol. 7, no. 4, pp. 294-305.

El-Halwagi, M.M., Gabriel, F., and Harell, D. 2003, Rigorous Graphical Targeting for Resource Conservation via Material Recycle/Reuse Networks', *Ind. Eng. Chem. Res.*, vol. 42, pp. 4319-4328.

7

AN ALGEBRAIC APPROACH TO THE TARGETING OF MASS EXCHANGE NETWORKS

As mentioned in Chapter Six, algebraic approaches have key advantages over graphical tools. While graphs provide visualization insights, they may become cumbersome as the problem size increases or when scaling problems become an issue. On the other hand, algebraic techniques can be effective in handling large problems as they can be conveniently carried out using spreadsheets or calculators. Additionally, major differences in composition scales do not constitute a computational problem. Finally, algebraic techniques can be readily integrated with other design tools including simulators. In this chapter, we present an algebraic approach to the targeting of mass exchange network that yields results equivalent to those provided by the graphical pinch analysis presented in Chapter Four. In particular, we focus on the identification of minimum load to be removed using external MSAs, excess capacity of process MSAs, and maximum removable load using process MSAs. For constructing the actual network of mass exchangers, the reader is referred to El-Halwagi and Manousiouthakis (1989a); El-Halwagi (1997).

7.1 THE COMPOSITION-INTERVAL DIAGRAM

In order to insure thermodynamic feasibility of mass exchange, we construct the composition-interval diagram (CID). On this diagram, $N_{sp}+1$ corresponding composition axes are generated. First, a composition axis, y, is established for the rich streams. This axis does not have to be drawn to scale. Each rich stream is represented as a vertical arrow whose tail corresponds to its supply composition and head corresponds to its target composition. Then, equation (4.17) is employed to each MSA construct, N_{sp} being corresponding composition scales for the process MSAs. Each process MSA is represented versus its composition axis as a vertical arrow extending between supply and target compositions. Next, horizontal lines are drawn at the heads and tails of the arrows. The vertical distance between two consecutive horizontal lines is referred to as a *composition interval*. The number of intervals is related to the number of process streams via

$$N_{int} \leq 2(N_R + N_{SP}) - 1 \tag{7.1}$$

with the equality applying in cases where no heads on tails coincide. An index k is used to designate composition intervals with $k=1$ being the top interval and $k=N_{int}$ being the bottom interval. Figure 7-1 provides a schematic representation of the CID. Mass exchange is thermodynamically (and practically) feasible from rich streams to MSAs within the same interval.

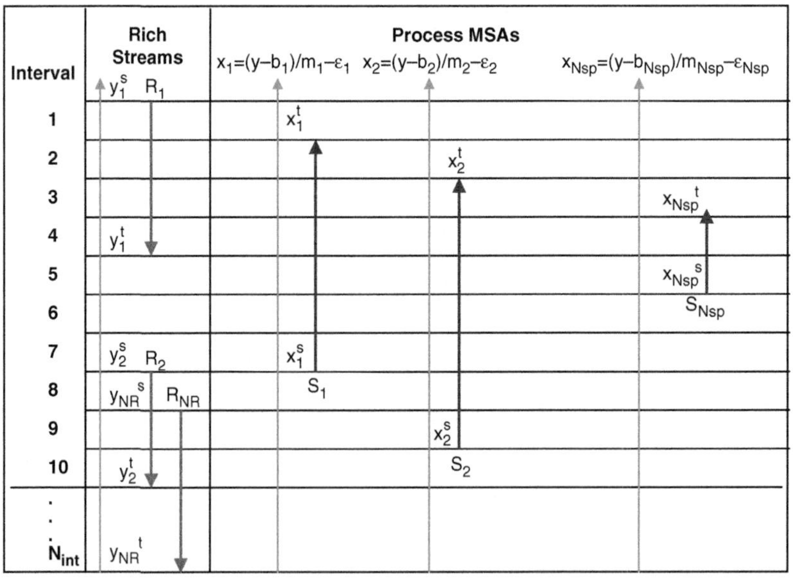

Interval	Rich Streams	Process MSAs $x_1=(y-b_1)/m_1-\varepsilon_1$	$x_2=(y-b_2)/m_2-\varepsilon_2$		$x_{Nsp}=(y-b_{Nsp})/m_{Nsp}-\varepsilon_{Nsp}$
	y_1^s R_1				
1		x_1^t			
2			x_2^t		
3					x_{Nsp}^t
4	y_1^t				
5					x_{Nsp}^s
6					S_{Nsp}
7	y_2^s R_2	x_1^s			
8	y_{NR}^s R_{NR}	S_1			
9			x_2^s		
10	y_2^t		S_2		
$\vdots$					
N_{int}	y_{NR}^t				

FIGURE 7-1   COMPOSITION INTERVAL DIAGRAM (EL-HALWAGI 1997)

Additionally, it is also feasible to transfer mass from a rich stream to lean stream that lies in an interval below it.

7.2 TABLE OF EXCHANGEABLE LOADS

In order to determine the exchangeable loads among the rich- and lean-streams in each interval, we construct the table of exchangeable loads (TEL). The exchangeable load of the ith rich stream that passes through the kth interval can be calculated from the following expression:

$$W_{i,k}^{R} = G_i(y_{k-1} - y_k) \tag{7.2}$$

where y_{k-1} and y_k are the rich-axis compositions of the transferable species that respectively correspond to the top and the bottom lines defining the kth interval. Similarly, maximum load that may be gained by the jth process MSA within interval k can be determined through the following equation:

$$W_{j,k}^{S} = L_j^{C}(x_{j,k-1} - x_{j,k}) \tag{7.3}$$

where $x_{j,k-1}$ and $x_{j,k}$ are the compositions on the jth lean-composition axis which respectively correspond to the higher and lower horizontal lines bounding the kth interval. Naturally, if a stream does not pass through an interval, its load within that interval is zero.

Once the individual loads of all rich and process MSAs have been determined for all composition intervals, one can also obtain the collective loads of the waste and the lean-streams. The collective load of the rich streams within the kth interval is calculated by summing the individual loads of the waste streams that pass through that interval, i.e.

$$W_{k}^{R} = \sum_{i \, \text{pass throught interval} \, k} W_{i,k}^{R} \cdot 1 \tag{7.4}$$

Similarly, the collective load of the lean-streams within the kth interval is evaluated as follows:

$$W_{k}^{S} = \sum_{j \, \text{passes through interval} \, k} W_{j,k}^{S} \cdot 2 \tag{7.5}$$

With the collective loads calculated, the next step is to exchange loads among rich- and lean-streams. This is accomplished through feasible mass balances as described in the next section.

7.3 MASS EXCHANGE CASCADE DIAGRAM

In constructing the CID, composition axes were mapped such that it is feasible to transfer mass from the rich streams to the lean-streams within the

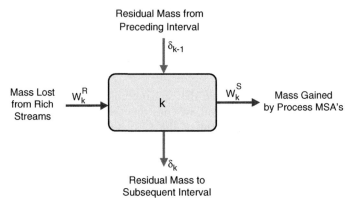

FIGURE 7-2 COMPONENT MATERIAL BALANCE AROUND A COMPOSITION INTERVAL

same interval or at any lower interval. Therefore, the following component material balance for the targeted species can be written for the kth interval:

$$W_k^R + \delta_{k-1} - W_k^S = \delta_k \qquad (7.6)$$

where δ_{k-1} and δ_k are the residual masses of the targeted species entering and leaving the kth interval with the residual mass entering the first interval (δ_0) being zero. Figure 7-2 illustrates the component material balance for the targeted species around the kth composition interval.

It is worth pointing out that δ_0 is zero because no waste streams exist above the first interval. In addition, thermodynamic feasibility is ensured when all the δ_k's are non-negative. Hence, a negative δ_k indicates that the capacity of the process lean-streams at that level is greater than the load of the waste streams. The most negative δ_k corresponds to the excess capacity of the process MSAs in removing the pollutant. Therefore, this excess capacity of process MSAs should be reduced by lowering the flowrate and/or the outlet composition of one or more of the MSAs. After removing the excess MSA capacity, one can construct a revised TEL in which the flowrates and/or outlet compositions of the process MSAs have been adjusted. Consequently, a revised cascade diagram can be generated. On the revised cascade diagram, the location at which the residual mass is zero corresponds to the mass exchange pinch composition. As expected, this location is the same as that with the most negative residual on the original cascade diagram. Since an overall material balance for the network must be realized, the residual mass leaving the lowest composition interval of the revised cascade diagram must be removed by external MSAs.

7.4 EXAMPLE ON CLEANING OF AQUEOUS WASTES

An organic pollutant is to be removed from two aqueous wastes. The data for the rich streams are given in Table 7-1.

TABLE 7-1 DATA OF RICH STREAMS FOR THE WASTEWATER CLEANING
EXAMPLE

Stream	Flowrate G_1 (kg/s)	Supply composition of pollutant y_i^s (mass fraction)	Target composition of pollutant y_i^t (mass fraction)
R_1	2.0	0.030	0.005
R_2	3.0	0.010	0.001

TABLE 7-2 DATA OF PROCESS LEAN-STREAMS FOR THE WASTEWATER
CLEANING EXAMPLE

Stream	Upper bound on flowrate (kg/s) L_j^C	Supply composition of pollutant (mass fraction) x_j^s	Target composition of pollutant (mass fraction) x_j^t
S_1	17.0	0.007	0.009
S_2	1.0	0.005	0.015

Two process MSAs are considered for separation. The stream data
for S_1 and S_2 are given in Table 7-2. The equilibrium data for the pollutant
in the two process MSAs are given by

$$y = 0.25x_1 \tag{7.7}$$

and

$$y = 0.50x_2 \tag{7.8}$$

The minimum allowable composition difference for both process MSAs
is taken as 0.001 (mass fraction of pollutant).

Determine the minimum load to be removed using an external MSA,
the pinch location, and excess capacity of the process MSAs.

SOLUTION

The CID for the problem is constructed as shown in Figure 7-3.
Then, the TEL is developed as shown in Table 7-3. Next, the mass exchange
cascade diagram is generated. As can be seen in Figure 7-4, the most negative residual mass is -0.006 kg/s. This value corresponds to the excess
capacity of process MSAs. Such excess capacity can be removed by reducing
the flowrates and/or outlet compositions of the process MSAs. If we decide
to eliminate this excess by decreasing the flowrate of S_1, the actual flowrate
of S_1 can be calculated as follows:

$$L_1^C = L_1^{\text{actual}} + L_1^{\text{excess}} \tag{7.9a}$$

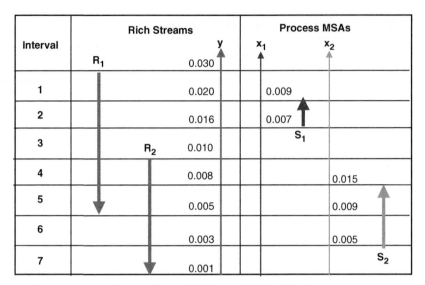

FIGURE 7-3   CID FOR THE WASTEWATER CLEANING EXAMPLE

TABLE 7-3 TEL FOR WASTEWATER CLEANING EXAMPLE

Interval	Load of R_1 (kg/s)	Load of R_2 (kg/s)	Load of $R_1 + R_2$ (kg/s)	Load of S_1 (kg/s)	Load of S_2 (kg/s)	Load of $S_1 + S_2$ (kg/s)
1	0.020	–	0.020	–	–	–
2	0.008	–	0.008	0.034	–	0.034
3	0.012	–	0.012	–	–	–
4	0.004	0.006	0.010	–	–	–
5	0.006	0.009	0.015	–	0.006	0.006
6	–	0.006	0.006	–	0.004	0.004
7	–	0.006	0.006	–	–	–

Multiplying both sides by the composition range for S_1, we get

$$L_1^C \times (0.009 - 0.007) = L_1^{actual} \times (0.009 - 0.007) + \text{Excess load} \qquad (7.9b)$$

i.e.,

$$17.0 \times (0.009 - 0.007) = L_1^{actual} \times (0.009 - 0.007) + 0.006 \qquad (7.9c)$$

Hence,

$$L_1^{actual} = 17.0 - \frac{0.006}{0.009 - 0.007} = 14.0 \, \text{kg/s} \qquad (7.9d)$$

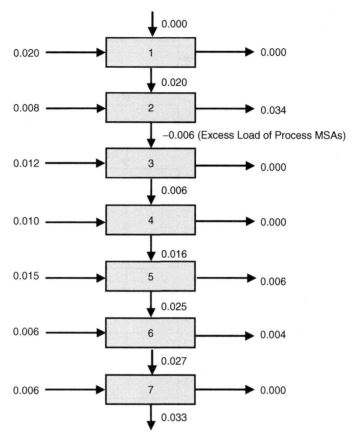

FIGURE 7-4   CASCADE DIAGRAM FOR THE WASTEWATER CLEANING EXAMPLE

Using the adjusted flowrate of S_1, we can now construct the revised cascade diagram as depicted by Figure 7-5. Hence, the revised mass exchange cascade diagram is generated as shown in Figure 7-5. On this diagram, the residual mass leaving the second interval is zero. Therefore, the mass exchange pinch is located on the line separating the second and the third intervals. As can be seen in Figure 7-3, this location corresponds to a y composition of 0.016 (which is equivalent to an x_1 composition of 0.007). Furthermore, Figure 7-5 shows the residual mass leaving the bottom interval as 0.039 kg/s. ■ ■ ■ This value is the amount of pollutant to be removed by external MSAs.

7.5 PROBLEMS

7.1 Using an algebraic procedure, synthesize an optimal MEN for the benzene recovery example described in Example 4-1.

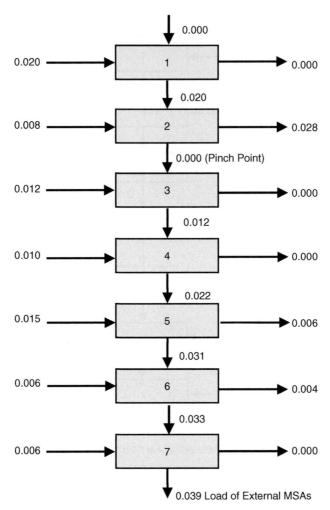

FIGURE 7-5   REVISED CASCADE DIAGRAM FOR THE WASTEWATER CLEANING EXAMPLE

7.2 Use the algebraic technique to solve the dephenolization case study described in Problem 4.8.

7.3 Resolve the dephenolization case study described in Problem 4.8 when the two waste streams are allowed to mix.

7.4 The techniques presented in this chapter can be generalized to tackle MENs with multiple pollutants. Consider the COG sweetening process addressed by the previous problem. Carbon dioxide often exists in COG in relatively large concentrations. Therefore, partial removal of CO_2 is sometimes desirable to improve the heating value of COG, and almost complete removal of CO_2 is required for gases undergoing low-temperature

■ **TABLE 7-4 STREAM DATA FOR COG-SWEETENING PROBLEM**

Rich streams				MSAs						
Stream	G_i (kg/s)	Supply mass fraction of CO_2	Target mass fraction of CO_2	Stream	L_j^C (kg/s)	Supply mass fraction of CO_2	Target mass fraction of CO_2	m_j for CO_2	b_j for CO_2	c_j $/kg
R_1	0.90	0.0600	0.0050	S_1	2.3	0.0100	?	0.35	0.000	0.00
R_2	0.10	0.1150	0.0100	S_2	?	0.0003	?	0.58	0.000	0.10

processing. The data for the CO_2-laden rich and lean-streams are given in Table 7-4.

Synthesize an MOC MEN which removes hydrogen sulfide and carbon dioxide simultaneously. *Hint*: See El-Halwagi and Manousiouthakis (1989a,b).

7.6 SYMBOLS

b_j intercept of equilibrium line of the jth MSA

G_i flowrate of the ith waste stream

i index of waste streams

j index of MSAs

k index of composition intervals

L_j flowrate of the jth MSA

L_j^C upper bound on available flowrate of the jth MSA

m_j slope of equilibrium line for the jth MSA

N_{int} number of composition intervals

N_{SP} number of process MSAs

R_i the ith waste stream

S_j the jth lean stream

$W_{i,k}^R$ the exchangeable lead of the ith rich stream which passes through the kth interval

$W_{j,k}^S$ the exchangeable load of the jth MSA passing through the kth interval

W_k^R the collective exchangeable load of the rich streams in interval k

W_k^S the collective exchangeable load of the MSAs in k

$x_{j,k-1}$ composition of key component in the jth MSA at the upper horizontal line defining the kth interval

$x_{j,k}$ composition of key component in the jth MSA at the lower horizontal line defining the kth interval

y_{k-1} composition of key component in the ith rich stream at the upper horizontal line defining the kth interval

y_k composition of key component in the ith rich stream at the lower horizontal line defining the kth interval

7.7 REFERENCES

El-Halwagi, M.M. 1997, *Pollution Prevention through Process Integration: Systematic Design Tools*, Academic Press, San Diego.

El-Halwagi, M.M. and Manousiouthakis, V. 1989a, 'Synthesis of mass exchange networks', *AIChE J.*, vol. 35, no. 8, 1233-1244.

El-Halwagi, M.M. and Manousiouthakis, V. 1989b, *Design and analysis of mass exchange networks with multicomponent targets*, AIChE Annu. Meet, San Francisco, November.

8
RECYCLE STRATEGIES USING PROPERTY INTEGRATION

The previous chapter dealt primarily with mass integration. A common feature of mass-integration techniques is that they are "chemo-centric"; namely they are based on tracking individual chemical species. Nonetheless, many material reuse problems are driven and governed by properties or functionalities of the streams and not by their chemical constituency. The following are some examples of property-based problems:

- The usage of material utilities (e.g., solvents) relies on their characteristics, such as equilibrium distribution coefficients, viscosity, and volatility without the need to chemically characterize these materials.
- Constraints on process units that can accept recycled/reused process streams and wastes are not limited to compositions of components but are also based on the properties of the feeds to processing units.
- The performance of process units depends on properties. For instance, a heat exchanger performs based on the heat capacities and heat transfer coefficients of the matched streams. The chemical identity of the components is only useful to the extent of determining the values of heat capacities and heat-transfer coefficients. Similar examples can be given for many other units (e.g., vapor pressure in condensers,

specific gravity in decantation, relative volatility in distillation, Henry's coefficient in absorption, density and head in pumps, density, pressure ratio, and heat-capacity ratio in compressors, etc.).

- Quantities of emission are dependent on properties of the pollutants (e.g., volatility, solubility, etc.).
- The environmental regulations involve limits on properties (e.g., pH, color, toxicity, TOC, BOD, ozone-depleting ability).
- Tracking numerous chemical pollutants is prohibitively difficult (e.g., complex hydrocarbons and ligno-cellulosic materials) while tracking properties is manageable.

Therefore, it is important to have a systematic design methodology which is based on properties and functionalities. In response, the paradigm of *property integration* has been introduced by El-Halwagi and co-workers. Property integration is a functionality-based, holistic approach to the allocation and manipulation of streams and processing units, which is based on the tracking, adjustment, assignment, and matching of functionalities throughout the process (El-Halwagi et al., 2004; Shelley and El-Halwagi 2000).

In this chapter, we focus on the problem of identifying rigorous targets for direct reuse in property-based applications through visualization techniques. The chapter also discusses the identification of interception tasks. First, the problem of direct recycle with a single-property constraints is addressed through a material recycle pinch diagram similar to the one presented in Chapter Three. Then, we deal with the problem of multiple properties.

8.1 PROPERTY-BASED MATERIAL RECYCLE PINCH DIAGRAM

The first problem we address is the property-based direct recycle when the constraints are based on one property. More details on the problem and its solution technique can be found in Kazantzi and El-Halwagi (2005). The problem can be stated as follows.

Consider a process with a number N_{sinks} of process sinks (units). Each sink, j, requires a feed with a given flowrate, G_j, and an inlet property, p_j^{in}, that satisfies the following constraint:

$$p_j^{min} \le p_j^{in} \le p_j^{max} \quad \text{where } j = 1, 2, \ldots, N_{sinks} \quad (8.1)$$

where p_j^{min} and p_j^{max} are the specified lower and upper bounds on admissible property to unit j.

The plant has a number $N_{sources}$ of process sources (e.g., process streams, wastes) that can be considered for possible reuse and replacement of the fresh material. Each source, i, has a given flowrate, W_i, and a given property, p_i.

Available for service is a fresh (external) resource whose property value is p_{Fresh} and can be purchased to supplement the use of process sources in sinks. The objective is to develop a non-iterative graphical procedure that determines the target for minimum usage of the fresh resource, maximum

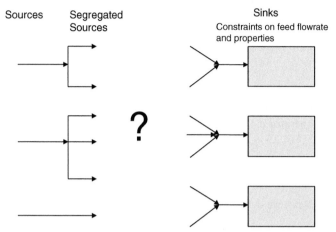

Sources Segregated Sources Sinks
Constraints on feed flowrate and properties

FIGURE 8-1 SCHEMATIC REPRESENTATION OF THE PROPERTY-BASED MATERIAL REUSE PROBLEM

material reuse, and minimum discharge to waste. The problem can be schematically represented through a source–sink allocation as shown in Figure 8-1. According to this representation, each source is allowed to split and be forwarded to all sinks. In particular, the objective here is to determine the optimum flowrate from each source to each sink so as to minimize the consumption of the fresh resource.

When several sources are mixed, it is necessary to evaluate the property of the mixture as a function of the flowrate and property of each stream. Consider the following mixing rule for estimating the resulting property of the mixture:

$$\overline{F} \times \psi(\overline{p}) = \sum_i F_i \times \psi(p_i) \qquad (8.2)$$

where $\psi(p_i)$ is the property-mixing operator, $\overline{p}$ is the property of the mixture and $\overline{F}$ is the total flowrate of the mixture which is given by:

$$\overline{F} = \sum_i F_i \qquad (8.3)$$

The property-mixing operators can be evaluated from first principles or estimated through empirical or semi-empirical methods. For instance, consider the mixing of two liquid sources whose flowrates are F_1 and F_2, volumetric flowrates are V_1 and V_2, and densities are ρ_1 and ρ_2. Suppose that the volumetric flowrate of the mixture is the sum of the volumetric flowrates of the two streams, i.e.,

$$\overline{V} = V_1 + V_2 \qquad (8.4)$$

Recalling the definition of density and designating the total flowrate of the mixture by $\overline{F}$, we get

$$\frac{\overline{F}}{\overline{\rho}} = \frac{F_1}{\rho_1} + \frac{F_2}{\rho_2} \tag{8.5}$$

Comparing equations (8.2) and (8.5), we can define the density-mixing operator as:

$$\psi(\rho_i) = \frac{1}{\rho_i} \tag{8.6}$$

Next, we define the property load of a stream as follows:

$$M_i^{\text{Source}} = W_i \times \psi(p_i) \tag{8.7}$$

For source i, let

$$y_i = \psi(p_i) \tag{8.8}$$

Therefore, equation (8.7) can be rewritten as

$$M_i^{\text{source}} = W_i \times y_i \tag{8.9}$$

We can rewrite the sink constraints given by equation (8.1) in terms of the property-mixing operator as:

$$\psi_j^{\min} \leq \psi_j^{\text{in}} \leq \psi_j^{\max} \tag{8.10}$$

In order to develop the targeting procedure, let us first start with the following special case for which the fresh source has superior properties compared to all other streams, and the sink constraint is given by:

$$\psi^{\text{Fresh}} \leq \psi_j^{\text{in}} \leq \psi_j^{\max} \tag{8.11}$$

where

$$\psi^{\text{Fresh}} = \psi(p_{\text{Fresh}}) \tag{8.12}$$

The property load of a sink is defined as the product of the flowrate times the property operator of the feed to the sink. Hence, we define the maximum property load for a sink as

$$M_j^{\text{Sink,max}} = G_j \times \psi_j^{\max} \tag{8.13}$$

For sink j, let

$$z_j^{\max} = \psi_j^{\max} \tag{8.14}$$

Therefore, equation (8.13) can be rewritten as

$$M_j^{\text{Sink,max}} = G_j \times z_j^{\max} \tag{8.15}$$

It is beneficial to observe the similarity between equations (8.9) and (8.15) with their equivalent expressions given by equations (3.60) and (3.58). Therefore, the property loads are analogous to the mass loads. Consequently, we can develop a targeting procedure similar to the one given by the material recycle pinch diagram. This tool is referred to as the *property-based material reuse pinch diagram* (Kazantzi and El-Halwagi 2005) and can be constructed through the following procedure:

1. **Sink Data:** Gather data on the flowrate and acceptable range of targeted property for each sink as in constraint (8.1). Using the admissible range of property value, calculate the maximum value of the property operator $\psi_j^{\max}$. Then, evaluate the maximum admissible property load ($M_j^{\text{Sink, max}}$) using equation (8.13). Rank the sinks in ascending order of $\psi_j^{\max}$.
2. **Sink Composite:** Using the required flowrate for each sink (G_j) and the calculated values of the maximum admissible loads ($M_j^{\text{Sink, max}}$), develop a representation for each sink in ascending order. Superposition is used to create a sink composite curve. This is shown in Figure 8-2.
3. **Fresh Line:** Use equation (8.12) to evaluate the property operator of fresh (ψ_{Fresh}). A locus of the fresh line is drawn starting from the origin with a slope of ψ_{Fresh}.

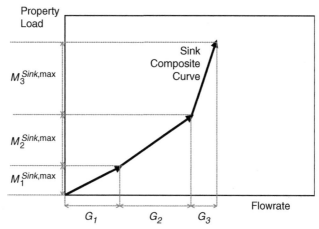

FIGURE 8-2   DEVELOPING SINK COMPOSITE CURVE

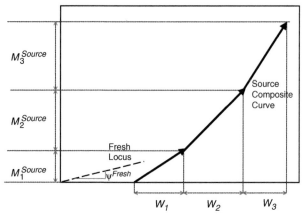

FIGURE 8-3   SOURCE COMPOSITE CURVE

4. **Source Data:** Gather data on the flowrate and property value for each process source. Using the functional form of the property operator, calculate the value of the property operator for each source (ψ_i). Rank the sources in ascending order of ψ_i. Also, calculate the property load of each source (M_i) using equation (8.7).

5. **Source Composite**: Using the flowrate for each source (G_i) and the calculated values of the property operator (ψ_i), develop a representation for each source in ascending order. Superposition is used to create a source composite curve. As mentioned in step 3, the locus of the fresh stream is represented by plotting a line with a slope of ψ_{Fresh}. The previous steps are represented in Figure 8-3.

6. **Pinch Diagram:** The source composite curve is slid on the fresh line pushing it to the left, while keeping it below the sink composite curve, until the two composites touch at the pinch point with the source composite completely below the sink composite in the overlapped region. Determine the minimum consumption of fresh resource and the minimum discharge of the waste as shown by the pinch diagram (Figure 8-4).

Similar to the design rules mentioned in Chapter Three, for the minimum usage of fresh resources, maximum reuse of the process sources and minimum discharge of waste, the following three design rules are needed:

- No property load should be passed through the pinch (i.e., the two composites touch)
- No waste should be discharged from sources below the pinch
- No fresh should be used in any sink above the pinch.

The targeting procedure identifies the targets for fresh, waste, and material reuse without commitment to the detailed design of the network matching the sources and sinks.

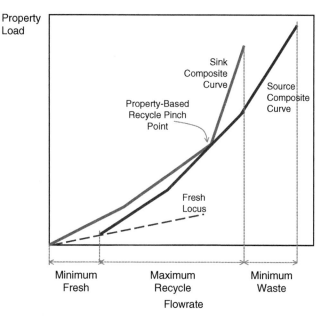

FIGURE 8-4   PROPERTY-BASED MATERIAL-REUSE PINCH DIAGRAM (KAZANTZI AND EL-HALWAGI 2005)

For other cases of property-based pinch diagrams (including the case when the property operators of the sources and the fresh are above those of the sinks or the case when the fresh source has a property operator that lies in the midst of property operators of process sources), the reader is referred to the paper by Kazantzi and El-Halwagi (2005).

8.2 PROCESS MODIFICATION BASED ON PROPERTY-BASED PINCH DIAGRAM

The property-based material-reuse pinch diagram and the aforementioned design rules can be used to guide the engineer in making process modifications to enhance material reuse; for instance, the observation that below the pinch there is deficiency of reusable sources, whereas above the pinch there is a surplus of sources. Therefore, sinks can be moved from below the pinch to above the pinch and sources can be moved from above to below the pinch to reduce the usage of fresh resources and the discharge of waste. Moving a sink from below to above the pinch can be achieved by increasing the upper bound on the property constraint of the sink given by equation (8.13). Conversely, moving a source from above to below the pinch can be accomplished by reducing the value of the property operator through changes in operating conditions or by adding an "interception" device (e.g., heat exchanger, separator, reactor, etc.) that can lower the value of the property operator. In such cases, the problem representation shown in Figure 8-1 can be extended to account for the addition of new "interception"

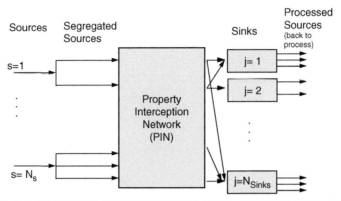

FIGURE 8-5 SCHEMATIC REPRESENTATION OF PROPERTY-BASED ALLOCATION AND INTERCEPTION (EL-HALWAGI ET AL., 2004)

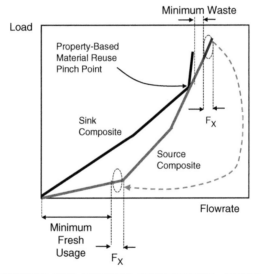

FIGURE 8-6 INSIGHTS FOR PROCESS MODIFICATION: MOVING A SOURCE ACROSS THE PINCH (KAZANTZI AND EL-HALWAGI 2005)

devices that serve to adjust the properties of the sources. Figure 8-5 is a schematic representation of the extended problem statement which incorporates stream allocation, interception, and process modification. Figure 8-6 illustrates an example of the pinch-based representation when a flowrate F_x is intercepted and its value of property operator is reduced to match that of the fresh. Compared to the nominal case of Figure 8-4, two benefits accrue as a result of this movement across the pinch: both the usage of fresh resource and the discharge of waste are reduced by F_x.

Another alternative is to reduce the load of a source below the pinch (again by altering operating conditions or adding an interception device). Consequently, the cumulative load of the source composite decreases and

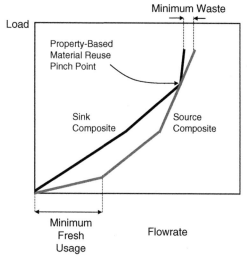

FIGURE 8-7 INSIGHTS FOR PROCESS MODIFICATION: REDUCING THE LOAD OF A SOURCE (KAZANTZI AND EL-HALWAGI 2005)

allows an additional reuse of process sources with the result of decreasing both the fresh consumption and waste discharge. Figure 8-7 demonstrates the revised pinch diagram for Figure 8-4, after the load of the process source (and therefore its slope) below the pinch has been decreased. When slid on the fresh line, an additional amount of the source above the pinch can be reused resulting in a net reduction in fresh and waste.

8.3 EXAMPLE ON SOLVENT RECYCLE IN METAL DEGREASING (SHELLEY AND EL-HALWAGI 2000; KAZANTZI AND EL-HALWAGI 2005)

A metal degreasing process presented in Figure 8-8 is considered here. Currently, a fresh organic solvent is used in the degreaser and the absorber. A reactive thermal processing and solvent regeneration system is used to decompose the grease and the organic additives, and regenerate the solvent from the degreaser. The liquid product of the solvent regeneration system is reused in the degreaser, while the gaseous product is passed through a condenser, an absorber, and a flare. The process produces two condensate streams: Condensate I from the solvent regeneration unit and Condensate II from the degreaser. The two streams are currently sent to hazardous waste disposal. Since these two streams possess many desirable properties that enable their possible use in the process, it is recommended that their recycle/reuse to be considered. The absorber and the degreaser are the two process sinks. The two process sources satisfy many properties required for the feed of the two sinks. An additional property should be examined; namely Reid Vapor Pressure (RVP), which is important in characterizing the volatility, makeup, and regeneration of the solvent.

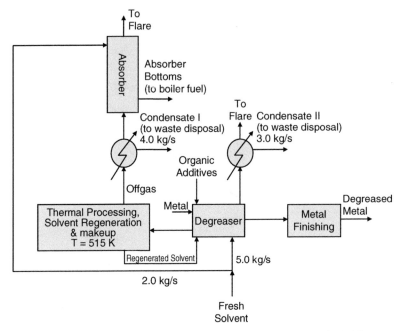

FIGURE 8-8   A DEGREASING PLANT (KAZANTZI AND EL-HALWAGI 2005)

TABLE 8-1 FLOWRATES AND BOUNDS ON PROPERTIES OF SINKS (KAZANTZI AND EL-HALWAGI 2005)

Sink	Flowrate (kg/s)	Lower bound on RVP (atm)	Upper bound on RVP (atm)
Degreaser	5.0	2.0	3.0
Absorber	2.0	2.0	4.0

The mixing rule for vapor pressure (RVP) is given by the following expression:

$$\overline{\text{RVP}}^{1.44} = \sum_{i=1}^{N_s} x_i \text{RVP}_i^{1.44} \tag{8.16}$$

Moreover, the inlet flowrates of the feed streams to the degreaser and absorber along with their constraints on the property (RVP) are given in Table 8-1.

The RVP for Condensate I is a function of the thermal regeneration temperature as follows:

$$\text{RVP}_{\text{Condensate I}} = 0.5e^{[(T-100)/175]} \tag{8.17}$$

where $RVP_{Condensate\ I}$ is the RVP of condensate I in atmosphere and T is the temperature of the thermal processing system in kelvin. The acceptable range of this temperature is 430 to 520 K. At present, the thermal processing system operates at 515 K leading to an RVP of 6.0. The data for Condensate I and Condensate II are given in Table 8-2.

    a.   What is the target for minimum fresh usage when direct recycle is used?

    b.   If the process is to be modified by manipulating the temperature of the condenser, what is the target for minimum fresh usage?

SOLUTION

    a.   The values of property loads for the sinks and the sources are calculated as shown in Tables 8-3 and 8-4. It is worth noting that all the sources satisfy the lower bound property constraints for the sinks. Therefore, we focus on the maximum admissible values of the property constraints for the sinks. Once the property loads are calculated, we construct the sink and source composite curves in ascending order of operators and slide the source composite on the fresh line until the two composites touch at the

TABLE 8-2 PROPERTIES OF PROCESS SOURCES AND FRESH (KAZANTZI AND EL-HALWAGI 2005)

Source	Flowrate (kg/s)	RVP (atm)
Process Condensate I	4.0	6.0
Process Condensate II	3.0	2.5
Fresh solvent	To be determined	2.0

TABLE 8-3 CALCULATING MAXIMUM PROPERTY LOADS FOR THE SINKS

Sink	Flowrate (kg/s)	Upper bound on RVP (atm)	Upper bound on $RVP^{1.44}$ ($atm^{1.44}$)	Maximum property load ($kg \cdot atm^{1.44}$/s)
Degreaser	5.0	3.0	4.87	24.35
Absorber	2.0	4.0	7.36	14.72

TABLE 8-4 CALCULATING PROPERTY LOADS FOR THE SOURCES

Source	Flowrate (kg/s)	RVP (atm)	$RVP^{1.44}$ ($atm^{1.44}$)	Property load ($kg \cdot atm^{1.44}$/s)
Process Condensate I	4.0	6.0	13.20	52.80
Process Condensate II	3.0	2.5	3.74	11.22

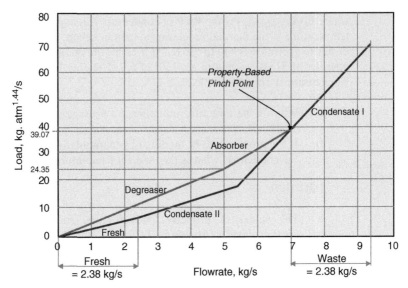

FIGURE 8-9   PROPERTY-BASED PINCH DIAGRAM FOR THE DEGREASING CASE STUDY (KAZANTZI AND EL-HALWAGI 2005)

pinch point, as shown in Figure 8-9. Thus, we can graphically determine the fresh consumption, which is 2.38 kg/s and the waste discharge, which is 2.38 kg/s as well. The fresh solvent consumption is, therefore, reduced by 66%.

b. As mentioned earlier, reuse can be enhanced through process modifications. For instance, if it is desired to completely eliminate the fresh consumption by reducing the load of Condensate I, then we reconstruct the pinch diagram with the two process sources only. Since Condensate I lies above the Absorber, it is necessary to decrease the slope of Condensate I, until it touches the sink composite curve with Condensate I completely below the sink composite curve. Figure 8-10 shows that the property load of Condensate I has to be reduced until it becomes $26.26 \, \text{kg} \cdot \text{atm}^{1.44}/\text{s}$. Recalling the definition of property load, we get:

$$26.26 = 4.0 \times \text{RVP}_{\text{Condensate I}}^{1.44} \qquad (8.18\text{a})$$

Hence,

$$\text{RVP}_{\text{Condensate I}} = 3.69 \, \text{atm} \qquad (8.18\text{b})$$

Substituting from equation (8.18b) into equation (8.17), we have

$$3.69 = 0.5e^{[(T-100)/175]} \qquad (8.19\text{a})$$

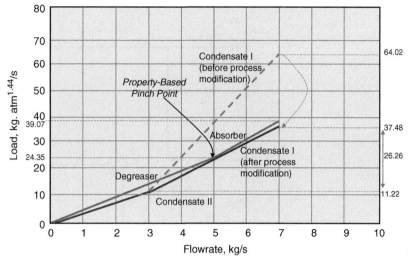

FIGURE 8-10 IDENTIFICATION OF PROCESS MODIFICATION TASK FOR THE DEGREASING CASE STUDY

i.e.,

$$T = 430 \, \text{K} \tag{8.19b}$$

Hence, in order to achieve this RVP, the temperature of the thermal processing system should be reduced to 430 K. The solution is shown by Figure 8-11.

8.4 CLUSTERING TECHNIQUES FOR MULTIPLE PROPERTIES

The previously presented pinch diagram is suitable for cases governed by a single dominant property. When the sink constraints and the process sources are characterized by multiple properties, it is necessary to address these properties simultaneously. Since properties (or functionalities) form the basis of performance of many units, it will be very insightful to develop design procedures based on key properties instead of key compounds. The challenge, however, is that while chemical components are conserved, properties are not. Therefore, the question is: whether or not it is possible to track these functionalities instead of compositions? Shelley and El-Halwagi (2000) have shown that it is possible to tailor conserved quantities, called clusters, that act as surrogate properties and enable the conserved tracking of functionalities instead of components. The basic mathematical expressions for clusters are given by Shelley and El-Halwagi (2000) and are summarized in the following section. Suppose that, we have N_{Sources} streams. Each stream, i, is characterized by N_p raw properties. The index for the properties is

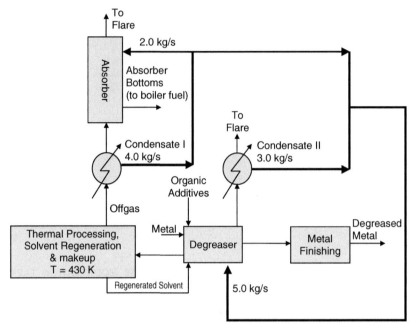

FIGURE 8-11   SOLUTION TO THE DEGREASING CASE STUDY WITH PROCESS MODIFICATION AND SOURCE REUSE (KAZANTZI AND EL-HALWAGI 2005)

designated by r. Consider the class of properties whose mixing rules for each raw property, r, are given by the following equation:

$$\psi_r(\overline{p}_r) = \sum_{i=1}^{N_{\text{Sources}}} x_i \psi_r(p_{r,i}) \tag{8.20}$$

Equation (8.20) is another form of equation (8.2) where x_i is the fractional contribution of the ith stream into the total flowrate of the mixture and $\psi_r(p_{r,i})$ is an operator on $p_{r,i}$ which can be normalized into a dimensionless operator by dividing it by a reference value

$$\Omega_{r,i} = \frac{\psi_r(p_{r,i})}{\psi_r^{\text{ref}}} \tag{8.21}$$

Then, an AUgmented Property (AUP) index for each stream, i, is defined as the summation of the dimensionless raw property operators:

$$\text{AUP}_i = \sum_{r=1}^{N_p} \Omega_{r,i} \quad i = 1, 2, \ldots, N_{\text{Sources}} \tag{8.22}$$

The cluster for property r in stream i, $C_{r,i}$, is defined as follows:

$$C_{r,i} = \frac{\Omega_{r,i}}{\text{AUP}_i} \qquad (8.23)$$

Property clusters are useful quantities that enable the conserved tracking of properties and the derivation of important design tools. In particular, property clusters possess two key characteristics: intra- and inter-stream conservation.

Intra-Stream Conservation: For any stream i, the sum of clusters must be conserved adding up to a constant (e.g., unity), i.e.,

$$\sum_{r=1}^{N_p} C_{r,i} = 1 \quad i = 1, 2, \ldots, N_{\text{Sources}} \qquad (8.24)$$

Figure 18-12a illustrates the ternary representation of equation (8.24) for three clusters of the ith stream.

Inter-Stream Conservation: When two or more streams are mixed, the resulting individual clusters are conserved. This is represented by consistent additive rules in the form of lever-arm rules, so that the mean cluster property of two or more streams with different property clusters can be easily

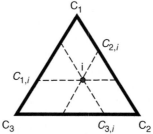

FIGURE 8-12a TERNARY REPRESENTATION OF INTRA-STREAM OF CLUSTERS WITHIN THE SAME STREAMS (SHELLEY AND EL-HALWAGI 2000). EACH TERNARY AXES RANGE FROM 0.0 TO 1.0

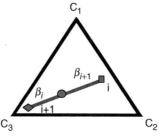

FIGURE 8-12b LEVER ARM ADDITION FOR CLUSTERS OF TWO STREAMS (SHELLEY AND EL-HALWAGI 2000)

determined graphically. The lever-arm additive rule for clusters can be expressed by:

$$\bar{C}_r = \sum_{i=1}^{N_{\text{Sources}}} \beta_i C_{r,i} \quad r = 1, 2, \ldots, N_p \tag{8.25}$$

where $\bar{C}_r$ is the mean cluster for the rth property resulting from adding the individual clusters of N_{Sources} streams and β_i represents the fractional lever arm of cluster, $C_{r,i}$, of stream i. The cluster arm is given by:

$$\beta_i = \frac{x_i \, \text{AUP}_i}{\overline{\text{AUP}}} \tag{8.26}$$

where

$$\overline{\text{AUP}} = \sum_{i=1}^{N_{\text{Sources}}} x_i \, \text{AUP}_i \tag{8.27}$$

Although the mixing of the original properties may be based on non-linear rules, the clusters are tailored to exhibit linear mixing rules in the cluster domain. When two sources (i and $i+1$) are mixed, the locus of all mixtures on the cluster ternary diagram is given by the straight line connecting sources i and $i+1$. Depending on the fractional contributions of the streams, the resulting mixture splits the mixing line in ratios β_i and β_{i+1}. Figure 8-12b illustrates the lever-arm addition for clusters of two streams, i and $i+1$. It looks similar to the lever-arm mixing rules for the three-component source–sink mapping diagram presented in Chapter Three. It is worth noting that equation (8.25) describes a revised lever-arm rule for which the arms are different from those for the mixing of masses. In addition to mass fractions, they also include ratios of the augmented properties of the mixed streams. The appearance of the augmented property index in the lever-arm rule allows one-to-one mapping from raw properties to property clusters and vice versa when streams are mixed.

It is instructive to verify the intra- and inter-stream conservation rules for the clusters as well as the expressions for the lever arms and augmented properties. First, we start by summing the individual clusters in the same stream (expressed by equation (8.23)) and substituting equation (8.22) to get:

$$\sum_{r=1}^{N_p} C_{i,s} = \frac{\sum_{r=1}^{N_p} \Omega_{r,i}}{\text{AUP}_i} = \frac{\text{AUP}_i}{\text{AUP}_i} = 1 \tag{8.28}$$

This proves the intra-stream conservation. Furthermore, if we mix streams (with each stream possessing a cluster of $C_{r,i}$ and a fractional flowrate contribution of x_i), we get a resulting cluster of $\overline{C}_r$. According to the definition given by equation (8.23), we have

$$\overline{C}_r = \frac{\overline{\Omega}_r}{\overline{\text{AUP}}} \tag{8.29}$$

Dividing both sides of equation (8.20) by ψ_r^{ref} and employing the definition given by equation (8.21), we get

$$\overline{\Omega}_r = \frac{\psi_r(\overline{p}_r)}{\psi_r^{\text{ref}}} = \frac{\sum\limits_{i=1}^{N_{\text{Sources}}} x_i \psi_r(p_{r,i})}{\psi_r^{\text{ref}}} = \sum_{i=1}^{N_{\text{Sources}}} x_i \Omega_{r,i} \tag{8.30}$$

Substituting from equation (8.30) into equation (8.29), we have:

$$\overline{C}_r = \sum_{i=1}^{N_{\text{Sources}}} \frac{x_i \Omega_{r,i}}{\overline{\text{AUP}}} \tag{8.31}$$

Substituting from equation (8.23) into equation (8.31), we obtain:

$$\overline{C}_r = \sum_{i=1}^{N_{\text{Sources}}} \frac{x_i \, \text{AUP}_i}{\overline{\text{AUP}}} C_{r,i} \tag{8.32}$$

Let us denote the cluster-mixing arm of the ith stream by

$$\beta_i = \frac{x_i \, \text{AUP}_i}{\overline{\text{AUP}}} \tag{8.33}$$

Therefore, equation (8.32) can be written as (see equation (8.25))

$$\overline{C}_r = \sum_{i=1}^{N_{\text{Sources}}} \beta_i C_{r,i} \quad r = 1, 2, \ldots, N_{\text{p}}$$

which is the inter-stream conservation rule. Now that we have defined the cluster arm, let us derive the expression for $\overline{\text{AUP}}$. Equation (8.24) can be applied to the mixture stream to give:

$$\sum_{r=1}^{N_{\text{p}}} \overline{C}_r = 1 \tag{8.34}$$

Substituting equation (8.25) into equation (8.34), we get

$$\sum_{r=1}^{N_{\text{p}}} \sum_{i=1}^{N_{\text{Sources}}} \beta_i C_{r,i} = \sum_{i=1}^{N_{\text{Sources}}} \sum_{r=1}^{N_{\text{p}}} \beta_i C_{r,i} = 1 \tag{8.35a}$$

Hence,

$$\sum_{i=1}^{N_{\text{Sources}}} \beta_i \sum_{r=1}^{N_{\text{p}}} C_{r,i} = 1 \tag{8.35b}$$

Substituting from equation (8.24) into equation (8.35b), we get

$$\sum_{i=1}^{N_{\text{Sources}}} \beta_i = 1 \tag{8.36}$$

which can be combined with equation (8.33) to give:

$$\frac{1}{\overline{\text{AUP}}} \sum_{i=1}^{N_{\text{Sources}}} x_i \, \text{AUP}_i = 1 \tag{8.37}$$

which can be rearranged to give the proof for equation (8.27):

$$\overline{\text{AUP}} = \sum_{i=1}^{N_{\text{Sources}}} x_i \, \text{AUP}_i$$

8.5 CLUSTER-BASED SOURCE–SINK MAPPING DIAGRAM FOR PROPERTY-BASED RECYCLE AND INTERCEPTION

With multiple properties involved, the overall problem for property integration involving the recycle, reuse, and interception of streams can be stated as follows (El-Halwagi et al., 2004): "Given a process with certain sources (streams) and sinks (units) along with their properties and constraints, it is desired to develop graphical techniques that identify optimum strategies for allocation and interception that integrate the properties of sources, sinks, and interceptors so as to optimize a desirable process objective (e.g., minimum usage of fresh resources, maximum utilization of process resources, minimum cost of external streams) while satisfying the constraints on properties and flowrate for the sinks".

Our objective is to develop visualization tools that systematically optimize a certain process objective (e.g., minimum usage of fresh resources, maximum utilization of process resources, minimum cost of external streams) while satisfying the constraints on properties and flowrate for the sinks. The solution strategies include a combination of allocation and interception. Allocation of sources involves the segregation and mixing of streams and their assignment to units throughout the process. Interception involves the use of processing units (typically new equipment) to adjust the properties of the various streams. If no capital investment is available for new interception devices, then a no/low cost solution will be based on the allocation of external sources and the recycle/reuse of internal sources to meet the constraints of the sinks.

In order to address the aforementioned problem, the following design decisions must be made:

- How to identify the geometrical shape of the feasibility region for each sink?
- To which sinks should the sources be allocated?
- Is there a need to use a fresh (external) source? If yes, how much and where should it be employed?
- Is there a need for segregation or mixing?
- Is interception required to modify the properties of the sources? What are the optimal interception tasks to reach process targets or objectives?

The three-property cluster source–sink mapping diagram is analogous to the ternary source–sink mapping diagram described in Chapter Three. However, constructing the boundaries of the feasibility region (BFR) for the sink is not as straightforward as in the case of composition-based case. Since the constraints are given in terms of properties while the cluster source–sink diagram is in the cluster domain, plotting the BFR requires derivation of construction rules. These rules have been derived by El-Halwagi et al. (2004) and are summarized as follows:

- The BFR can be accurately represented by no more than **six** linear segments.
- When extended, the linear segments of the BFR constitute three convex hulls (cones) with their heads lying on the three vertices of the ternary cluster diagram. This observation is shown by Figure 8-13a.
- The six points defining the BFR can be determined before constructing the BFR. These six points are characterized by the following values

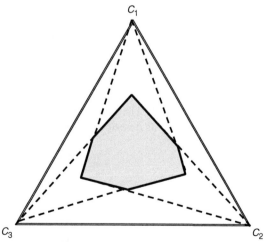

FIGURE 8-13a   THE BFR IS BOUNDED BY THREE CONES EMANATING FROM THE CLUSTER VERTICES (EL-HALWAGI ET AL., 2004)

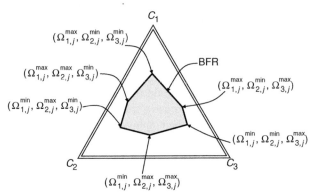

FIGURE 8-13b THE SIX POINTS DEFINING THE BOUNDARIES OF THE FEASIBILITY REGION OF A SINK j (EL-HALWAGI ET AL., 2004)

of dimensionless operators Ω's for the sink constraints: $(\Omega_{i,s}^{min}, \Omega_{j,s}^{min}, \Omega_{k,s}^{max})$, $(\Omega_{i,s}^{min}, \Omega_{j,s}^{max}, \Omega_{k,s}^{max})$, $(\Omega_{i,s}^{min}, \Omega_{j,s}^{max}, \Omega_{k,s}^{min})$, $(\Omega_{i,s}^{max}, \Omega_{j,s}^{max}, \Omega_{k,s}^{min})$, $(\Omega_{i,s}^{max}, \Omega_{j,s}^{min}, \Omega_{k,s}^{min})$, and $(\Omega_{i,s}^{max}, \Omega_{j,s}^{min}, \Omega_{k,s}^{max})$. This observation is shown by Figure 8-13b.

8.6 PROPERTY-BASED DESIGN RULES FOR RECYCLE AND INTERCEPTION

With the rigorous determination of the BFR, we can now proceed and describe the design rules. In particular, the following visualization techniques and revised lever-arm rules that can be systematically used for mixing points and property modification tasks. We also present the optimality conditions for selecting values of the augmented property index of a sink. For the derivation of these rules, the reader is referred to El-Halwagi et al. (2004).

Source Prioritization Rule: When two sources (i and $i+1$) are mixed to satisfy the property constraints of a sink with source i *being more expensive than* $i+1$, minimizing the mixture cost is achieved by selecting the minimum feasible value of x_i.

It is important to note that x_i cannot be directly visualized on the ternary cluster diagram. Instead, the lever arms on the ternary cluster diagram represent another quantity, β_i. The two terms are related through the Augmented Property Index (AUP) as described earlier, see equation (8.33):

$$\beta_i = \frac{x_i \, \mathrm{AUP}_i}{\overline{\mathrm{AUP}}}$$

Using equation (8.27), let us rewrite equation (8.33) in the case of mixing two sources (i and $i+1$):

$$\beta_i = \frac{x_i \, \mathrm{AUP}_i}{x_i \, \mathrm{AUP}_i + (1 - x_i) \, \mathrm{AUP}_{i+1}} \tag{8.38a}$$

Rearranging, we get

$$x_i = \frac{\beta_i \, \text{AUP}_{i+1}}{\beta_i \, \text{AUP}_{i+1} + (1 - \beta_i) \, \text{AUP}_i} \tag{8.38b}$$

Next, let us take the first derivative of x_i with respect to β_i, we get:

$$\frac{dx_i}{d\beta_i} = \frac{\text{AUP}_{s+1}[\beta_i \, \text{AUP}_{i+1} + (1 - \beta_i) \, \text{AUP}_i] - \beta_i \, \text{AUP}_{i+1}[\text{AUP}_{i+1} - \text{AUP}_i]}{[\beta_i \, \text{AUP}_{i+1} + (1 - \beta_i) \, \text{AUP}_i]^2}$$

$$\tag{8.39}$$

Rearranging and simplifying, we get

$$\frac{dx_i}{d\beta_i} = \frac{\text{AUP}_i \, \text{AUP}_{i+1}}{[\beta_i \, \text{AUP}_{i+1} + (1 - \beta_i) \, \text{AUP}_i]^2} \tag{8.40}$$

With both AUP_i and AUP_{i+1} being non-negative, the right-hand side of equation (8.40) is also non-negative. Therefore x_i as a function of β_i is monotonically increasing. From this we can state that following rule (El-Halwagi et al., 2004):

Lever Arm Source Prioritization Rule: On a ternary cluster diagram, minimization of the cluster arm of a source corresponds to minimization of the flow contribution of that source. In other words, **minimum β_i corresponds to minimum x_i.**

For instance, consider the case of a fresh (external) resource (F) whose flowrate is to be minimized by recycling the maximum possible flowrate of a process streams (e.g., waste W) as shown by Figure 8-14. The straight line

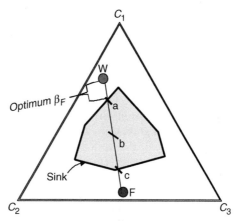

FIGURE 8-14 DETERMINATION OF OPTIMAL MIXING POINT FOR A FRESH RESOURCE AND A PROCESS SOURCE (EL-HALWAGI ET AL., 2004)

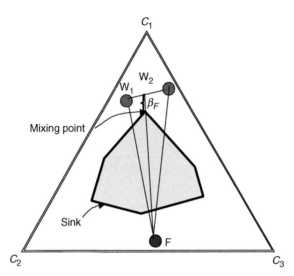

FIGURE 8-15   DETERMINATION OF FRESH ARM WHEN TWO PROCESS SOURCES ARE USED (EL-HALWAGI ET AL., 2004)

connecting the two sources represents the locus for any mixture of W and F. The resulting mixture splits the total mixing arm in the ratios of β_F to β_W. The intersection of the mixing line with the feasibility region of the sink gives the line segment representing all feasible mixtures. This is shown in Figure 8-14 by the segment connecting points a and c. The question is what should be the optimum mixing point (e.g., a, b, or c)? Our objective is to minimize the mixture cost. Based on the lever-arm and prioritization rules, the optimal mixing point corresponds to the minimum feasible β_F. This is point a as shown in Figure 8-14. It is worth mentioning that this is a necessary condition only. For sufficiency, values of the augmented property index and flowrate should match as well. This issue is explained in detail by El-Halwagi et al. (2004).

The same concept can be generalized when more than two sources are mixed. For instance, consider the mixing of a fresh resource (F) with two process streams (W_1 and W_2). As can be seen from Figure 8-15, the mixing region is defined by the triangle connecting points F, W_1, and W_2. Any mixture of the three sources can be represented by a point in that triangle. It is worth noting that the line connecting represents all possible mixtures of W_1 and W_2. For a given mixing point (e.g., the mixing point shown in Figure 8-15), one can graphically determine the fresh arm (β_F) and use it to calculate the flowrate of the fresh source according to equation (8.33).

In case the target for recycling a process stream (W) is not met, more of the process stream may be recycled by adjusting its properties via an interception device (e.g., separation, reaction, etc.). Our objective is to alter the properties of W such that the use of F is minimized. The logical question is what should be the task of the interception device in changing the properties (and consequently the cluster values) of the process stream? The cluster

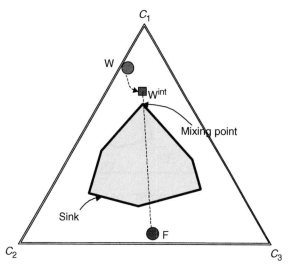

FIGURE 8-16 TASK IDENTIFICATION FOR THE INTERCEPTION SYSTEM (EL-HALWAGI ET AL., 2004)

diagram can rigorously answer this question. As shown by Figure 8-16, for the selected mixing point and the desired value of β_F, the fresh arm can be drawn to determine the desired location of the intercepted internal stream (W^{int}). Additionally, equation (8.38b) may be employed to determine the desired value of the augmented property index for W^{int} (since the values of the augmented property index are known for F and the mixing point of the sink). Once the ternary cluster value for W^{int} and its augmented property index are determined, equation (8.23) which defines the cluster equation is solved backwards to determine the dimensionless property operators for the intercepted stream. Next, equation (8.21) is solved to calculate the raw properties of W^{int}. This is the minimum extent of interception (as measured on the cluster domain) to achieve maximum recycle of W or minimum usage of the fresh since additional interception will still lead to the same target of minimum usage, but will result in a mixing point inside the sink and not just on the periphery of the sink. Once the task for the interception system is defined, conventional process synthesis techniques can be employed to develop the design and operating parameters for the interception system. The same procedure can be repeated for various mixing points resulting in the task identification of the locus for minimum extent of interception (Figure 8-17).

8.7 DEALING WITH MULTIPLICITY OF CLUSTER-TO-PROPERTY MAPPING (EL-HALWAGI ET AL., 2004)

Consider a certain point whose coordinates on the ternary cluster diagram are given by $C_1^{sink}, C_2^{sink}, C_3^{sink}$ This cluster point may correspond to multiple

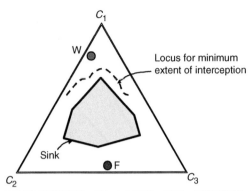

FIGURE 8-17 DEVELOPMENT OF LOCUS FOR MINIMUM INTERCEPTION (EL-HALWAGI ET AL., 2004)

combinations of property points[1]. In other words, as a result of the non-linear mapping from the property domain to the cluster domain, it is possible to have multiple property points (n_{Multiple}) from the feasible property domain that result in the same value of clusters, i.e.,

$$(C_1^{\text{sink}}, C_2^{\text{sink}}, C_3^{\text{sink}}) \equiv (p_{1,m}, p_{2,m}, p_{3,m}) \quad \text{where } m = 1, 2, \ldots n_{\text{Multiple}} \tag{8.41}$$
$$\text{and} \quad (p_{1,m}, p_{2,m}, p_{3,m}) \text{ are feasible for the considered sink.}$$

For each property combination ($p_{1,m}$, $p_{2,m}$, $p_{3,m}$), the corresponding augmented property index is designated by AUP_m. The set of all these multiple feasible values of AUP_m is defined as:

$$\text{SET\_AUP}_{(C_1^{\text{sink}}, C_2^{\text{sink}}, C_3^{\text{sink}})}^{\text{Feasible}} = \{\text{AUP}_m | m = 1, 2, \ldots n_{\text{Multiple}}\} \tag{8.42}$$

Looking at the minimum and maximum values of AUP_m, the range for the values of AUP_m is given by the following interval:

$$\text{INTERVAL\_AUP}_m = [\text{Argmin AUP}_m, \text{Argmax AUP}_m] \tag{8.43}$$

where Argmin AUP_m and Argmax AUP_m are the lowest and highest values of $\text{AUP}_m \in \text{SET\_AUP}_{(C_1^{\text{sink}}, C_2^{\text{sink}}, C_3^{\text{sink}})}^{\text{Feasible}}$

As a result of such multiplicity, three conditions must be satisfied to insure feasibility of feeding sources (or mixtures of sources) into a sink:

1. The cluster value for the source (or mixture of sources) must be contained within the feasibility region of the sink on the cluster ternary diagram.

---

[1]Points on the BFR are exceptions. Each point on the BFR has a one-to-one mapping to the property domain. The reason for such one-to-one mapping stems from the uniqueness of the six vertices on the BFR. Mixtures of unique points also correspond to unique points.

2. The values of the augmented property index for the source (or mixture of sources) and the sink must match.
3. The flowrate of the source (or mixture of sources) must lie within the acceptable feed flowrate range for the sink.

Suppose that two sources i and $i+1$ are mixed in a ratio of x_i to x_{i+1} such that the resulting mixture has a cluster value which matches that of a feasible sink cluster, i.e.,

$$(C_1^{\text{mixture}}, C_2^{\text{mixture}}, C_3^{\text{mixture}}) = (C_1^{\text{sink}}, C_2^{\text{sink}}, C_3^{\text{sink}}) \qquad (8.44)$$

Equation (8.44) corresponds to the abovementioned second condition for insuring feasibility of feeding sources into a sink. As stated earlier, this is a necessary but not sufficient condition for satisfying the sink constraint. In addition to satisfying the previous condition and satisfying the flowrate (third condition), sufficiency is guaranteed when the value of the augmented property for the mixture matches a feasible value of the augment property for the sink, i.e.,

$$\text{AUP}_{\text{mixture}} = \text{AUP}_m \in \text{SET\_AUP}^{\text{Feasible}}_{(C_1^{\text{sink}}, C_2^{\text{sink}}, C_3^{\text{sink}})} \qquad (8.45)$$

So, which of the n_{Multiple} feasible values of the augmented properties in the set $\text{SET\_AUP}^{\text{Feasible}}_{(C_1^{\text{sink}}, C_2^{\text{sink}}, C_3^{\text{sink}})}$ should be selected?

Recalling the Source Prioritization Rule and designating source i to be more expensive than source $i+1$, minimizing x_i results in minimizing cost of the mixture. Consequently, we should select an AUP_m, which minimizes x_i. This can be determined by establishing the relationship between AUP_m and x_i.

Let us denote the numerical values of the augmented properties for sources i and $i+1$ as AUP_i and AUP_{i+1}, respectively. These are constants. According to equation (8.27), we can describe the augmented property of the mixture in terms of the individual augmented properties. Substituting into equation (8.45), we get

$$x_i \text{AUP}_i + (1 - x_i)\text{AUP}_{i+1} = \text{AUP}_m \qquad (8.46a)$$

or

$$\text{AUP}_m = x_i(\text{AUP}_i - \text{AUP}_{i+1}) + \text{AUP}_{i+1} \qquad (8.46b)$$

Hence,

$$x_i = \frac{\text{AUP}_m - \text{AUP}_{i+1}}{\text{AUP}_i - \text{AUP}_{i+1}} \qquad (8.47)$$

Therefore,

$$\frac{\partial x_i}{\partial \text{AUP}_m} = \frac{1}{\text{AUP}_i - \text{AUP}_{i+1}} \qquad (8.48)$$

which is monotonically increasing if $AUP_i > AUP_{i+1}$ and monotonically decreasing if $AUP_i < AUP_{i+1}$. Therefore, to minimize x_i (and consequently the cost), we should select

$$AUP_m^{optimum} = \text{Argmin } AUP_m \quad \text{if} \quad AUP_i > AUP_{i+1} \qquad (8.49a)$$

$$AUP_m^{optimum} = \text{Argmax } AUP_m \quad \text{if} \quad AUP_i < AUP_{i+1} \qquad (8.49b)$$

These results can be shown graphically in Figures 8-18a and b. According to equation (8.46b), the relationship between AUP_m and x_s is represented by a straight line whose slope is $AUP_s - AUP_{s+1}$. If $AUP_s > AUP_{s+1}$, the slope is positive (Figure 8-18a) which corresponds to equation (8.49a). On the other hand, when $AUP_s < AUP_{s+1}$, the slope is negative (Figure 8-18b) which corresponds to equation (8.49b).

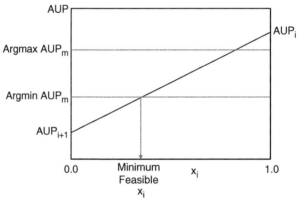

FIGURE 8-18a SELECTION OF OPTIMUM AUP$_m$ WHEN AUP$_i$ > AUP$_{i+1}$ (EL-HALWAGI ET AL., 2004)

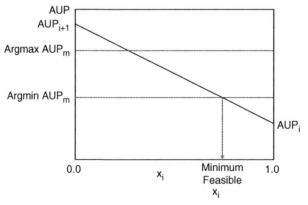

FIGURE 8-18b SELECTION OF OPTIMUM AUP$_m$ WHEN AUP$_i$ < AUP$_{i+1}$ (EL-HALWAGI ET AL., 2004)

As described by equation (8.46a), the AUP of the sink should match that of the mixture (first condition for feasibility). If no possible mixture can have an AUP matching that selected for the sink (e.g., Argmax AUP_m as per equation (8.49a)), then we systematically decrease the value of the sink's AUP starting with Argmax AUP_m until we get the highest value of $AUP_m \in SET\_AUP^{Feasible}_{(C^{sink}_1, C^{sink}_2, C^{sink}_3)}$ which matches that of the mixture. A similar procedure is adopted for the conditions of equation (8.49b) by systematically increasing the value of the sink's AUP starting with Argmin AUP_m until we get the highest value of $AUP_m \in SET\_AUP^{Feasible}_{(C^{sink}_1, C^{sink}_2, C^{sink}_3)}$ which matches that of the mixture.

Now that the foregoing rules and tools have been developed, it is useful to proceed to a case study to illustrate the applicability of these rules and tools.

8.8 PAPERMAKING AND FIBER RECYCLE EXAMPLE (EL-HALWAGI ET AL., 2004)

Figure 8-19 is a schematic representation of the papermaking process. A Kraft pulping process is used to yield digested pulp using a chemical-pulping process. The digested pulp is passed to a bleaching system to produce bleached pulp (fiber I). The plant also purchases an external pulp (fiber II). Two types of paper are produced through two papermaking machines (sinks I and II). Paper machine I employs 100 ton/h of fiber I. On the other hand, a mixture of fibers I and II (16.4 and 23.6 ton/h, respectively) is fed to paper machine II. Because of the occasional malfunctions of the process operation, a certain amount of partly and completely manufactured paper is rejected. These waste fibers are referred to as *reject*. The reject is typically passed through a hydro-pulper and a hydro-sieve with the net result of producing an

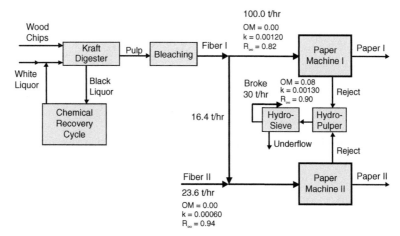

FIGURE 8-19 SCHEMATIC REPRESENTATION OF THE PULP AND PAPER PROCESS (EL-HALWAGI ET AL., 2004)

underflow, which is burnt, and an overflow (referred to as *broke*) which goes to waste treatment. It is worth noting that the broke contains fibers that may be partially recycled for papermaking.

The case study is aimed at providing optimal solutions to the following design questions:

a. Direct Recycle and Reallocation: What is the optimal allocation of the three fiber sources (fiber I, fiber II, and broke) for a direct recycle/reuse situation (no new equipment)?

b. Interception of Broke: To maximize the use of process resources and minimize wasteful discharge (broke), how should the properties of broke be altered so as to achieve its maximum recycle?

The performance of the paper machines and, consequently, the quality of the produced papers rely on three primary properties (Biermann 1996; Brandon 1981; and Willets 1958):

- **Objectionable Material (OM)**: this refers to the undesired species in the fiber (expressed as mass fraction).
- **Absorption Coefficient (k)**: which is an intensive property that provides a measure of absorptivity of light into the fibers (black paper has a high value of k). Hemicellulose and cellulose have very little absorption of light in the visible region. However, lignin has a high absorbance. Therefore, light absorbance is mostly attributed to lignin. The light absorption coefficient is a very useful property in determining the opacity of the fibers.
- **Reflectivity (R_∞)**: which is defined as the reflectance of an infinitely thick material compared to an absolute standard, which is Magnesium Oxide (MgO).

The mixing rules for OM and k are linear (Brandon 1981), i.e.,

$$\overline{\text{OM}} = \sum_{s=1}^{N_s} x_s \text{OM}_s \qquad (8.50)$$

$$\overline{k}(\tfrac{m^2}{g}) = \sum_{s=1}^{N_s} x_s k_s(\tfrac{m^2}{g}) \qquad (8.51)$$

On the other hand, a non-linear empirical mixing rule for R_∞ is developed using data from Willets (1958):

$$\overline{R}_\infty^{5.92} = \sum_{s=1}^{N_s} x_s R_{\infty_s}^{5.92} \qquad (8.52)$$

Tables 8-5 and 8-6 describe the constraints for the two sinks, while Table 8-7 provides the data on the properties of the sources.

TABLE 8-5 CONSTRAINTS FOR PAPER MACHINE I (SINK I)

Property	Lower bound	Upper bound
OM (mass fraction)	0.00	0.02
k (m^2/gm)	0.00115	0.00125
R_∞	0.80	0.90
Flowrate (ton/h)	100	105

TABLE 8-6 CONSTRAINTS FOR PAPER MACHINE II (SINK II)

Property	Lower bound	Upper bound
OM (mass fraction)	0.00	0.00
k (m^2/gm)	0.00070	0.00125
R_∞	0.85	0.90
Flowrate (ton/h)	40	40

TABLE 8-7 PROPERTIES OF FIBER SOURCES

Source	OM (mass fraction)	k (m^2/gm)	R_∞	Maximum available flowrate (ton/h)	Cost ($/ton)
Broke	0.08	0.00130	0.90	30	0
Fiber I	0.00	0.00120	0.82	∞	210
Fiber II	0.00	0.00060	0.94	∞	400

SOLUTION

To transform the problem from the property domain to the cluster domain, let us arbitrarily select the following reference values of the raw properties:

$$OM^{ref} = 0.01 \tag{8.53a}$$

$$k^{ref} = 0.001\,\text{m}^2/\text{gm} \tag{8.53b}$$

$$R_\infty^{ref} = 1.0 \tag{8.53c}$$

First, we use the values of the properties for the three usable sources (broke, fiber I, and fiber II) from Table 8-7 and map them to the cluster domain. As an illustration, consider the broke. The following are the values of the property operators for the broke:

$$\psi_{OM}^{Broke} = 0.08 \tag{8.54a}$$

$$\psi_{k}^{Broke} = 0.0013\,\text{m}^2/\text{gm} \tag{8.54b}$$

$$\psi_{R_\infty}^{Broke} = 0.9^{5.92} = 0.5359 \tag{8.54c}$$

Let us also calculate the values of the property operators for the reference properties. Hence

$$\psi_{OM}^{Broke} = 0.01 \tag{8.55a}$$

$$\psi_k^{Broke} = 0.001 \, m^2/gm \tag{8.55b}$$

$$\psi_{R_\infty}^{Broke} = 1.0^{5.92} = 1.0 \tag{8.55c}$$

Using the definition of equation (8.21), we get

$$\Omega_{OM}^{Broke} = \frac{0.08}{0.01} = 8.0000 \tag{8.56a}$$

$$\Omega_k^{Broke} = \frac{0.0013}{0.001} = 1.3000 \tag{8.56b}$$

$$\Omega_{R_\infty}^{Broke} = \frac{0.5359}{1.0000} = 0.5359 \tag{8.56c}$$

According to equation (8.22), we have

$$AUP^{Broke} = 8.0000 + 1.3000 + 0.5359 = 9.8359 \tag{8.57}$$

Therefore, the cluster values for broke can be calculated as follows:

$$C_{OM}^{Broke} = \frac{8.0000}{9.8359} = 0.8133 \tag{8.58a}$$

$$C_k^{Broke} = \frac{1.3000}{9.8359} = 0.1322 \tag{8.58b}$$

$$C_{R_\infty}^{Broke} = \frac{0.5359}{9.8359} = 0.0545 \tag{8.58c}$$

This cluster point for the broke is represented in Figure 8-20. Similarly, the sink constraints defined by Tables 8-5 and 8-6 are transformed into feasibility regions on the ternary cluster diagram. The rules derived in Table 8-8 illustrate the six points defining the BFR for sink I.

Figure 8-20 shows the results. While the feasibility region for sink I is a two-dimensional zone, it is a line segments for sink II as it lies on the zero cluster line for the objectionable materials (the base of the triangle).

As has been mentioned in the theoretical analysis (equations (8.41–8.43)), each cluster point within the feasibility region of the sink may correspond to multiple combinations of property points. For each property combination, the corresponding augmented property index is calculated. Consequently, the range for the values of AUP shown in Figure 8-21 refer to the interval

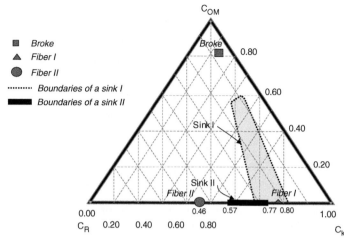

FIGURE 8-20 REPRESENTATION OF THREE USABLE SOURCES AND THE FEASIBILITY REGIONS FOR SINKS I AND II (EL-HALWAGI ET AL., 2004)

TABLE 8-8 VERTICES OF THE BFR FOR SINK I (EL-HALWAGI ET AL., 2004)

Characteristic dimensionless operators	Corresponding values of raw properties	Corresponding values of clusters	AUP$_{\text{Sink}}$
$(\Omega_{\text{OM,Sink I}}^{\min}, \Omega_{k,\text{Sink I}}^{\min}, \Omega_{R_\infty,\text{Sink I}}^{\max})$	0.00 0.00115 0.90	0.00 0.68 0.32	1.69
$(\Omega_{\text{OM,Sink I}}^{\min}, \Omega_{k,\text{Sink I}}^{\max}, \Omega_{R_\infty,\text{Sink I}}^{\max})$	0.00 0.00125 0.90	0.00 0.70 0.30	1.79
$(\Omega_{\text{OM,Sink I}}^{\min}, \Omega_{k,\text{Sink I}}^{\max}, \Omega_{R_\infty,\text{Sink I}}^{\min})$	0.00 0.00125 0.80	0.00 0.82 0.18	1.52
$(\Omega_{\text{OM,Sink I}}^{\max}, \Omega_{k,\text{Sink I}}^{\max}, \Omega_{R_\infty,\text{Sink I}}^{\min})$	0.02 0.00125 0.80	0.57 0.36 0.07	3.52
$(\Omega_{\text{OM,Sink I}}^{\max}, \Omega_{k,\text{Sink I}}^{\min}, \Omega_{R_\infty,\text{Sink I}}^{\min})$	0.02 0.00115 0.80	0.58 0.34 0.08	3.42
$(\Omega_{\text{OM,Sink I}}^{\max}, \Omega_{k,\text{Sink I}}^{\min}, \Omega_{R_\infty,\text{Sink I}}^{\max})$	0.02 0.00115 0.90	0.54 0.31 0.15	3.69

designating [Argmin AUP$_m$, Argmax AUP$_m$]. Figure 8-21 shows these intervals for selected points, primarily on the boundaries of the feasibility region of the sink because of the importance that boundary points play according to the lever-arm visualization tools. For instance, the cluster point

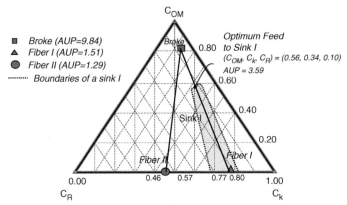

FIGURE 8-21 DETERMINATION OF OPTIMAL MIXING POINT FOR SINK I (EL-HALWAGI ET AL., 2004)

($C_{OM} = 0.56$, $C_k = 0.34$, $C_R = 0.10$) corresponds to multiple combinations of feasible properties that yield this value of the cluster. As an example, let us consider two property combinations: (OM = 0.01877, $k = 0.00115$, $R_\infty = 0.83475$) and (OM = 0.02000, $k = 0.001225$, $R_\infty = 0.84371$). Both yield the same value of the cluster ($C_{OM} = 0.557$, $C_k = 0.341$, $C_R = 0.102$) but the first one has an AUP of 3.37 while the second one has an AUP of 3.59.

With the mapping of the problem data from the property domain to the cluster domain completed, we can now readily solve the property integration problem.

Our first objective is to maximize the recycle of broke since it is the internal source (process stream for free). Any mixture of broke, fiber I, and fiber II is contained in the triangle connecting the three sources. Let us start with sink I. From Figure 8-21, it can be seen that the optimum mixing point is the intersection of the broke-fiber I line with the feasibility region of sink I. This optimal mixing point, whose coordinates are ($C_{OM} = 0.56$, $C_k = 0.34$, $C_R = 0.10$), provides the longest relative arm for the broke and the shortest relative arm for fiber I. The cluster coordinates for the broke are ($C_{OM} = 0.81$, $C_k = 0.13$, $C_R = 0.06$) with an AUP_{broke} of 9.84. The value of $AUP_{Fiber\ I}$ is 1.51. Hence, equation (8.49b) applies and for $AUP_{Sink\ I\ @\ (C_{OM}=0.56,\ C_k=0.34,\ C_R=0.10)} = 3.59$, the relative arm for fiber I can be calculated from the graph:

$$\beta_{Fiber\ I} = \frac{\text{Arm for fiber I}}{\text{Total arm connecting broke to fiber I}} \qquad (8.59)$$

These arms can be measured from the graph. Equivalently, we can use one of the cluster coordinates because of similitude. Hence,

$$\beta_{Fiber\ I} = \frac{0.81 - 0.56}{0.81 - 0.00} = 0.31 \qquad (8.60)$$

Hence, according to equation (8.26):

$$x_{\text{Fiber I}} = \frac{3.59}{1.51}(0.31) = 0.74 \qquad (8.61)$$

Alternatively, equation (8.47) can be used to calculate $x_{\text{Fiber I}}$:

$$x_{\text{Fiber I}} = \frac{3.59 - 9.84}{1.51 - 9.84} = 0.75 \qquad (8.62)$$

The slight discrepancy between the calculated values of equations (8.61) and (8.62) is within the graphical accuracy of the ternary diagram.

Using the value of fractional contribution of fiber I from equation (8.62) and for a total flowrate of 100.0 ton/h fed to sink I, we get:

$$\text{Optimum flowrate of fiber I} = 0.75 \times 100 = 75.0 \text{ ton/h} \qquad (8.63)$$

And through material balance around sink I, the flowrate of recycled broke can be obtained:

$$\text{Maximum flowrate of directly recycled broke} = 100.0 - 75.0 = 25.0 \text{ ton/h} \qquad (8.64)$$

We can now map the cluster of the optimal mixing point back to the property domain. For the optimal mixing point, whose coordinates are ($C_{\text{OM}} = 0.56$, $C_k = 0.34$, $C_R = 0.10$), and whose AUP is 3.59, we can use the clustering equations to back calculate the equivalent properties to be:

$$\text{OM}_{\text{sink I}} = 0.02 \qquad (8.65a)$$

$$k_{\text{sink I}} = 0.001225 \text{ m}^2/\text{gm} \qquad (8.65b)$$

$$R_{\infty, \text{ sink I}} = 0.844 \qquad (8.65c)$$

Next, we proceed to sink II. Since the boundaries for this sink lie on the zero-OM cluster line, no broke can be recycled to the sink and the only feasible mixture is between fibers I and II. The values of the augmented property indices for fibers I and II are $\text{AUP}_{\text{Fiber I}} = 1.51$ and $\text{AUP}_{\text{Fiber II}} = 1.29$. Since fiber II is more expensive than fiber I, but $\text{AUP}_{\text{Fiber I}} > \text{AUP}_{\text{Fiber II}}$, then equation (8.49b) applies and we should select the highest feasible value for $\text{AUP}_{\text{Sink II}}$ at the optimum mixing point. Therefore, we start with the shortest arm for fiber II (as close as possible to the cluster location of fiber I). Figure 8-22 is represented on the zero-OM cluster line which is the horizontal base of the cluster triangle. As can be seen from Figure 8-22, the AUP range for sink II at the shortest arm (at $C_k = 0.77$) is [1.63, 1.63] which is higher than either AUP for fibers I and II. Therefore, there can be

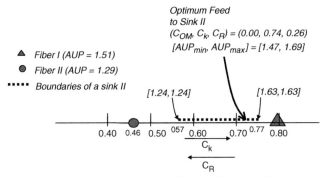

FIGURE 8-22 DETERMINATION OF OPTIMAL MIXING POINT FOR SINK II (EL-HALWAGI ET AL., 2004)

no feasible mixture of fibers I and II that matches the AUP of the sink at that point. Therefore, we systematically increase the arm of fiber II (i.e., we move to the left closer to fiber II) and for each feasible mixing point we start with Argmax $AUP_{sink\ II}$ and keep decreasing till we get a feasible answer. Hence, the optimum answer is the one shown in Figure 8-22 with coordinates of ($C_{OM} = 0.00$, $C_k = 0.74$, $C_R = 0.26$) and an AUP of 1.47.

The relative arm for fiber II can be calculated from the graph:

$$\beta_{Fiber\ II} = \frac{\text{Arm for fiber II}}{\text{Total arm connecting fiber II to fiber I}} \tag{8.66}$$

These arms can be measured from the graph leading to:

$$\beta_{Fiber\ II} = \frac{0.795 - 0.740}{0.795 - 0.464} = 0.17 \tag{8.67}$$

Hence, according to equation (8.33):

$$x_{Fiber\ II} = \frac{1.47}{1.29}(0.17) = 0.19 \tag{8.68}$$

Alternatively, equation (8.47) can be used to calculate $x_{Fiber\ II}$:

$$x_{Fiber\ II} = \frac{1.47 - 1.51}{1.29 - 1.51} = 0.18 \tag{8.69}$$

Again, the slight discrepancy between the calculated values of equations (8.68) and (8.69) is within the accuracy of reading from the ternary diagram.

Based on the definition of fractional contribution and the value of fractional contribution of fiber II from equation (8.68) and a total flowrate of 40.0 ton/h fed to sink II, we get:

$$\text{Optimum flowrate of fiber II} = 0.19 \times 40 = 7.6 \text{ ton/h} \tag{8.70}$$

And through material balance around sink II, the flowrate of fiber I can be obtained:

$$\text{Optimum flowrate of fiber I} = 40.0 - 7.6 = 32.4 \text{ ton/h} \qquad (8.71)$$

We can now map the cluster of the optimal mixing point back to the property domain. For the optimal mixing point, whose coordinates are ($C_{OM} = 0.00$, $C_k = 0.74$, $C_R = 0.26$), and whose AUP is 1.47, we can use the clustering equations to back calculate the equivalent properties to be:

$$OM_{\text{sink II}} = 0.00 \qquad (8.72a)$$

$$k_{\text{sink II}} = 0.001086 \text{ m}^2/\text{gm} \qquad (8.72b)$$

$$R_{\infty,\text{ sink II}} = 0.850 \qquad (8.72c)$$

The optimum direct recycle and reallocation of sources solution is shown in Figure 8-23. It illustrates the revised flowrates of the sources and the new properties entering the sinks. It is useful to compare the cost of raw materials before and after recycle.

$$\text{The initial cost of raw materials before recycle/reallocation} =$$
$$116.4 \text{ ton of fiber I} \times \$210/\text{ton} + 23.6 \text{ ton of fiber II} \times \$400/\text{ton} = \$33,884/\text{h} \qquad (8.73)$$

$$\text{The cost of raw materials after recycle/reallocation} =$$
$$107.4 \text{ ton of fiber I} \times \$210/\text{ton} + 7.6 \text{ ton of fiber II} \times \$400/\text{ton} = \$25,594/\text{h} \qquad (8.74)$$

Therefore, direct recycle and reallocation of sources result in a 24.5% reduction in cost of raw materials.

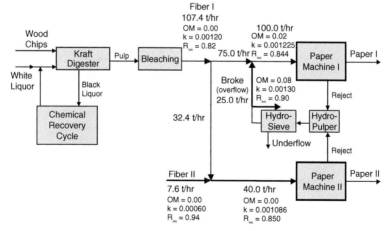

FIGURE 8-23   OPTIMUM DIRECT-RECYCLE SOLUTION

SOLUTION TO PART B

Part b deals with the interception of the broke. First, what is the minimum target for fresh fiber usage. Since sink II requires zero OM, it is not possible to intercept the broke and reach absolute zero content of OM. Therefore, we focus our attention on recycling the broke to sink I by substituting fiber I. As Figure 8-24 illustrates, it is theoretically possible to recover all the broke, adjust its properties to be used in lieu of fiber I, and recycle 30 ton/h. This intercepted recycle leads to a reduction of fiber I usage by 30 ton/h.

To recycle all of the process resources (the broke), how should the properties of broke be altered so as to achieve its maximum recycle? As can be seen from Figure 8-16, once a mixing point is selected, one can determine the minimum extent of interception to adjust the properties of the broke. Multiple candidate mixing points can be selected as mentioned earlier (Figures 8-16 and 8-17). For instance, let us select the same mixing point as in the case of direct recycle. The coordinates for the selected mixing point are ($C_{OM} = 0.56$, $C_k = 0.34$, $C_R = 0.10$), and the optimum AUP of the sink is 3.59. Maximum recycle of broke is total recycle, i.e. 30 ton/h. Hence, minimum usage of fiber I in sink I is 70 ton/h leading to a fractional contribution of:

$$x_{Fiber\ I} = 70/100 = 0.70 \tag{8.75}$$

Recalling that $AUP_{Fiber\ I} = 1.51$ and substituting this value and equation (8.75) into equation (8.33), we get

$$\beta_{Fiber\ I} = \frac{0.70}{3.59}(1.51) = 0.29 \tag{8.76}$$

(a) Overall Papermaking Balance Before Property Integration

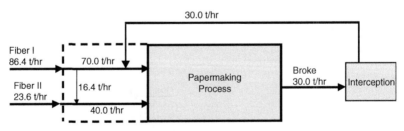

(b) Overall Papermaking Balance After Property Integration

FIGURE 8-24   TARGETING FOR MINIMUM FRESH FIBER USAGE

But, the relative arm for fiber I is defined as:

$$\beta_{\text{Fiber I}} = \frac{\text{Arm for fiber I}}{\text{Total arm connecting intercepted broke to fiber I}} \tag{8.77}$$

These arms can be measured from the graph. Equivalently, one can use the cluster coordinates based on equation (8.25) or because of similitude. Hence,

$$\beta_{\text{Fiber I}} = \frac{C_{\text{OM}}^{\text{Intercepted}} - 0.56}{C_{\text{OM}}^{\text{Intercepted}} - 0.00} \tag{8.78}$$

Equating equations (8.77) and (8.78), we get:

$$C_{\text{OM}}^{\text{Intercepted}} = 0.79 \tag{8.79}$$

Similarly,

$$\beta_{\text{Fiber I}} = \frac{C_k^{\text{Intercepted}} - 0.34}{C_k^{\text{Intercepted}} - 0.80} = 0.29 \rightarrow C_k^{\text{Intercepted}} = 0.15 \tag{8.80}$$

and, according to equation (8.24),

$$C_R^{\text{Intercepted}} = 0.06 \tag{8.81}$$

Using equation (8.46a), we get

$$3.59 = 0.7 \times 1.51 + 0.3 \times \text{AUP}_{\text{broke}}^{\text{Intercepted}} \rightarrow \text{AUP}_{\text{broke}}^{\text{Intercepted}} = 8.44 \tag{8.82}$$

Therefore, the minimum interception task entails bringing the broke to an intercepted cluster point of (0.79, 0.15, 0.06) and an $\text{AUP}_{\text{broke}}^{\text{Intercepted}} = 8.44$. Using the clustering equations to map back to the property domain for the intercepted broke, we get

$$\text{OM}^{\text{Intercepted}} = 0.067 \tag{8.83a}$$

$$k^{\text{Intercepted}} = 0.0013 \text{ m}^2/\text{gm} \tag{8.83b}$$

$$R_\infty^{\text{Intercepted}} = 0.90 \tag{8.83c}$$

This analysis has identified the minimum interception task for the broke without committing to the specific nature of the interception device. Later, conventional process synthesis techniques can be used to screen candidate interception techniques and to select optimum system. In essence,

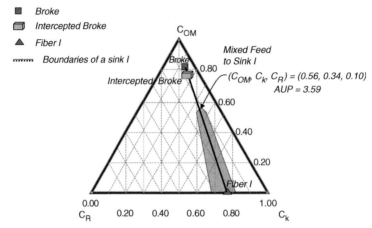

FIGURE 8-25   DETERMINATION OF MINIMUM EXTENT OF INTERCEPTION FOR THE BROKE (EL-HALWAGI ET AL., 2004)

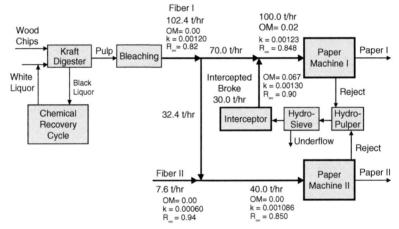

FIGURE 8-26   OPTIMAL SOLUTION WITH INTERCEPTION AND REALLOCATION (EL-HALWAGI ET AL., 2004)

we have targeted for the performance of the interception device. This is an important linkage between property integration and process synthesis. It is worth pointing out that the same procedure can be carried out for other mixing points or for the second sink.

Figure 8-25 shows the interception task for the broke stream while Figure 8-26 represents the revised flowsheet following interception and reallocation.

8.9 RELATIONSHIP BETWEEN CLUSTERS AND MASS FRACTIONS

Since compositions represent a special class of properties, it is useful to think of mass integration as a special case of property integration. Analogy has been shown between the material recycle pinch diagram and the property-based pinch diagram and between the source–sink ternary diagram and its cluster-based analogue. Therefore, it is useful to illustrate the relationship between mass fractions and clusters. Let us designate the mass fraction of component r in stream i by $y_{r,i}$. When a number N_{sources} of streams are mixed, we get the following balance equations:

Overall material balance (see equation (8.3)):

$$\bar{F} = \sum_{i=1}^{N_{\text{Sources}}} F_i$$

Component material balance:

$$\bar{F}\,\bar{y}_r = \sum_{i=1}^{N_{\text{Sources}}} F_i\, y_{r,i} \tag{8.84}$$

Diving both sides of equation (8.84) by $\bar{F}$, we get

$$\bar{y}_r = \sum_{i=1}^{N_{\text{Sources}}} x_i\, y_{r,i} \tag{8.85}$$

where

$$x_i = \frac{F_i}{\bar{F}} \tag{8.86}$$

Equation (8.85) is analogous with equation (8.20) with the mass fraction being the property of choice and the operator on the mass fraction corresponding to the mass fraction itself, i.e.,

$$p_{r,i} = y_{r,i} \tag{8.87}$$

and

$$\psi_r(p_{r,i}) = y_{r,i} \tag{8.88}$$

Let us choose a reference value of 1.0 for all mass fractions. Therefore,

$$\Omega_{r,i} = y_{r,i} \tag{8.89}$$

Substituting from equation (8.89) into equation (8.22) and recalling that the sum of mass fractions in a stream is 1.0, we get:

$$\text{AUP}_i = 1.0 \tag{8.90}$$

Substituting from equation (8.90) into equation (8.23), we get,

$$C_{r,i} = y_{r,i} \tag{8.91}$$

Therefore, when the property of choice is the mass fraction, the cluster of the property is also the mass fraction.

Additional observations may also be noted. For instance, comparing equations (8.25) and (8.85), we get

$$\beta_i = x_i \tag{8.92}$$

Also, when equations (8.90) and (8.92) are substituted in equation (8.26), we get that the AUP for a sink or a mixture is always equal to 1.0.

The foregoing observations reiterate the earlier statement that mass integration may be regarded as a special case of property integration.

8.10 ADDITIONAL READINGS

Recent work has expanded the concept of property integration. The concept of property-based clusters and componentless design is covered by Shelley and El-Halwagi (2000). This chapter also addresses property integration techniques for the recovery and recycle of volatile organic compounds. Graphical techniques for property integration can address up to three properties. Recent work on the subject includes property-based process modification (Kazantzi et al., 2004a and b) as well as property-based recycle and interception (El-Halwagi et al., 2004; Kazantzi and El-Halwagi 2004; Gabriel et al., 2003a, b; Glasgow et al., 2001). Algebraic approaches for property integration have been developed to address any number of properties (Qin et al., 2004). Furthermore, an algebraic technique has been developed by Foo et al. (2006) and Kazantzi et al. (2004c) to solve the property-based pinch diagram via a cascade analysis. Property-based clusters have been used to develop reverse problem formulations as well as integrated process and product design (Eden et al., 2002, 2004, 2005). Property-based modeling and its impact on process and product design has been discussed by Gani and Pistikopoulos (2002). Unsteady state processes involving property-based scheduling and operation have been addressed by Grooms et al. (2005).

8.11 PROBLEMS

8.1 Consider the microelectronics manufacturing facility (Gabriel et al., 2003a, Kazantzi and El-Halwagi 2005) represented by Figure 8-27. The Wafer Fabrication (Wafer Fab) Section and the combined Chemical and Mechanical Processing (CMP) Section are identified as the sinks of the problem that both accept ultra pure water (UPW) as their feed. There are also two main process sources that are available for reuse to the sinks, i.e., the 50% Spent Rinse and the 100% Spent Rinse. We are interested in reusing them as feed to the sinks, in order to reduce the ultra pure water consumption. The main characteristic that we consider here, in order to evaluate the reuse of the rinse streams to the sinks, is resistivity (R), which constitutes an index of the ionic content of aqueous streams.

The mixing rule for resistivity is the following (Gabriel et al., 2003a):

$$\frac{1}{\overline{R}} = \sum_{i=1}^{N_s} \frac{x_i}{R_i} \tag{8.93}$$

Moreover, the inlet flowrates of the feed streams to the Wafer Fabrication and the CMP Sections along with their constraints on resistivity are given in Table 8-9, whereas the source flowrates and property values are given in Table 8-10.

Using direct recycle, what is the target for fresh (ultra pure) water usage?

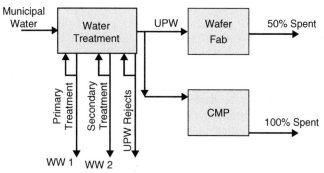

FIGURE 8-27   A MICROELECTRONICS MANUFACTURING FLOWSHEET (GABRIEL ET AL., 2003a)

TABLE 8-9 FLOWRATES AND BOUNDS ON PROPERTIES OF SINKS

Sink	Flowrate (gal/min)	Lower bound on R (kΩ/cm)	Upper bound on R (kΩ/cm)
Wafer Fab	800	16,000	20,000
CMP	700	10,000	18,000

TABLE 8-10 PROPERTIES OF PROCESS SOURCES AND FRESH

Source	Flowrate (gal/min)	R (kΩ/cm)
50% Spent	1000	8000
100% Spent	1000	2000
UPW (fresh)	To be determined	18,000

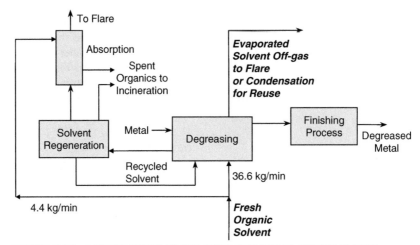

FIGURE 8-28 A DEGREASING PLANT (SHELLEY AND EL-HALWAGI 2000)

8.2 Consider the metal degreasing process (Shelley and El-Halwagi 2000) shown in Figure 8-28. The process uses a fresh organic solvent in the absorption column and the degreaser. The organic solvent contains numerous components. It is used in the absorption column to capture lights which escape from the solvent regeneration unit. The organic solvent is also used in the degreaser to degrease the metal parts. Currently, the off-gas volatile organic compounds (VOCs) that evaporate from the degreasing process are flared leading to economic loss and environmental pollution. In this problem, it is desired to explore the possibility of condensing and reusing the off-gas VOCs instead of flaring, thereby optimizing the usage of fresh solvent.

Three properties are examined to determine the suitability of a solvent for usage in the absorber or the degreaser; sulfur content, density, and Reid vapor pressure (RVP). The fresh organic solvent has the following properties:

$$S_{fresh} = 0.1 \ (\text{wt\%}) \tag{8.94}$$

$$\rho_{fresh} = 610 \ \text{kg/m}^3 \tag{8.95}$$

$$RVP_{fresh} = 2.1 \ \text{atm} \tag{8.96}$$

The sinks have the following constraints on the properties and flowrates:

$$0.0 \leq S_{degreaser}(\text{wt\%}) \leq 1.0 \tag{8.97}$$

$$555 \leq \rho_{degreaser}(\text{kg/m}^3) \leq 615 \tag{8.98}$$

$$2.1 \leq RVP_{degreaser}(\text{atm}) \leq 4.0 \tag{8.99}$$

$$36.6 \leq F_{\text{degreaser}}(\text{kg/min}) \leq 36.8 \tag{8.100}$$

$$0.0 \leq S_{\text{absorber}}(\text{wt\%}) \leq 0.1 \tag{8.101}$$

$$530 \leq \rho_{\text{absorber}}(\text{kg/m}^3) \leq 610 \tag{8.102}$$

$$1.5 \leq \text{RVP}_{\text{absorber}}(\text{atm}) \leq 2.5 \tag{8.103}$$

$$4.4 \leq F_{\text{absorber}}(\text{kg/min}) \leq 6.2 \tag{8.104}$$

Experimental data are available for the degreaser off-gas condensate. Samples of the off-gas were taken then condensed at various condensation temperatures and the three properties as well as flowrate of the condensate were measured as shown in Figures 8-29–32.

The following mixing rules can be used to evaluate the properties resulting from mixing several streams (Shelley and El-Halwagi 2000):

$$\overline{S}(\text{wt\%}) = \sum_{s=1}^{N_s} x_s \, S_s(\text{wt\%}) \tag{8.105}$$

$$\frac{1}{\overline{\rho}} = \sum_{s=1}^{N_s} \frac{x_s}{\rho_s} \tag{8.106}$$

$$\overline{\text{RVP}}^{1.44} = \sum_{s=1}^{N_s} x_s \text{RVP}_s^{1.44} \tag{8.107}$$

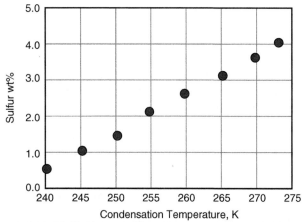

FIGURE 8-29 EXPERIMENTAL DATA FOR SULFUR CONTENT OF THE CONDENSATE VERSUS CONDENSATION TEMPERATURE (SHELLEY AND EL-HALWAGI 2000)

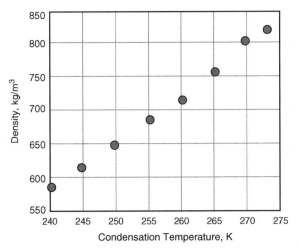

FIGURE 8-30 EXPERIMENTAL DATA FOR DENSITY OF THE CONDENSATE VERSUS CONDENSATION TEMPERATURE (SHELLEY AND EL-HALWAGI 2000)

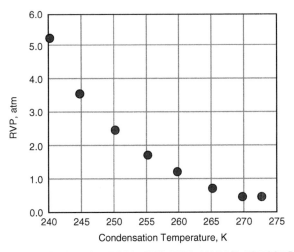

FIGURE 8-31 EXPERIMENTAL DATA FOR REID VAPOR PRESSURE OF THE CONDENSATE VERSUS CONDENSATION TEMPERATURE (SHELLEY AND EL-HALWAGI 2000)

The objective of this problem is to minimize the total annualized cost (TAC) of the condensation system along with the fresh solvent cost, i.e.,

$$\text{Minimize TAC of the solution} = \text{TAC of the condensation system} + \text{TAC of fresh solvent} \quad (8.108)$$

where the TAC of condensation system can be estimated to be:

$$\text{TAC of condensation system (\$/year)} = 27,000 + 1070(273 - T)^{1.8} \quad (8.109)$$

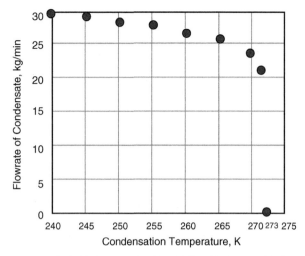

FIGURE 8-32   EXPERIMENTAL DATA FOR CONDENSATE FLOWRATE VERSUS CONDENSATION TEMPERATURE (SHELLEY AND EL-HALWAGI 2000)

where T is the condensation temperature in kelvin. The cost of the fresh solvent is \$0.08/kg.

8.12 SYMBOLS

Argmax	highest value of an element in a set
Argmin	lowest value of an element in a set
AUP	augmented property index
AUP_m	augmented property index for point m in a sink
AUP_i	augmented property index for source s
$\overline{\text{AUP}}$	augmented property index of mixture
C	cluster
$C_{r,i}$	cluster of property r in source i
$\overline{C}$	cluster of mixing
F_i	flowrate of source i, ton/h
i	index for sources
INTERVAL_AUP_m	the interval bounding the range for the AUP at sink point m
j	index for sinks
k	absorption coefficient
m	a feasible point satisfying sink constraints
n_{Multiple}	number of multiple property points leading to the same value of the cluster
N_{Sources}	number of sources
p_i	i^{th} property
$p_{i,m}$	i^{th} property for point m in the sink
r	index for properties or clusters
R_∞	reflectivity

$\text{SET\_AUP}^{\text{Feasible}}_{(C_1^{\text{sink}}, C_2^{\text{sink}}, C_3^{\text{sink}})}$ set of all feasible AUP's for property combinations yielding the same value of clusters $(C_1^{\text{sink}}, C_2^{\text{sink}}, C_3^{\text{sink}})$

W internal process stream

X_i fractional contribution of the ith stream into the total flowrate of the mixture

Subscripts

k	absorption coefficient
m	a feasible point satisfying sink constraints
OM	objectionable materials
R	reflectivity

Superscripts

Feasible	a feasible point in a sink
int	intercepted value
ref	reference value

Greek

β_i	mixing arm of stream i on the ternary cluster diagram
ψ_r	operator used in the mixing formula for the rth property
$\Omega_{r,j}$	normalized, dimensionless operator for the rth property of the ith source

8.13 REFERENCES

Biermann, C.J. 1996, *Handbook of Pulping and Papermaking*, Academic Press, San Diego, CA, pp. 32-40, 158-159, 193-208, 251-252, 502-515.

Brandon, C.E. 1981, 'Properties of paper' in *Pulp and Paper Chemistry and Chemical Technology*, 3rd Edition, volume III, ed. James P. Casey, John Wiley & Sons, New York, pp. 1739-1746, 1819-1886.

Eden, M.R., Jørgensen, S.B., Gani, R., and El-Halwagi, M.M. 2002, 'Property integration – A new approach for simultaneous solution of process and molecular design problems', *Comp. Aided Chem. Eng.*, vol. 10, pp. 79-84.

Eden, M.R., Jørgensen, S.B., Gani R., and El-Halwagi, M.M. 2004, 'A novel framework for simultaneous separation process and product design', *Chem. Eng. & Proc.*, vol. 43, no. 5, pp. 595-608.

El-Halwagi, M.M., Glasgow, I.M., Eden, M.R., and Qin, X. 2004, 'Property integration: Componentless design techniques and visualization tools', *AIChE J.*, vol. 50, no. 8, 1854-1869.

Eljack, F.T., Abdelhady, A.F., Eden, M.R., Gabriel, F., Qin, X., and El-Halwagi, M.M. 2005, 'Targeting optimum resource allocation using reverse problem formulations and property clustering techniques', *Comp. Chem. Eng.*, vol. 29, 2304-2317.

Foo, D.C.Y., Kazantzi, V., El-Halwagi, M.M., and Manan, Z. 2006, 'Surplus diagram and cascade analysis technique for targeting property-based material reuse network', *Chem. Eng. Sci.*, (in press).

Gabriel, F.B.; Harell, D.A., Dozal, E., and El-Halwagi, M.M. 2003a, Pollution Targeting via Functionality Tracking, *AIChE Spring Meeting*, New Orleans.

Gabriel, F., Harell, D., Kazantzi, V., Qin, X., and El-Halwagi, M. 2003b, 'A Novel Approach to the Synthesis of Property Integration Networks', *AIChE* Annual Meeting, San Francisco, November.

Gani, R., and Pistikopoulos, E. 2002, 'Property modeling and simulation for product and process design', *Fluid Phase Equilib.*, 194-197, pp. 43-59.

Glasgow I.M., Eden M.R., Shelley M.D., Krishnagopalan G., and El-Halwagi M.M. 2001, 'Property Integration for Process Optimization', *AIChE* Annual Meeting 2001, Reno NV, USA.

Grooms, D., Kazantzi, V., and El-Halwagi, M.M. 2005, 'Scheduling and operation of property-interception networks for resource conservation', *Comp. Chem. Eng.*, vol. 29, 2318-2325.

Kazantzi, V., and El-Halwagi, M.M. 2005, 'Targeting material reuse via property integration', *Chem. Eng. Prog.*, vol. 101, no. 8, 28-37.

Kazantzi, V., Harell, D., Gabriel, F., Qin X., and El-Halwagi, M.M. 2004a, 'Property-based integration for sustainable development', *Proceedings of European Sympoisum on Computer-Aided Process Engineering 14 (ESCAPE 14)*, eds. A. Barbosa-Povoa and H. Matos, pp. 1069-1074, Elsevier.

Kazantzi, V., Qin, X., Gabriel, F., Harell, D., and El-Halwagi, M.M. 2004b, 'Process modification through visualization and inclusion techniques for property based integration', *Proceedings of the Sixth Foundations of Computer Aided Design (FOCAPD)*, eds. C.A. Floudas and R. Agrawal, *CACHE Corp.* pp. 279-282.

Kazantzi, V., Foo, D.C.Y., Almutlaq, A., Manan, Z., and El-Halwagi, M.M. 2004c, 'Resource conservation and waste minimization for property networks using cascade analysis', *AIChE* Annual Meeting, Austin, TX November 9.

Kazantzi, V., Harell, D., Gabriel, F., Qin, X., and El-Halwagi, M. 2003, 'Property-based integration for sustainable design', *AIChE* Annual Meeting, San Francisco, November.

Qin, X.F., Gabriel, Harell, D., and El-Halwagi, M. 2004, 'Algebraic techniques for property integration via componentless design', *Ind. Eng. Chem.*, vol. 43, 3792-3798.

Shelley, M.D. and El-Halwagi, M.M. 2000, 'Component-less design of recovery and allocation systems: a functionality-based clustering approach', *Comp. Chem. Eng.*, vol. 24, 2081-2091.

Willets, W.R. 1958, 'Titanium Pigments' in '*Paper Loading Materials* TAPPI Monograph Series – No. 19', Technical Association of the Pulp and Paper Industry, New York, NY, pp. 96-114.

9 ■ HEAT INTEGRATION

Up to this chapter, attention has been given to mass integration. As mentioned in Chapter One, there are two main commodities handled in the process: mass and energy. Both contribute to the overall performance of the process and both affect the capital and operating costs of the process. Heat is one of the most important energy forms in the process. The current chapter provides an overview of heat integration. First, the problem of synthesizing networks of heat exchangers is discussed while highlighting the analogy with synthesizing networks of mass exchangers. Next, targeting procedures are presented with the objective of minimizing heating and cooling utilities while maximizing heat exchange among the process streams.

9.1 SYNTHESIS OF HEAT EXCHANGE NETWORKS (HENs)

In a typical process, there are normally several hot streams that must be cooled and several cold streams that must be heated. The usage of external cooling and heating utilities (e.g., cooling water, refrigerants, steam, heating oils, etc.) to address all the heating and cooling duties is not cost effective. Indeed, integration of heating and cooling tasks may lead to significant cost reduction. The key concept is to transfer heat from the process hot streams to the process cold streams before the external utilities are used.

The result of this heat integration is the simultaneous reduction of heating and cooling duties of the external utilities.

The problem of synthesizing HENs can be stated as follows:

Given a number N_H of process hot streams (to be cooled) and a number N_C of process cold streams (to be heated), it is desired to synthesize a cost-effective network of heat exchangers that can transfer heat from the hot streams to the cold streams. Given also are the heat capacity (flowrate x specific heat) of each process hot stream, $FC_{P,u}$; its supply (inlet) temperature, T_u^s; and its target (outlet) temperature, T_u^t, where $u = 1, 2, \ldots, N_H$. In addition, the heat capacity, $fc_{P,v}$, supply and target temperatures, t_u^s and t_v^t, are given for each process cold stream, where $v = 1, 2, \ldots, N_C$. Available for service are N_{HU} heating utilities and N_{CU} cooling utilities whose supply and target temperatures (but not flowrates) are known. Figure 9-1 is a schematic representation of the HEN problem statement.

For a given system, the synthesis of HENs entails answering several questions:

- Which heating/cooling utilities should be employed ?
- What is the optimal heat load to be removed/added by each utility?
- How should the hot and cold streams be matched (i.e., stream pairings)?
- What is the optimal system configuration (e.g., how should the heat exchangers be arranged? Is there any stream splitting and mixing ?)?

Numerous methods have been developed for the synthesis of HENs. These methods have been reviewed by Shenoy (1995); Linnhoff (1993); Gundersen and Naess (1988); and Douglas (1988). One of the key advances in synthesizing HENs is the identification of minimum utility targets ahead of designing the network using the thermal pinch analysis. This technique is presented in the following section.

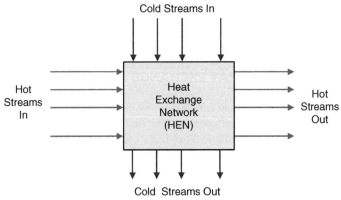

FIGURE 9-1   SYNTHESIS OF HENs

9.2 HEAT EXCHANGE PINCH DIAGRAM

Let us consider a heat exchanger for which the thermal equilibrium relation governing the transfer of the heat from a hot stream to a cold stream is simply given by

$$T = t \tag{9.1}$$

By employing a minimum heat exchange driving force of ΔT^{min}, one can establish a one-to-one correspondence between the temperatures of the hot and the cold streams for which heat transfer is feasible, i.e.,

$$T = t + \Delta T^{min} \tag{9.2}$$

This expression ensures that the heat-transfer considerations of the second law of thermodynamics are satisfied. For a given pair of corresponding temperatures (T, t) it is thermodynamically and practically feasible to transfer heat from any hot stream whose temperature is greater than or equal to T to any cold stream whose temperature is less than or equal to t. It is worth noting the analogy between equations (9.2) and (4.16). Thermal equilibrium is a special case of mass exchange equilibrium with T, t, and ΔT^{min} corresponding to y_i, x_j, and ε_j, respectively, while the values of m_j and b_j are one and zero, respectively. Table 9-1 summarizes the analogous terms in MENs and HENs. Similar to the role of ε_j in cost optimization, ΔT^{min} can be used to trade off capital versus operating costs as shown in Figure 9-2.

In order to accomplish the minimum usage of heating and cooling utilities, it is necessary to maximize the heat exchange among process streams. In this context, one can use a very useful graphical technique referred to as the "thermal pinch diagram". This technique is primarily based on the work of Linnhoff and co-workers (e.g., Linnhoff and Hindmarsh (1983); Umeda et al. (1979); and Hohmann (1971)). The first step in constructing the thermal-pinch diagram is creating a global representation for all the hot streams by plotting the enthalpy exchanged by each process hot stream versus its

TABLE 9-1 ANALOGY BETWEEN MENs AND HENs

MENs	HENs
Transferred commodity: Mass	Transferred commodity: Heat
Donors: rich streams	Donors: hot streams
Recipient: lean streams	Recipient: cold streams
Rich composition: y	Hot temperature: T
Lean composition: x	Cold temperature: t
Slope of equilibrium: m	Slope of equilibrium: 1
Intercept of equilibrium: b	Intercept of equilibrium: 0
Driving force: ε	Driving force: ΔT^{min}

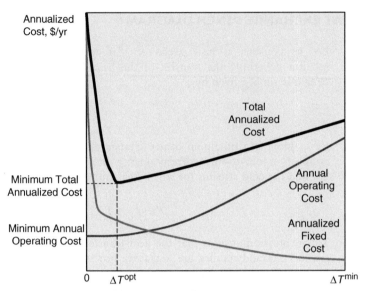

FIGURE 9-2   ROLE OF MINIMUM APPROACH TEMPERATURE IN TRADING OFF CAPITAL VERSUS OPERATING COSTS

temperature[1]. Hence, a hot stream losing sensible heat[2] is represented as an arrow whose tail corresponds to its supply temperature and its head corresponds to its target temperature. Assuming constant heat capacity over the operating range, the slope of each arrow is equal to $F_u C_{P,u}$. The vertical distance between the tail and the head of each arrow represents the enthalpy lost by that hot stream according to the following expression:

Heat lost from the uth hot stream

$$HH_u = F_u C_{P,u}(T_u^s - T_u^t) \qquad \text{where } u = 1, 2, \ldots, N_H \qquad (9.3)$$

Note that any stream can be moved up or down while preserving the same vertical distance between the arrow head and tail and maintaining the same supply and target temperatures. Similar to the graphical superposition described in Chapter Four, one can create a hot composite stream using the diagonal rule. Figures 9-3a and b illustrate this concept for two hot streams.

Next, a cold-temperature scale, t, is created in one-to-one correspondence with the hot temperature scale, T, using equation (9.2). The enthalpy of each cold stream is plotted versus the cold temperature scale, t.

---

[1]In most HEN literature, the temperature is plotted versus the enthalpy. However, in this chapter enthalpy is plotted versus temperature in order to draw the analogy with MEN synthesis. Furthermore, when there is a strong interaction between mass and energy objectives, the enthalpy expressions become non-linear functions of temperature and composition. In such cases, it is easier to represent enthalpy on the vertical axis.

[2]Whenever there is a change in phase, the latent heat should also be included.

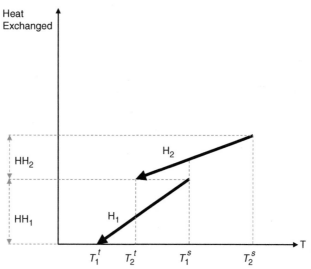

FIGURE 9-3a   REPRESENTING HOT STREAMS

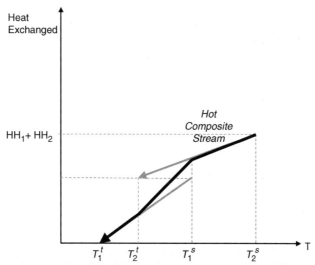

FIGURE 9-3b CONSTRUCTING A HOT COMPOSITE STREAM USING
SUPERPOSITION (DASHED LINE REPRESENTS COMPOSITE LINE)

The vertical distance between the arrow head and tail for a cold stream is
given by
Heat gained by the vth cold stream

$$HC_v = f_v c_{p,v}(t_v^t - t_v^s) \qquad \text{where } v = 1, 2, \ldots, N_C \qquad (9.4)$$

In a similar manner to constructing the hot-composite line, a cold composite stream is plotted (see Figures 9-4a and b for a two-cold-stream example).

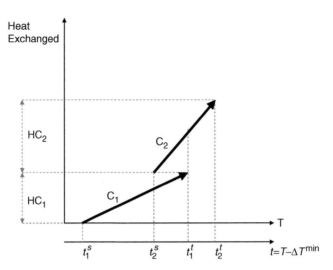

FIGURE 9-4a   REPRESENTING COLD STREAMS

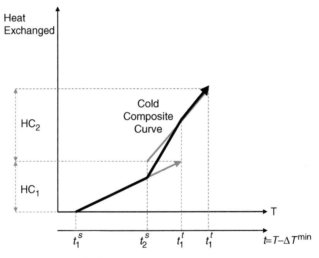

FIGURE 9-4b CONSTRUCTING A COLD COMPOSITE STREAM USING SUPERPOSITION (DASHED LINE REPRESENTS COMPOSITE LINE)

Next, both composite streams are plotted on the same diagram (Figure 9-5). On this diagram, thermodynamic feasibility of heat exchange is guaranteed when at any heat exchange level (which corresponds to a horizontal line), the temperature of the cold composite stream is located to the left of the hot composite stream (i.e., temperature of the hot is higher than or equal to the cold temperature plus the minimum approach temperature). Hence, for a given set of corresponding temperatures, it is thermodynamically and practically feasible to transfer heat from any hot stream to any cold stream.

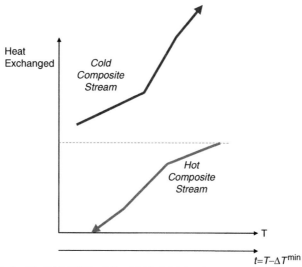

FIGURE 9-5 PLACEMENT OF COMPOSITE STREAMS WITH NO HEAT INTEGRATION

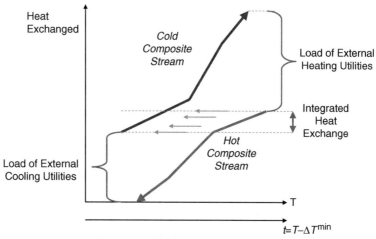

FIGURE 9-6 PARTIAL HEAT INTEGRATION

The cold composite stream can be moved up and down which implies different heat exchange decisions. For instance, if we move the cold composite stream upwards in a way that leaves no horizontal overlap with the hot composite stream, then there is no integrated heat exchange between the hot composite stream and the cold composite stream as seen in Figure 9-5. When the cold composite stream is moved downwards so as to provide some horizontal overlap, some integrated heat exchange can be achieved (Figure 9-6). However, if the cold composite stream is moved downwards such that a portion of the cold is placed to the right of the hot composite stream, thereby creating infeasibility (Figure 9-7). Therefore, the optimal

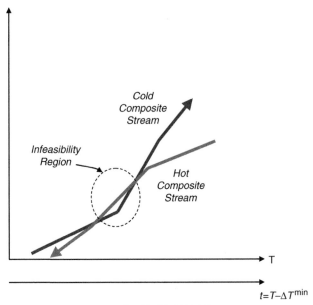

FIGURE 9-7   INFEASIBLE HEAT INTEGRATION

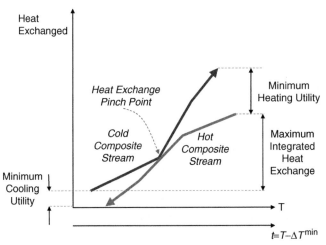

FIGURE 9-8   THERMAL PINCH DIAGRAM

situation is constructed when the cold composite stream is slid vertically until it touches the rich composite stream while lying completely to the left of the hot composite stream at any horizontal level. Therefore, the cold composite stream can be slid down until it touches the hot composite stream. The point where the two composite streams touch is called the "thermal pinch point". As Figure 9-8 shows, one can use the pinch diagram to determine the minimum heating and cooling utility requirements. Again, the cold composite line cannot be slid down any further; otherwise,

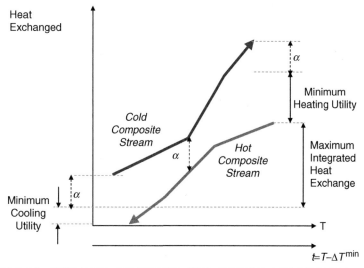

FIGURE 9-9 PENALTIES ASSOCIATED WITH PASSING HEAT THROUGH THE PINCH

portions of the cold composite stream would be the right of the hot composite stream, causing thermodynamic infeasibility. On the other hand, if the cold composite stream is moved up (i.e., passing heat through the pinch), less heat integration is possible, and consequently, additional heating and cooling utilities are required. Therefore, for a minimum utility usage the following design rules must be observed:

- No heat should be passed through the pinch.
- Above the pinch, no cooling utilities should be used.
- Below the pinch, no heating utilities should be used.

The first rule is illustrated by Figure 9-9. The passage of a heat flow through the pinch (α) results in a double penalty: an increase of α in both heating utility and cooling utility. The second and third rules can be explained by noting that above the pinch there is a surplus of cooling capacity. Adding a cooling utility above the pinch will replace a load that can be removed (virtually for no operating cost) by a process cold stream. A similar argument can be made against using a heating utility below the pinch.

EXAMPLE 9-1 UTILITY MINIMIZATION IN A CHEMICAL PLANT

Consider the chemical processing facility illustrated in Figure 9-10. The process has two adiabatic reactors. The intermediate product leaving the first reactor (C_1) is heated from 420 to 490 K before being fed to the second reactor. The off-gases leaving the reactor (H_1) at 460 K are cooled to 350 K prior to being forwarded to the gas-treatment unit. The product leaving the bottom of the reactor is fed to a separation network. The product stream

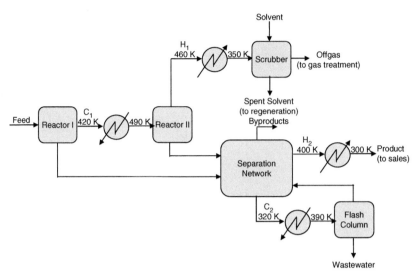

FIGURE 9-10   SIMPLIFIED FLOWSHEET FOR THE CHEMICAL PROCESSING FACILITY

TABLE 9-2 STREAM DATA FOR THE CHEMICAL PROCESS

Stream	Flowrate × specific heat (kW/K)	Supply temperature (K)	Target temperature (K)	Enthalpy change (kW)
H_1	300	460	350	33,000
H_2	500	400	300	50,000
C_1	600	420	490	42,000
C_2	200	320	390	14,000

leaving the separation network (H_2) is cooled from 400 to 300 prior to sales. A byproduct stream (C_2) is heated from 320 to 390 K before being fed to a flash column. Stream data are given in Table 9-1.

In the current operation, the heat exchange duties of H_1, H_2, C_1, and C_2 are fulfilled using the cooling and heating utilities. Therefore, the current usage of cooling and heating utilities are 83,000 and 56,000 kW, respectively.

The objective of this case study is to use heat integration via the pinch diagram to identify the target for minimum heating and cooling utilities. A value of $\Delta T^{\min} = 10$ K is used.

SOLUTION

Figures 9-11–9-13 illustrate the hot composite stream, the cold composite stream and the pinch diagram, respectively. As can be seen from Figure 9-13a, the two composite streams touch at 430 K on the hot scale (420 K on the cold scale). This designates the location of the heat exchange pinch point. The minimum heating and cooling utilities are 33,000 and

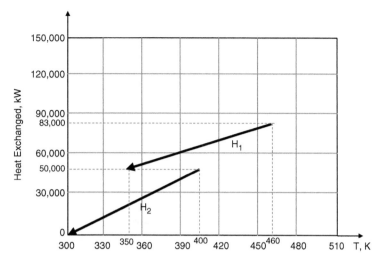

FIGURE 9-11a   HOT STREAMS EXAMPLE 9-1

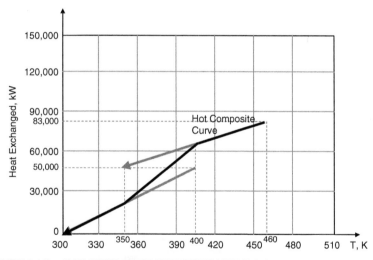

FIGURE 9-11b   HOT COMPOSITE STREAM EXAMPLE 9-1

60,000 kW, respectively. Therefore, the potential reduction in utilities can be calculated as follows:

$$\text{Target for percentage savings in heating utility} = \frac{56,000 - 33,000}{56,000} \times 100\% = 41\%$$

$$(9.5)$$

$$\text{Target for percentage savings in cooling utility} = \frac{83,000 - 60,000}{83,000} \times 100\% = 28\%$$

$$(9.6)$$

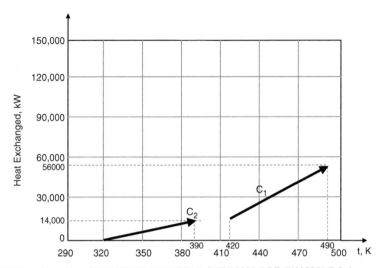

FIGURE 9-12a   REPRESENTING THE COLD STREAMS FOR EXAMPLE 9-1

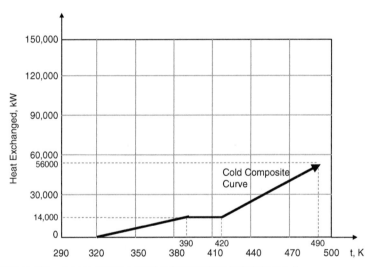

FIGURE 9-12b   COLD COMPOSITE STREAM FOR EXAMPLE 9-1

Once the minimum operating cost is determined, a network of heat exchangers can be synthesized[3]. The trade off between capital and operating costs can be established by iteratively varying ΔT^{min} until the minimum total annualized cost is attained.

[3]Constructing the HEN with minimum number of units and minimum heat transfer area is analogous to constructing a MEN. The design starts from the pinch following two matching criteria relating number of streams and heat capacities. A detailed discussion on this issue can be found in Linnhoff and Hindmarsh (1983); Douglas (1988); Shenoy (1995); and Smith (1995).

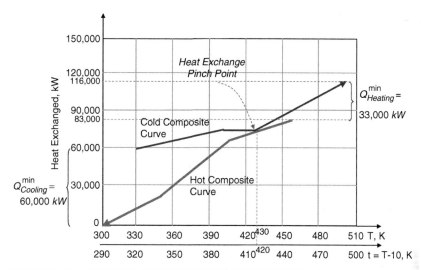

FIGURE 9-13a THERMAL PINCH DIAGRAM FOR EXAMPLE 9-1

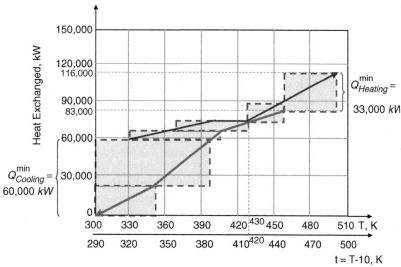

FIGURE 9-13b MATCHING OF HOT AND COLD STREAMS

As mentioned before, there are several techniques for configuring the actual network of heat exchangers that satisfies the utility targets. One technique is to match streams that lie at the same temperature level by transferring heat horizontally. Figure 9-13b is an illustration of this approach with the dotted boxes representing the horizontal heat transfers. Each box may represent an actual heat exchanger or more (in case of multiple streams in the box). It is worth noting that the first and the last two boxes involve the use of heating and cooling utilities, respectively.

9.3 MINIMUM UTILITY TARGETING THROUGH AN ALGEBRAIC PROCEDURE

The temperature-interval diagram (TID) is a useful tool for ensuring thermodynamic feasibility of heat exchange. It is a special case of the CID described in Chapter Seven in which only two corresponding temperature scales are generated: hot and cold. The scale correspondence is determined using equation (9.2). Each stream is represented as a vertical arrow whose tail corresponds to its supply temperature, while its head represents its target temperature. Next, horizontal lines are drawn at the heads and tails of the arrows. These horizontal lines define a series of temperature intervals $z = 1, 2, \ldots, n_{\text{int}}$. Within any interval, it is thermodynamically feasible to transfer heat from the hot streams to the cold streams. It is also feasible to transfer heat from a hot stream in an interval z to any cold stream which lies in an interval below it.

Next, we construct a table of exchangeable heat loads (TEHL) to determine the heat exchange loads of the process streams in each temperature interval. The exchangeable load of the uth hot stream (losing sensible heat) which passes through the zth interval is defined as

$$HH_{u,z} = F_u C_{P,u}(T_{z-1} - T_z) \tag{9.7}$$

where T_{z-1} and T_z are the hot-scale temperatures at the top and the bottom lines defining the zth interval. On the other hand, the exchangeable capacity of the vth cold stream (gaining sensible heat) which passes through the zth interval is computed through

$$HC_{v,z} = f_v C_{P,v}(t_{z-1} - t_z) \tag{9.8}$$

where t_{z-1} and t_z are the cold-scale temperatures at the top and the bottom lines defining the zth interval.

Having determined the individual heating loads and cooling capacities of all process streams for all temperature intervals, one can also obtain the collective loads (capacities) of the hot (cold) process streams. The collective load of hot process streams within the zth interval is calculated by summing up the individual loads of the hot process streams that pass through that interval, i.e.,

$$HH_z^{\text{Total}} = \sum_{\substack{u \text{ passes through interval } z \\ \text{where } u=1,2,\ldots,N_H}} HH_{u,z} \tag{9.9}$$

Similarly, the collective cooling capacity of the cold process streams within the zth interval is evaluated as follows:

$$HC_z^{\text{Total}} = \sum_{\substack{v \text{ passes through interval } z \\ \text{and } v=1,2,\ldots,N_C}} HC_{v,z} \tag{9.10}$$

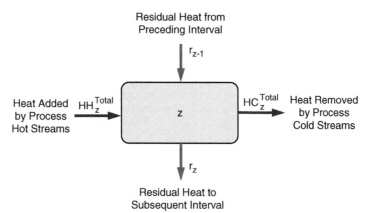

FIGURE 9-14 HEAT BALANCE AROUND TEMPERATURE INTERVAL

As has been mentioned earlier, within each temperature interval, it is thermodynamically as well as technically feasible to transfer heat from a hot process stream to a cold process stream. Moreover, it is feasible to pass heat from a hot process stream in an interval to any cold process stream in a lower interval. Hence, for the zth temperature interval, one can write the following heat-balance equation:

$$r_z = HH_z^{\text{Total}} - HC_z^{\text{Total}} + r_{z-1} \qquad (9.11)$$

where r_{z-1} and r_z are the residual heats entering and leaving the zth interval. Figure 9-14 illustrates the heat balance around the zth temperature interval.

r_0 is zero, since no process streams exist above the first interval. In addition, thermodynamic feasibility is ensured when all the r_z's are non-negative. Hence, a negative r_z indicates that residual heat is flowing upwards, which is thermodynamically infeasible. All negative residual heats can be made non-negative if a hot load equal to the most negative r_z is added to the problem. This load is referred to as the minimum heating utility requirement, $Q_{\text{Heating}}^{\min}$. Once this hot load is added, the cascade diagram is revised. A zero residual heat designates the thermal pinch location. The load leaving the last temperature interval is the minimum cooling utility requirement, $Q_{\text{Cooling}}^{\min}$.

9.4 CASE STUDY REVISITED USING THE ALGEBRAIC PROCEDURE

We now solve the chemical-plan case study described earlier using the algebraic cascade diagram. The first step is the construction of the TID (Figure 9-15). Next, the TEHLs for the process hot and cold streams are developed (Tables 9-3 and 9-4). Figures 9-16 and 9-17 show the cascade-diagram calculations. The results obtained from the revised

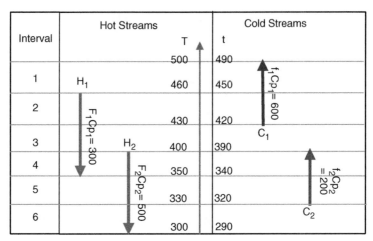

FIGURE 9-15   TEMPERATURE INTERVAL DIAGRAM FOR EXAMPLE 9-1

TABLE 9-3 TEHL FOR PROCESS HOT STREAMS

Interval	Load of H_1 (kW)	Load of H_2 (kW)	Total load (kW)
1	–	–	–
2	9000	–	9000
3	9000	–	9000
4	15,000	25,000	40,000
5	–	10,000	10,000
6	–	15,000	15,000

TABLE 9-4 TEHL FOR PROCESS COLD STREAMS

Interval	Capacity of C_1 (kW)	Capacity of C_2 (kW)	Total capacity (kW)
1	24,000	–	24,000
2	18,000	–	18,000
3	–	–	–
4	–	10,000	10,000
5	–	4000	4000
6	–	–	–

cascade diagram are identical to those obtained using the graphical pinch approach.

As mentioned earlier, for minimum utility usage no heat should be passed through the pinch. Let us illustrate this point using the cascade diagram. Suppose that we use $Q_{\text{Heating}}^{\text{extra}}$ kW more than the minimum heating

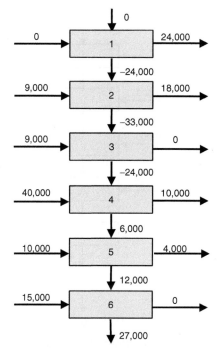

FIGURE 9-16 CASCADE DIAGRAM FOR EXAMPLE 9-1

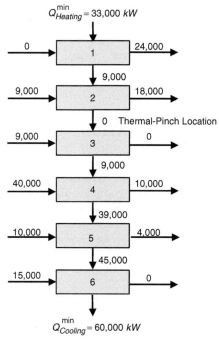

FIGURE 9-17 REVISED CASCADE DIAGRAM FOR EXAMPLE 9-1

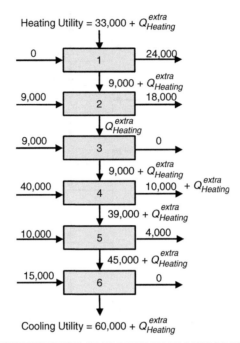

FIGURE 9-18   CONSEQUENCES OF PASSING HEAT ACROSS THE PINCH

utility. As can be seen from Figure 9-18, this additional heating utility passes down through the cascade diagram in the form of an increased residual heat load. At the pinch, the residual load becomes $Q_{\text{Heating}}^{\text{extra}}$. The net effect is not only an increase in the heating utility load, but also an equivalent increase in the cooling utility load.

9.5 SCREENING OF MULTIPLE UTILITIES USING THE GRAND COMPOSITE REPRESENTATION

So far, construction of the heat exchange pinch diagram started by maximizing the heat exchange among the process hot and cold streams and minimizing the external heating and cooling utilities. In many cases, multiple utilizes are available for service. These utilities must be screened so as to determine which one(s) should be used and the task of each utility. In order to minimize the cost of utilities, it may be necessary to stage the use of utilities such that at each level the use of the cheapest utility ($/kJ) is maximized while insuring its feasibility. A convenient way of screening multiple utilities is the *grand composite curve* (GCC).

The GCC may be directly constructed from the cascade diagram. To illustrate the procedure for constructing the GCC, let us consider the cascade diagram shown in Figure 9-19a. The residual heat loads are shown leaving the temperature intervals. Suppose that r_4 is the most negative

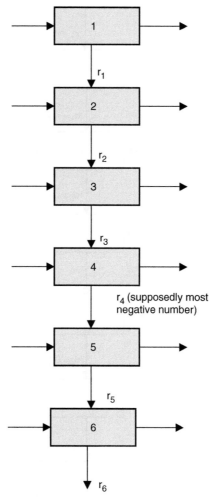

FIGURE 9-19a   CASCADE DIAGRAM

residual. As mentioned before, this infeasibility and all other infeasibilities are removed by adding the absolute value of r_4 to the top of the cascade diagram. This value is also the minimum heating utility. The residual loads are re-calculated with the load leaving the last temperature interval being the minimum cold utility.

Each residual heat corresponds to a hot temperature and a cold temperature. In order to have a single-temperature representation, we use an adjusted temperature scale which is calculated as the arithmetic average of the hot and the cold temperature, i.e.,

$$\text{Adjusted temperature} = \frac{T + t}{2} \qquad (9.12)$$

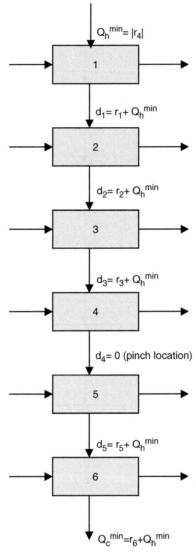

FIGURE 9-19b REVISED CASCADE DIAGRAM

Given the relationship between the hot and cold temperature (described by equation 9.2), we get:

$$\text{Adjusted temperature} = T - \frac{\Delta T^{\min}}{2} \qquad (9.13a)$$

$$= t + \frac{\Delta T^{\min}}{2} \qquad (9.13b)$$

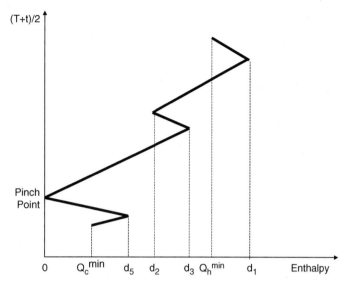

FIGURE 9-20a   CONSTRUCTION OF THE GCC

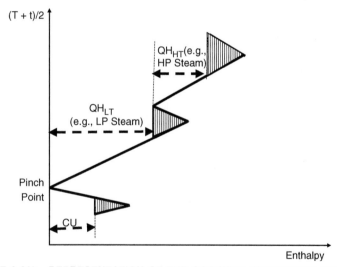

FIGURE 9-20b   REPRESENTATION OF THE GCC WITH INTEGRATED POCKETS AND OPTIMAL PLACEMENT OF UTILITIES

Next, we represent the adjusted temperature versus the residual enthalpy as shown in Figure 9-20a. This representation is the GCC. The pinch point corresponds to the zero-residual point. Additionally, the top and bottom residuals represent the minimum heating and cooling utilities. The question is how to distribute these loads over the multiple utilities? Any time the enthalpy representation is given by a line drawn from left to right, it corresponds to a surplus of heat in that interval. Conversely, when an enthalpy line is drawn from right to left, it corresponds to deficiency in heat in that interval. A heat surplus may be used to satisfy a heat residual below it. Therefore, the shaded regions (referred to as "*pockets*") shown in Figure 9-20b are fully integrated by transferring heat from process hot streams to process cold streams. Then, we represent each utility based on its temperature. The adjusted temperature of a heating utility is given by equation (9.13a) while that for a cooling utility is given by equation (9.13b). We start with the cheapest utility and maximize its use by filling the enthalpy gap (deficiency) at that level. Then, we move up for heating utilities and down for cooling utilities and continue to fill the enthalpy gaps by the cheapest utility at that level. Figure 9-20b is an illustration of this concept by screening low-and high-pressure steam where the low-pressure steam is cheaper ($/kJ) than the high-pressure steam. It is worth noting that the sum of the heating loads of the low- and high-pressure steams is equal to the minimum heating utility (the value of the top heat residual).

EXAMPLE 9-2 UTILITY SELECTION

Consider the stream data given in Table 9-5. Available for service are two heating utilities: a high pressure (HP) steam and a very high pressure (VHP) steam whose temperatures are 450 and 660°F, respectively. The VHP steam is more expensive than the HP. Also available for service is a cooling utility whose temperature is 100°F. The minimum approach temperature is taken as 10°F. Figures 9-21–9-23 represent the temperature-interval diagram, cascade diagram, and the GCC. As can be seen from Figure 9-22, the minimum heating requirement is 90 MM Btu/h. In order to maximize the use of the HP steam, we represent the HP on the GCC (a horizontal line at $450 - 10/2 = 445°F$). The deficit below this line is 50 MM Btu/h. Therefore, the duty of the HP steam is 50 MM Btu/h and the rest of the heating requirement (40 MM Btu/h) will be provided by the VHP steam.

TABLE 9-5 STREAM DATA FOR EXAMPLE 9-2

Stream	Flowrate × specific heat MMBtu/(h °F)	Supply temperature (°F)	Target temperature (°F)	Enthalpy change MMBtu/h
H_1	0.5	650	150	250.0
H_2	2.0	550	500	100.0
C_1	0.9	490	640	−135.0
C_2	1.5	360	490	−195.0

9.6 PROBLEMS

9.1 A plant has two process hot streams (H_1 and H_2), two process cold streams (C_1 and C_2), a heating utility (HU_1), and a cooling utility (CU_1). The problem data are given in Table 9-6. A value of $\Delta T^{\min} = 10°F$ is used. Using graphical and algebraic techniques, determine the minimum heating and cooling requirements for the problem.

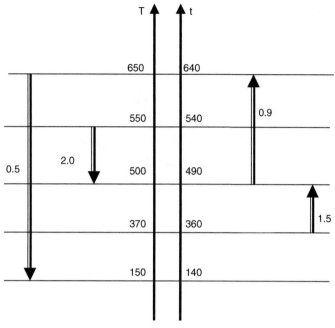

FIGURE 9-21   THE TEMPERATURE INTERVAL DIAGRAM FOR EXAMPLE 9-2

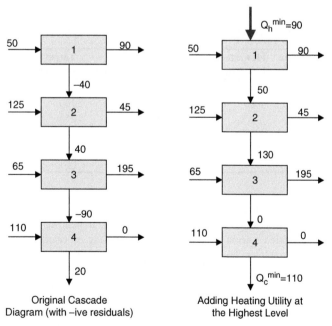

FIGURE 9-22   THE CASCADE AND REVISED CASCADE DIAGRAMS FOR EXAMPLE 9-2

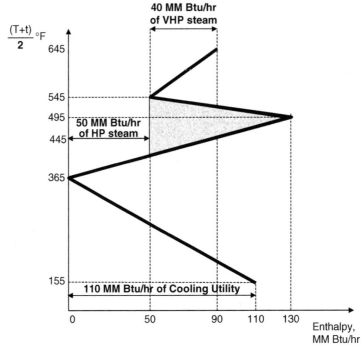

FIGURE 9-23   THE GCC FOR EXAMPLE 9-2

TABLE 9-6 STREAM DATA FOR PROBLEM 9.1 (DOUGLAS 1988)

Stream	Flowrate × specific heat (Btu/h °F)	Supply temperature (°F)	Target temperature (°F)	Enthalpy change (10^3 Btu/h)
H_1	1000	250	120	130
H_2	4000	200	100	400
HU_1	?	280	250	?
C_1	3000	90	150	−180
C_2	6000	130	190	−360
CU_1	?	60	80	?

9.2 Consider a process that has two process hot streams (H_1 and H_2), two process cold streams (C_1 and C_2), a heating utility (HU_1, which is a saturated vapor that loses its latent heat of condensation), and a cooling utility (CU_1). The problem data are given in Table 9-7. The cost of the heating utility is $\$4/10^6$ kJ added, and the cost of the coolant is $\$7/10^6$ kJ. A value of $\Delta T^{min} = 10$ K is used. Employ graphical, algebraic, and optimization techniques to determine the minimum heating and cooling requirements for the process.

9.3 Consider the pharmaceutical processing facility illustrated in Figure 9-24 (El-Halwagi 1997). The feed mixture (C_1) is first heated to 550 K, then fed to

TABLE 9-7 STREAM DATA FOR PROBLEM 9.2 (PAPOULIAS AND GROSSMANN 1983)

Stream	Flowrate × specific heat (kW/°C)	Supply temperature (°C)	Target temperature (°C)
H_1	10.55	249	138
H_2	8.79	160	93
HU_1	?	270	270
C_1	7.62	60	160
C_2	6.08	116	260
CU_1	?	38	82

TABLE 9-8 STREAM DATA FOR PHARMACEUTICAL PROCESS (EL-HALWAGI 1997)

Stream	Flowrate × specific heat (kW/°C)	Supply temperature (K)	Target temperature (K)	Enthalpy change (kW)
H_1	10	520	330	1900
H_2	5	380	300	400
HU_1	?	560	520	?
C_1	19	300	550	−4750
C_2	2	320	380	−120
CU_1	?	290	300	?

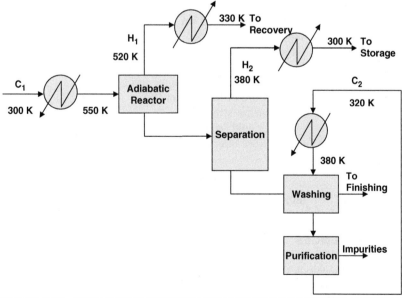

FIGURE 9-24 SIMPLIFIED FLOWSHEET FOR THE PHARMACEUTICAL PROCESS (EL-HALWAGI 1997)

an adiabatic reactor where an endothermic reaction takes place. The off-gases leaving the reactor (H_1) at 520 K are cooled to 330 K prior to being forwarded to the recovery unit. The mixture leaving the bottom of the reactor is separated into a vapor fraction and a slurry fraction. The vapor fraction (H_2) exits the separation unit at 380 K and is to be cooled to 300 prior to storage. The slurry fraction is washed with a hot immiscible liquid at 380 K. The wash liquid is purified and recycled to the washing unit. During purification, the temperature drops to 320 K. Therefore, the recycled liquid (C_2) is heated to 380 K. Two utilities are available for service, HU_1 and CU_1. The cost of the heating and cooling utilities ($\$/10^6$ kJ) are 3 and 5, respectively. Stream data are given in Table 9-8.

A value of $\Delta T^{min} = 10$ K is used. Using the pinch analysis, determine the minimum heating and cooling utilities for the process.

9.7 SYMBOLS

$C_{P,u}$	specific heat of hot stream u [kJ/(kg K)]
$C_{P,v}$	specific heat of cold stream v [kJ/(kg K)]
f	flowrate of cold stream (kg/s)
F	flowrate of hot stream (kg/s)
$HC_{v,z}$	cold load in interval z
$HH_{u,z}$	hot load in interval z
N_C	number of process cold streams
N_{CU}	number of cooling utilities
N_H	number of process hot streams
N_{HU}	number of process cold streams
r_z	residual heat leaving interval z
t	temperature of cold stream (K)
t_v^s	supply temperature of cold stream v (K)
t_v^t	target temperature of cold stream v (K)
T_v^s	supply temperature of hot stream u (K)
T_v^t	target temperature of hot stream u (K)
T	temperature of hot stream (K)
u	index for hot streams
v	index for cold streams
Z	temperature interval

Greek

ΔT^{min}	minimum approach temperature (K)

9.8 REFERENCES

Douglas, J.M. 1988, *Conceptual Design of Chemical Processes*, McGraw Hill, New York. pp. 216-288.

El-Halwagi, M.M. 1997, *Pollution Prevention through Process Integration: Systematic Design Tools*, Academic Press, San Diego.

Gundersen, T., and Naess, L. 1988, 'The synthesis of cost optimal heat exchanger networks: An industrial review of the state of the art', *Comput. Chem. Eng.* vol. 12, no. 6, 503-530.

Hohmann, E.C. 1971, *Optimum networks for heat exchanger*, Ph.D. Thesis, University of Southern California, Los Angeles.

Linnhoff, B. 1993, 'Pinch analysis – A state of the art overview,' *Trans. Inst. Chem. Eng. Chem. Eng. Res. Des.* vol. 71, Part A5, 503-522.

Linnhoff, B. and Hindmarsh, E. 1983, 'The pinch design method for heat exchanger networks,' *Chem. Eng. Sci.* vol. 38, no. 5, 745-763.

Papoulias, S.A. and Grossmann, I.E. 1983, 'A structural optimization approach in process synthesis. II. Heat recovery networks" *Comput. Chem. Eng.* vol. 7, no. 6, 707-721.

Shenoy, U.V. 1995, *Heat Exchange Network Synthesis: Process Optimization by Energy and Resource Analysis*, Gulf Publ. Co., Houston, TX.

Smith, R. 1995, *Chemical Process Design*, McGraw Hill, New York.

Umeda, T., Itoh, J., and Shiroko, K. 1979, 'A thermodynamic approach to the synthesis of heat integration systems in chemical processes', *Comp. Chem. Eng.* vol. 3, 273-282.

10

COMBINED HEAT AND POWER INTEGRATION

Chapter Nine addressed heat integration and the selection of heating and cooling utilities. In heat integration, thermal energy is the focus. Energy integration is a more general concept which provides a holistic view to the generation, allocation, transformation, and exchange of all forms of energy including heat and work (or heat rate and power when the system is studied on a per unit-time basis). This chapter discusses key issues in integrating heat and power which is commonly referred to as combined heat and power (CHP).

10.1 HEAT ENGINES

A heat engine is a device which uses energy in the form of heat to provide work. Because of the second law of thermodynamics, it is impossible to convert all the input heat into useful work. Consequently, a heat engine discharges heat. A heat engine has three elements:

- A heat source (or hot reservoir) which provides the heat input to the engine. The temperature of the heat source is referred to as T_H.
- A heat sink (or cold reservoir) which receives the discharged heat from the engine. The temperature of the heat sink is referred to as T_L.
- An object on which work is done.

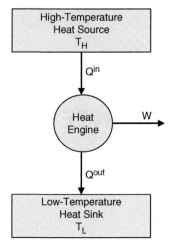

FIGURE 10-1 A HEAT ENGINE

The concept of a heat engine is widely used in many applications such as the automobile engine (internal combustion engine) where heat is generated during one part of the cycle and is used in another part of the cycle to produce useful work. Another application of heat engines is in the power plants with steam turbines where fuel is burned to generate heat which is added to water to produce steam which is let down through a turbine to produce useful work. Heat engines may be modeled through thermodynamic cycles such as Brayton, Otto, Diesel, Stirling, Rankine, and Carnot.

Figure 10-1 is a schematic representation of a heat engine. The energy balance around the heat engine can be written as follows:

$$Q^{\text{in}} = W + Q^{\text{out}} \tag{10.1}$$

The efficiency of the heat engine provides a measure of how much work can be extracted from the added heat, i.e.,

$$\eta = \frac{W}{Q^{\text{in}}} \tag{10.2}$$

As mentioned earlier, the second law of thermodynamics prevents the engine from transforming all the input heat to useful work. Consequently, the efficiency of the heat engine cannot reach 100%. The most efficient heat engine cycle is the Carnot cycle and it sets the limiting value on the fraction of the added heat which can be used to produce useful work. The Carnot cycle is assumed to be reversible and isentropic (constant entropy). The Carnot efficiency is given by:

$$\eta_{\text{Carnot}} = 1 - \frac{T_{\text{L}}}{T_{\text{H}}} \tag{10.3}$$

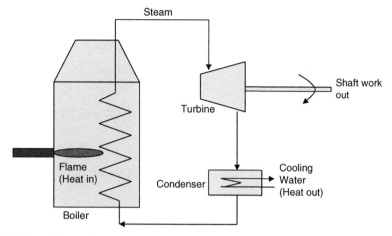

FIGURE 10-2   A POWER PLANT WITH A STEAM TURBINE

where T_L and T_H are the absolute temperatures of the heat sink and source, respectively.

A common example of a heat engine is a power plant with a steam turbine. As shown in Figure 10-2, the power plant includes a boiler which corresponds to the high-temperature heat source where a fuel is burned to transfer heat to water, thereby producing high-pressure steam. The high-pressure steam is directed towards a turbine where the pressure energy of the steam is converted into rotational energy in the form of a shaft work. The shaft work can be converted to electric energy through an electric generator. The steam leaving the turbine is cooled/condensed and the condensate is returned back to the boiler to be heated, transformed into steam, and the cycle continues. The net effect of the system is that the heat from the boiler is converted into work while the discharged heat (with the steam exiting the turbine) is transferred to the condenser which serves as the low-temperature heat sink.

EXAMPLE 10-1 EFFICIENCY AND WORK OF A HEAT ENGINE
Consider a heat engine that approximates the performance of a Carnot heat engine. The heat engine draws $40,000\,kW$ of heat from a hot reservoir at $833\,K$ and discharges heat to a heat sink at $500\,K$. Determine the following:

- Efficiency of the heat engine
- Generated work rate (power)
- Rate of heat discharge to heat sink

SOLUTION
According to equation (10.3), the efficiency is calculated as follows

$$\eta_{Carnot} = 1 - \frac{500}{833} = 0.4 \tag{10.4}$$

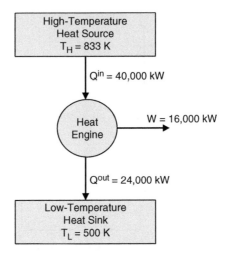

FIGURE 10-3   HEAT ENGINE OF EXAMPLE 10-1

Generated power is determined through equation (10.2):

$$W = 0.4 \times 40,000 = 16,000 \, \text{kW} \qquad (10.5)$$

Performing the energy balance given by equation (10.1), we get the rate of heat discharge at 500 K to be:

$$Q^{\text{out}} = 40,000 - 16,000 = 24,000 \, \text{kW} \qquad (10.6)$$

■ ■ ■   Figure 10-3 is a schematic representation of this heat engine.

10.2 HEAT PUMPS

A heat pump is a device which uses external energy in the form of external work to extract heat from a low-temperature heat source to a high-temperature heat sink. Consequently, a heat pump uses work to force the flow of heat from low to high temperature. A heat pump has three elements:

- A heat source (or cold reservoir) which provides the heat input to the heat pump. The temperature of the heat source is referred to as T_L.
- A heat sink (or hot reservoir) which receives the discharged heat from the heat pump. The temperature of the heat pump is referred to as T_H.
- An external source which delivers work.

The concept of a heat pump is widely used in many applications such as the refrigeration where heat is generated extracted from a low-temperature source (e.g., object to be cooled) and is discharged in another

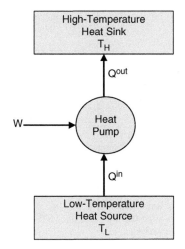

FIGURE 10-4   A HEAT PUMP

FIGURE 10-4   A HEAT PUMP

part of the systems (e.g., the surroundings) by exerting external work on the system.

Figure 10-4 is a schematic representation of a heat pump. The energy balance around the heat pump can be written as follows:

$$Q^{in} + W = Q^{out} \tag{10.7}$$

The coefficient of performance (COP) of the heat pump provides a measure of how much heat is extracted from the cold reservoir to how much work is added, i.e.,

$$COP = \frac{Q^{in}}{W} \tag{10.8a}$$

or

$$COP = \frac{Q^{in}}{Q^{out} - Q^{in}} \tag{10.8b}$$

The ideal limit for the COP of a heat pump is that of the Carnot cycle which is given by:

$$COP_{Carnot} = \frac{T_L}{T_H - T_L} \tag{10.9}$$

where T_L and T_H are the absolute temperatures of the cold reservoir (heat source) and hot reservoir (heat sink), respectively.

There are two main types of a heat pump: vapor compression and absorption cycle. A schematic representation of the vapor-compression heat pump is given by Figure 10-5. The main components in such a

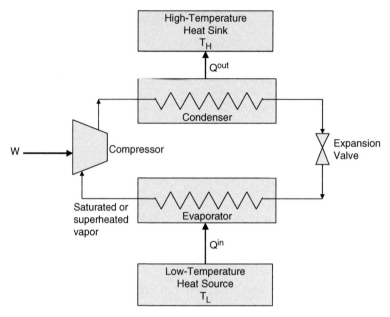

FIGURE 10-5   A VAPOR-COMPRESSION HEAT PUMP

vapor-compression heat pump system are two heat exchangers (a condenser and an evaporator), a compressor, and an expansion. Another important element of the vapor-compression heat pump is a volatile liquid which is referred to as the working fluid or the refrigerant is circulated through the four components. A saturated or superheated vapor leaving the evaporator is fed to the compressor where an external energy (e.g., mechanical, electric) is added. As a result of the compression the pressure and temperature of the vapor increase. The hot, compressed vapor is fed to a condenser where it condenses and releases heat to a hot reservoir or heat sink. Then, the working fluid is expanded through an expansion valve and fed to the evaporator. The temperature of the working fluid fed to the evaporator is maintained below the temperature of the heat source (or cold reservoir). Therefore, heat flows from the heat source to the working fluid in the evaporator resulting in the working fluid becoming a saturated or superheated vapor (which is the original state we started) and the cycle continues.

A common form of the vapor-compression heat pump used in refrigeration is the Brayton cycle shown in Figure 10-6. In this cycle, the external work is reduced by expanding the vapor in a turbine instead of an expansion valve. The work generated in the turbine is used to supplement the work exerted on the compressor.

The second type of a heat pump is the absorption cycle heat pump. Compared to vapor compression heat pumps where the major source of work is typically mechanical, the absorption heat pumps are driven primarily by thermal energy. The source of heat may be a fossil fuel (e.g., gas) or a high-pressure steam or waste heat from the process. The absorption cycle heat

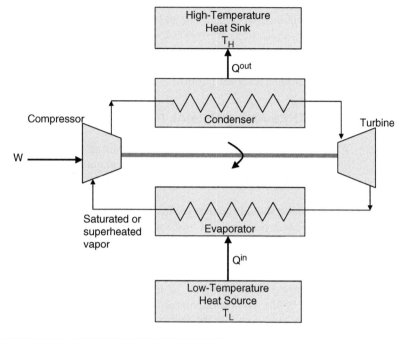

FIGURE 10-6 BRAYTON CYCLE HEAT PUMP

pump uses two fluids: a working fluid and an absorbent. While in vapor compression heat pumps, the compressor is used to create the pressure difference to circulate the working fluid, the absorption cycle heat pumps uses the absorbent to circulate the working fluid. This is done through three main steps. First, the low-pressure vapor from the evaporator is absorbed in the liquid absorbent. The dissolution of the working fluid in the absorbent is exothermic and heat must be removed from the solution. Next, the pressure of the solution is increased using a pump which consumes a modest amount of mechanical or electric energy. Then, the vapor is stripped or released in a stripper (or generator) where heat is added to the solution to induce stripping. The absorbent is returned to the absorber, the high-pressure vapor is fed to the condenser and the rest of the cycle is similar to the vapor-compression heat pump.

Because the primary driver of the absorption cycle is heat added to the stripper (Q_2), the COP for the absorption cycle is defined as:

$$\text{COP} = \frac{Q_2}{Q^{\text{in}}} \tag{10.10}$$

The most common fluids used for the working fluid and absorbent are:

- water (working fluid) and lithium bromide (absorbent): this is normally used for applications above 273 K (e.g., air conditioning);

- ammonia (working fluid) and water (absorbent): this is normally used for applications above 273 K (e.g., refrigeration).

Figures 10-7a and b illustrate an overview and a detailed sketch for the absorption cycle heat pumps.

10.3 HEAT ENGINES AND THERMAL PINCH DIAGRAM

When heat engines are used, they have a strong interaction with the heat integration of the process. Since the heat engine extracts and discharges heat, the following questions should be addressed:

- Where should the heat engine be placed on the temperature scale?
- What is the source of heat for the engine?
- Where should the heat leaving the engine be discharged?
- How should the heat engine be interfaced with the HEN?

To answer these questions, let us first consider the cascade diagram for the HEN of the process. Constructing the cascade diagram was described in Chapter Nine. Suppose the cascade diagram of the HEN is represented by Figure 10-8 and involves five temperature intervals with the pinch location corresponding to the temperature between intervals three and four (i.e., lower temperature for interval three which is upper temperature for interval four).

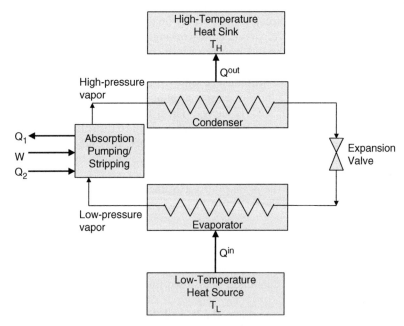

FIGURE 10-7a   OVERALL SKETCH OF THE ABSORPTION CYCLE HEAT PUMP

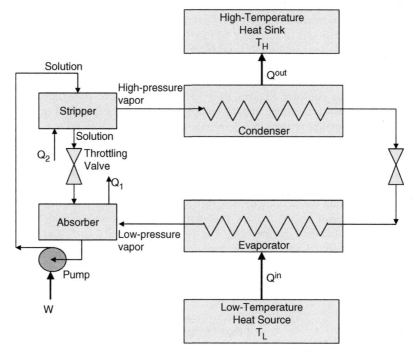

FIGURE 10-7b   DETAILED SKETCH OF THE ABSORPTION CYCLE HEAT PUMP

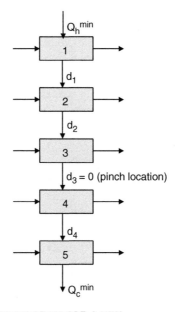

FIGURE 10-8   CASCADE DIAGRAM FOR A HEN

The residual heat for the kth interval is designated by d_k. The pinch location corresponds to zero heat residual. The minimum heating and cooling utilities are designated by Q_H^{min} and Q_C^{min}.

Townsend and Linnhoff (1983a,b) identified several important observations for the placement of heat engines. In order to discuss these observations, let us consider Figures 10-9–10-11. First, Figure 10-9 illustrates the placement of the heat engine completely above the pinch. For the HEN, the region above the pinch has deficit in heat which is equal to minimum heating utility. Meanwhile, the heat engine has a heat surplus in the form of heat discharge to the low-temperature sink which is an indication of a thermodynamic waste. Therefore, it is beneficial to match the heat deficit of the HEN with the heat discharge from the engine. This can be achieved by placing the heat exchanger above the pinch. In this case, heat is extracted for the heat engine from a high-temperature heat source which is hotter than the required temperature for the heating utility and the discharged heat is selected to match the minimum heating utility. Therefore, the process requirement for the heating utility serves as the heat sink for the heat engine. Consequently, the discharged heat from the heat engine is fully utilized

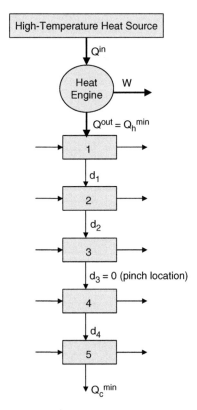

FIGURE 10-9   PLACEMENT OF THE HEAT ENGINE ABOVE THE PINCH

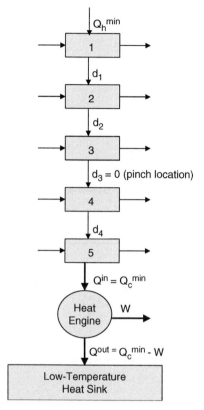

FIGURE 10-10 PLACEMENT OF THE HEAT ENGINE BELOW THE PINCH

to provide the heating utility for the process. This way, the heat discharged from the heat engine is no longer wasted and is effectively utilized in the HEN. So, while the efficiency of a heat engine is thermodynamically limited, such efficiency can be boosted if the heat engine is augmented with the HEN. When the efficiency is considered for the integrated heat engine and HEN, Figure 10-8 shows no losses for the heat engine. All the heat extracted from the hot source is transformed either to useful work or discharged as heat which is used to provide heating utility.

Figure 10-10 shows the placement of the heat engine below the pinch. For the HEN, the region below the pinch has a net surplus in heat. Therefore, a sensible proposition is to place the heat engine below the pinch such that the HEN surplus (required cooling utility) is used as the heat input to the heat engine. Consequently, two benefits accrue: the heat surplus below the pinch is removed by the heat engine and this extracted heat is partially transformed to useful work.

Finally, let us consider placing the heat engine across the pinch where heat is extracted from above the pinch and discharged below the pinch. As

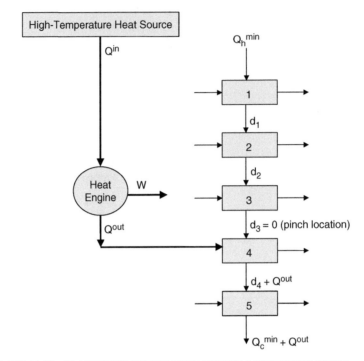

FIGURE 10-11   PLACEMENT OF THE HEAT ENGINE ACROSS THE PINCH

can be seen from Figure 10-11, there is no benefit from this arrangement. No reduction is achieved in heating or cooling utilities, and the heat engine performs in a similar way to the case when it is operated separately from the HEN.

The above discussion illustrates the need to appropriately place the heat engine. As observed by Townsend and Linnhoff (1983a,b), *the heat engine should be placed above the pinch or below the pinch (but not across the pinch).*

EXAMPLE 10-2 PLACEMENT OF A HEAT ENGINE

Consider the heat engine of Example 10-1. Suppose that this heat engine is used in the same process described by Example 9-1. What is the benefit for integrating the heat engine with the HEN?

SOLUTION

The TID and the cascade diagram for Example 9-1 are shown by Figures 10-12 and 10-13. The heat is discharged from the heat engine at 500 K which is hot enough to provide a portion of the needed heating utility for the HEN. Therefore, the heating utility is reduced from 33,000 kW to 9000 kW as shown in Figure 10-14.

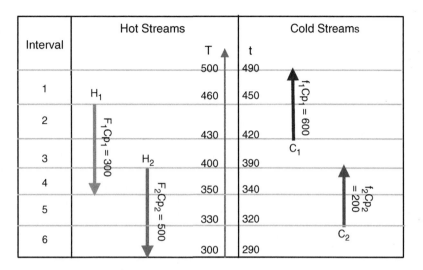

FIGURE 10-12 TEMPERATURE INTERVAL DIAGRAM FOR THE HEN OF
EXAMPLE 9-1

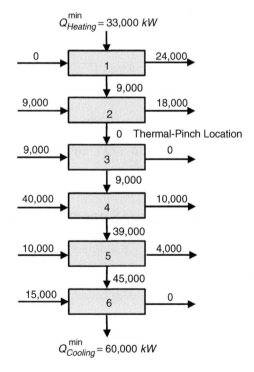

FIGURE 10-13 CASCADE DIAGRAM FOR THE HEN OF EXAMPLE 9-1

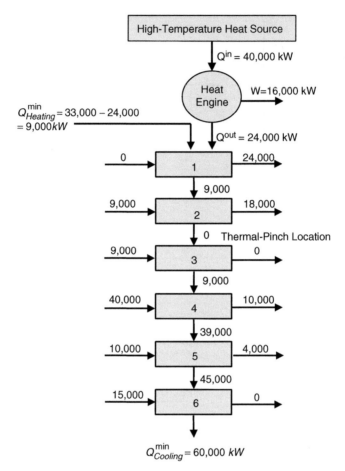

FIGURE 10-14 INTEGRATED HEAT ENGINE AND HEN FOR EXAMPLES 9-1 AND 10-1

10.4 HEAT PUMPS AND THERMAL PINCH DIAGRAM

Heat pumps provide an efficient way of elevating the quality of heat. Therefore, they have the potential to be integrated with the HEN. The challenge is in proper placement of the heat pump. To illustrate the role of placing the heat pump, let us consider the following three cases:

- Placing the heat pump above the pinch: As can be seen in Figure 10-15, this arrangement does not provide any benefit. While the heating utility is reduced by W, an equivalent amount of work must be added to the heat pump thereby nullifying the benefit.
- Placing the heat pump below the pinch: In this case, there are two penalties: work is exerted on the heat pump without a benefit and the

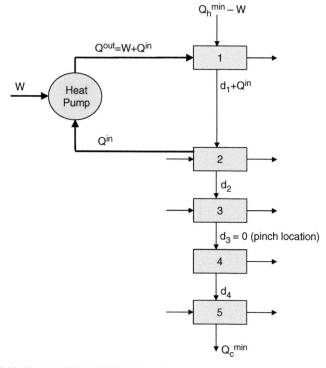

FIGURE 10-15   PLACING THE HEAT PUMP ABOVE THE PINCH

cooling utility increases by a similar amount. This scenario is illustrated by Figure 10-16.

- Placing the heat pump across the pinch: This arrangement is shown by Figure 10-17. It offers two benefits: a reduction in heating and cooling utilities. For a work, W, exerted on the heat pump, the heating utility is reduced by $W+Q^{in}$ and the cooling utility is reduced by Q^{in}.

The above discussion illustrates the need to appropriately place the heat pumps. As observed by Townsend and Linnhoff (1983a,b), *the heat pump should be placed across the pinch (not above the pinch or below the pinch).*

10.5 COGENERATION TARGETING

The previous sections illustrate the benefits of combining heat and power. In this regard, steam plays a major role in the process. Steam is a convenient medium in capturing and delivering heat. Steam may also have non-heating uses in the process (e.g., stripping agent, blanketing, etc.). Each application of steam requires a certain pressure (and/or temperature). For instance, as was seen in Chapter Nine, the grand composite curve provides a method for utility

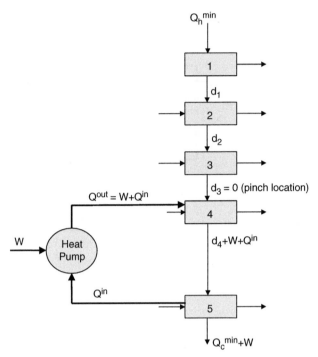

FIGURE 10-16   PLACING THE HEAT PUMP BELOW THE PINCH

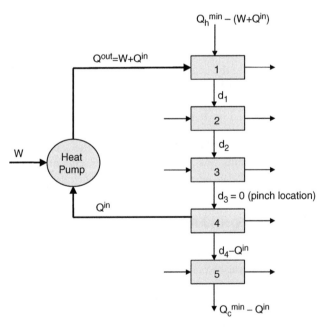

FIGURE 10-17   PLACING THE HEAT PUMP ACROSS THE PINCH

selection including steam levels. In addition to using steam for heating and non-heating purposes, it may be advantageous to generate the steam at higher pressures than the required levels then pass the stem through steam turbines to generate electric power (or shaft work). This way, the steam plays a dual role: it provides the process requirements and produces power. This is often referred to as "*cogeneration*" and when the process requirements are in the form of heating demands, it is referred to as combined heat and power.

This section provides a shortcut method to identifying the cogeneration target for a process. More details on this method can be found in Harell and El-Halwagi (2003); Harell (2004); and Harell et al. (2005). The problem is stated as follows:

Consider a process with:

- A number of combustible wastes and byproducts
- Specific heating and cooling demands
- Non-heating steam demands (e.g., tracing, blanketing, stripping, etc.).

It is desired to identify a target for power cogeneration which makes effective use of the combustible wastes while fulfilling all process demands for heating and non-heating steam.

First, we start by considering the combustible wastes that can be burned in existing boilers and industrial furnaces provide heat typically in the form of steam. Depending on the specific boiler or industrial furnace, the steam is generated at a specific pressure and is fed to the appropriate steam header. The steam demands can be determined through heat and mass integration. Heat integration (e.g., grand composite analysis) can be used to determine the heating steam requirements and their levels. Mass integration can be used to determine the non-heating steam demands. The result of the mass integration analysis is the identification of process supply of steam (from the combustible wastes) and the process demand of steam (for non-heating purposes). The supplies and demands are determined in terms of quantities and levels of the steam headers. Those supplies and demands are now known both in terms of quantities and pressures (or header level). A typical industrial process has several pressure levels in its steam system. Consider a process with the following headers: very high pressure (VHP), high pressure (HP), medium pressure (MP), and low pressure (LP). Depending on the supply and demand to each header, the net balance of steam for each header will be positive (surplus) or negative (deficit). For instance, Figure 10-18 shows a case where there is a surplus of steam at the VHP and HP levels, while there is a deficit of steam at the MP and LP levels. In order to satisfy the steam deficits in the MP and LP levels, steam may be let down from the surplus headers and/or supplied from fired boilers. These decisions will be made later.

When steam is let down from higher pressure to lower pressure in a steam turbine, the power generation can be obtained from a Mollier diagram as illustrated in Figure 10-19a. For a given pair of inlet pressure and temperature and for a given outlet pressure of the turbine, the isentropic enthalpy change in the turbine can be determined as:

$$\Delta H^{\text{isentropic}} = H^{\text{in}} - H^{\text{out}}_{\text{is}} \qquad (10.11)$$

where $\Delta H^{\text{isentropic}}$ is the specific isentropic enthalpy change in the turbine, H^{in} is the specific enthalpy of the steam at the inlet temperature and pressure of the turbine and $H^{\text{out}}_{\text{is}}$ is the specific isentropic enthalpy at the

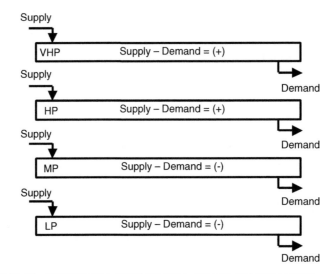

FIGURE 10-18   HEADER BALANCE (HARELL AND EL-HALWAGI 2003)

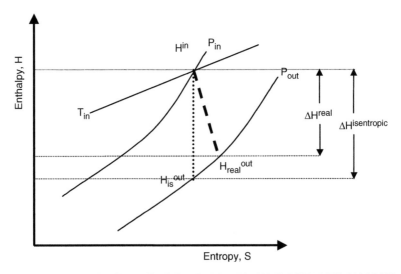

FIGURE 10-19a   TURBINE POWER REPRESENTED ON THE MOLLIER DIAGRAM

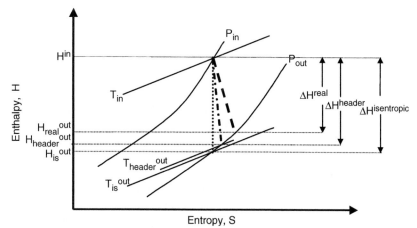

FIGURE 10-19b   APPROXIMATION OF TURBINE PERFORMANCE USING HEADER ENTHALPIES (HARELL AND EL-HALWAGI 2003)

outlet pressure of the turbine. In reality, a turbine cannot extract all the energy expressed by the isentropic expansion. Therefore, an isentropic efficiency term is defined as follows:

$$\eta_{is} = \frac{\Delta H^{real}}{\Delta H^{isentropic}} \qquad (10.12)$$

where η_{is} is the isentropic efficiency and ΔH^{real} is the actual specific enthalpy difference across the turbine. Based on equations (10.11) and (10.12), the specific work, w, produced by the turbine is expressed as:

$$w = \Delta H^{real} = \eta_{is}(H^{in} - H_{is}^{out}) \qquad (10.13)$$

Depending on the flowrate of steam passing through the turbine, $\dot{m}$, the power produced by the turbine, W, is given by:

$$W = \dot{m}\eta_{is}(H^{in} - H_{is}^{out}) \qquad (10.14)$$

When the objective is to target cogeneration potential for a process, it is advantageous to determine this target without detailed calculations. Equation (10.14) requires individual turbine calculations to evaluate the outlet enthalpy at constant pressure and the flowrate of steam passing through the turbine. As such, it is useful to develop expressions for estimating turbine power using easily determined terms. Since turbines are placed between steams headers with known temperature and pressures. Consequently, the specific enthalpies of these steam headers are also known. Harell and El-Halwagi (2003) proposed an approximation of the turbine power to be based on the header enthalpies. They introduced the term "*extractable energy*" which is based on the header levels that the turbine actually operates between, rather than the isentropic conditions at the outlet pressure.

$$e_{\text{Header}} = \eta_{\text{Header}} H_{\text{Header}} \tag{10.15}$$

where e_{Header} is the extractable energy for a given header, η_{Header} is an efficiency term and H_{Header} is the specific enthalpy at a given set of conditions for the header. Let the extractable power, E_{Header}, of a header be defined as:

$$E_{\text{Header}} = \dot{m} \eta_{\text{Header}} H_{\text{Header}} = \dot{m} e_{\text{Header}} \tag{10.16}$$

The advantage of this definition is that it does not involve detailed turbine calculations such as isentropic outlet enthalpy.

Then, the power generation expression can be rewritten as the difference between the inlet and outlet extractable power:

$$W = E^{\text{in}} - E^{\text{out}} \tag{10.17}$$

where E^{in} is the extractable power at the header conditions feeding the inlet steam to the turbine and E^{out} is the extractable power at the header conditions receiving the outlet steam from the turbine. The representation of this form of power evaluation on a Mollier diagram is shown by Figure 10-19b. As can be seen from this figure, the header efficiency is given by:

$$\eta_{\text{header}} = \frac{\Delta H^{\text{real}}}{H^{\text{in}} - H^{\text{out}}} \tag{10.18}$$

Based on the concept of extractable energy, Harell and El-Halwagi (2003) developed a graphical targeting procedure that identifies cogeneration targets and can serve as the basis for developing feasible turbine network designs which meet the predicted target. According to this method, the extractable power expressed by equation (10.16) is plotted for each header versus the net flowrate of the header. First, the extractable power is plotted versus the steam flowrate for each surplus header in ascending order of pressure levels. The result of this superposition is the development of a *surplus composite line*. Similarly, the extractable power for the deficit headers is plotted in ascending order leading to the *deficit composite line*. Consider the case shown in Figure 10-20 which is based on Figure 10-18. There are two surplus headers (VHP and HP) and two deficit headers (MP and LP). The extractable power at the VHP, HP, MP, and LP levels are referred to as $E1$, $E2$, $E3$, and $E4$, respectively, and the steam flowrates of these headers are referred to as $M1$, $M2$, $M3$, and $M4$, respectively. The construction of the surplus composite line is shown by Figure (10-20). The deficit composite line can be constructed in a similar manner.

The next step is to represent the two composite lines on the same diagram (Figure 10-21). Hence, the cogeneration potential of the system can be

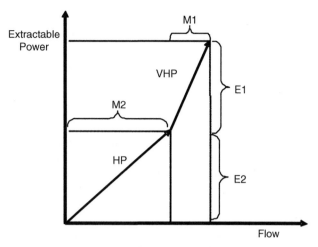

FIGURE 10-20 CONSTRUCTION OF THE EXTRACTABLE POWER SURPLUS COMPOSITE CURVE (HARELL AND EL-HALWAGI 2003)

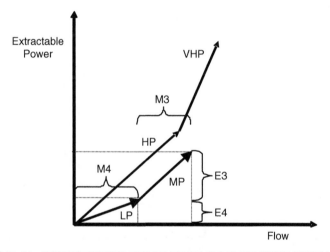

FIGURE 10-21 CONSTRUCTION OF THE EXTRACTABLE POWER SURPLUS AND DEFICIT COMPOSITE CURVES (HARELL AND EL-HALWAGI 2003)

evaluated by shifting the deficit composite line to the right and up until it is directly below the terminal point of the surplus line. As illustrated by Figure 10-22, the vertical distance (of the "jaw") between the two terminal points of the surplus and the deficit composite lines is the *target for cogeneration potential*. This diagram which is referred to as the *"extractable power cogeneration targeting pinch diagram"*. In order to guarantee that the cogeneration target is feasible, higher pressure surplus headers must be directly above lower pressure deficit headers. By letting down steam from the higher-pressure headers, to the lower-pressure headers, the deficit is removed while power can be generated by virtue of the difference between the extractable powers. Therefore, both steam demands (heating and non-heating) are satisfied while power is cogenerated. The steam flowrate of the portion of the surplus composite line which does not overlap with the deficit composite line represents excess steam. Since there is no header demand for this excess, it can be used for power generation (not cogeneration) by letting it down through a condensing turbine, used for other process purposes, or simply vented.

EXAMPLE 10-3 COGENERATION TARGET (HARELL ET AL., 2005)

Consider a process with four steam headers: VHP, HP, MP, and LP. The temperature and pressure of these steam headers are given in Table 10-1. For this case study, the power efficiency for any header is taken as 70% (i.e., $\eta = 0.7$).

As a result of mass and heat integration, a number of steam streams is supplied to the headers. The results are given in Table 10-2.

By plotting the extractable power versus flow rate, the results are shown by Figure 10-23. Next the deficit composite line is shifted up and to the right such that its terminal point is vertically alighted with the terminal point of the surplus composite line (Figure 10-24). As can be seen

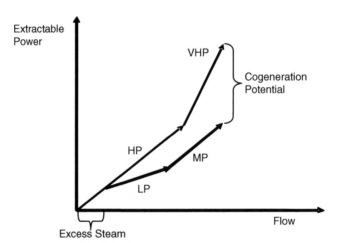

FIGURE 10-22 EXTRACTABLE POWER COGENERATION TARGETING PINCH DIAGRAM (HARELL AND EL-HALWAGI 2003)

TABLE 10-1 STEAM HEADER DATA (HARELL ET AL., 2005)

Header	Pressure (psia)	Temperature (°F)	Specific enthalpy (Btu/lb)
VHP	600	805	1414
HP	160	600	1327
MP	80	395	1226
LP	60	300	1181

TABLE 10-2 SURPLUS/DEFICIT STEAM FLOWRATES AND EXTRACTABLE POWER (HARELL ET AL., 2005)

Surplus header	Pressure (psia)	Flowrate (lb/h)	Net enthalpy difference per hour (MMBtu/h)	Extractable power (MMBtu/h)
VHP	600	119,519	169	118.3
HP	160	101,676	135	94.5
Deficit headers				
MP	80	110,114	−135	94.5
LP	60	60,019	−71	49.7

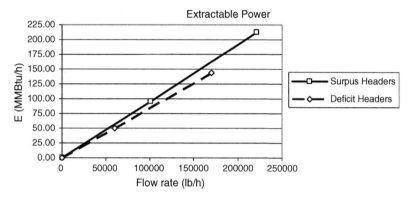

FIGURE 10-23 UNSHIFTED EXTRACTABLE POWER VERSUS FLOWRATE PLOT (HARELL ET AL., 2005)

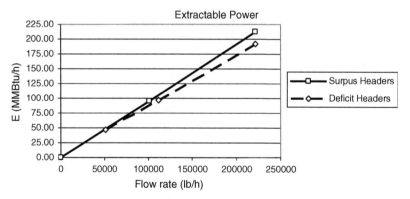

FIGURE 10-24 SHIFTED EXTRACTABLE POWER VERSUS FLOWRATE PLOT (HARELL ET AL., 2005)

from Figure 10-24, the cogeneration target is 21.1 MMBtu/h, and there is an excess of approximately 51,000 lb/h of steam being generated within the process. The excess steam in this case should be let down through a condensing turbine because it primarily comes from a process source.

■ ■ ■ This additional power is considered to be generation (and not cogeneration) since it does not contribute to satisfying other process needs.

10.6 ADDITIONAL READINGS

Several methods have been used to assess the cogeneration opportunities of a process. Dhole and Linnhoff (1992) proposed to use exergy analysis to estimate cogeneration opportunities. Exergy is a measure of the useful work available in a heat source. Raissi (1994) developed the TH-shaftwork targeting model. This model is based on an observation that the specific enthalpy at the turbine outlet minus the specific enthalpy of saturated water is relatively constant regardless of the outlet conditions. The TH-shaftwork target combines this observation with the observation that specific power can be approximated by a linear function of the outlet saturation temperature (Klemes et al., 1997). Mathematical-programming techniques have also been made in the area of modeling and optimization of turbine network systems. The turbine hardware model (Mavromatis, 1996; Mavromatis and Kokossis, 1998) is based on the Willans line (commonly used in turbine modeling to represent steam consumption versus rated power of the turbine) and utilizes typical maximum efficiency plots and rules of thumb to target cogeneration potential. The turbine hardware model also incorporates complex turbines by modeling them as sets of simpler turbines.

10.7 PROBLEMS

10.1   Consider a heat engine that approximates the performance of a Carnot heat engine. The heat engine draws 3000 kW of heat from a hot reservoir at 800 K and discharges heat to a heat sink at 600 K. Determine the following:
   Efficiency of the heat engine
   Generated work rate (power)
   Rate of heat discharge to heat sink
10.2   Consider the heat engine of Problem 10.1. Suppose that this heat engine is used in the same process described by Problem 9.3. What is the benefit for integrating the heat engine with the HEN?
10.3   Resolve Example 10-3 when the net surplus flowrate of the VHP header is 150,000 lb/h.

10.8 REFERENCES

Dhole V.R., and Linnhoff, B. 1992, *Comp Chem Eng*, vol. 17, S101-S109.
Harell, D.A. 2004, *Resource Conservation and Allocation via Process Integration*, Ph.D. Dissertation, Texas A&M University.
Harell D., and El-Halwagi M. 2003, *Design Techniques and Software Development for Cogeneration Targeting with Mass and Energy Integration*, AIChE Spring Meeting, New Orleans, March.

Harell, D.A., El-Halwagi, M.M., and Spriggs, H.D. 2005, 'Targeting Cogeneration and Waste Utilization through Process Integration', submitted.

Klemes, J., Dhole, V. R., Raissi, K., Perry, L., and Puigjaner, S.J. 1997, 'Targeting and design methodology for reduction of fuel, power, and CO_2 on total sites', *J. Applied Thermal Eng.*, vol. 17, 993.

Linnhoff, B. 1993, 'Pinch analysis – A state of the art overview', *Trans. Inst. Chem. Eng. Chem. Eng. Res. Des.* vol. 71, Part A5, 503-522.

Linnhoff, B., Townsend, Boland, D.W.D., Hewitt, G.F., Thomas, B.E.A., Guy, A.R., and Marsland, R.H. 1982, *User Guide on Process Integration for the Efficient Use of Energy*, ICHemE Press, Rugby, Warwickshire, UK.

Mavromatis, S.P. 1996, *Conceptual Design and Operation of Industrial Steam Turbine Networks*, Ph.D. Thesis, University of Manchester Institute of Science and Technology, Manchester, UK.

Mavromatis, S. P. and Kokossis, A. C. 1998, 'Conceptual optimization of utility networks for operational variation-I targets and level optimization', *Chem. Eng. Sci.*, vol. 53, 1585.

Raissi, K. 1994, *Total Site Integration*, Ph.D. Thesis, University of Manchester Institute of Science and Technology, Manchester, UK.

Townsend, D.W. and Linnhoff, B. 1983a, 'Heat and power networks in process design. I. Criteria for placement of heat engines and heat pumps in process networks', *AIChE J.*, vol. 29, pp. 742-747.

Townsend, D.W. and Linnhoff, B. 1983b, 'Heat and power networks in process design. II. Design procedure for equipment selection and process matching', *AIChE J.*, vol. 29, pp. 747-754

11

OVERVIEW OF OPTIMIZATION

Optimization is one of the most powerful tools in process integration. Optimization involves the selection of the "best" solution from among the set of candidate solutions. The degree of goodness of the solution is quantified using an *objective function* (e.g., cost) which is to be minimized or maximized. The search process is undertaken subject to the system model and restrictions which are termed *constraints*. Hence, the purpose of optimization is to **maximize (or minimize)** the value of a function (called **objective function**) subject to a number of restrictions (called **constraints**). These constraints are in the form of equality and inequality expressions. Examples of equality constraints include material and energy balances, process modeling equations, and thermodynamic requirements. On the other hand, the nature of inequality constraints may be environmental (e.g., the quantity of certain pollutants should be below specific levels), technical (e.g., pressure, temperature or flowrate should not exceed some given values) or thermodynamic (e.g., the state of the system cannot violate second law of thermodynamics). The principles of optimization theory and algorithms are covered by various books (e.g., Diwekar 2003; Tawarmalani and Sahinidis 2003; Floudas and Pardalos, 2001; Edgar and Himmelblau 2001; Floudas 1999; Grossmann 1996). Additionally, for a retrospective and future prospective on optimization, the reader is referred to recent publications (e.g., Biegler and Grossmann

2005; Grossmann and Biegler 2005; Floudas et al., 2004). This chapter presents an overview of using mathematical techniques to formulate optimization problems. Additionally, the use of optimization software to solve and analyze optimization programs will be presented.

11.1 MATHEMATICAL PROGRAMMING

Over the past few decades, significant progress has been made in the field of *mathematical programming* which deals with the formulation, solution, and analysis of optimization problems or mathematical programs. The analysis and solution of an optimization problem may involve graphical, algebraic, or computer-aided tools. As such the word "programming" does not necessarily entail the use of computers or computer coding although in solving complex optimization problems, the use of computer-aided tools is becoming a common practice. A mathematical program (or a mathematical programming model or an optimization problem) can be represented by:

$$\text{min (or max)} : f(x_1, x_2 \dots, x_N) \tag{P11.1}$$

Subject to:

Inequality constraints

$$g_1(x_1, x_2, \dots, x_N) \leq 0$$

$$g_2(x_1, x_2, \dots, x_N) \leq 0$$

$$\vdots$$

$$g_m(x_1, x_2, \dots x_N) \leq 0$$

Equality constraints

$$h_1(x_1, x_2, \dots, x_N) = 0$$

$$h_2(x_1, x_2, \dots, x_N) = 0$$

$$\vdots$$

$$h_E(x_1, x_2, \dots x_N) = 0 \quad E \leq N$$

The variables x_1, x_2,…,x_n are referred to as *decision or optimization variables*. In a vector notation, a mathematical program can be expressed as:

$$\min \text{ (or max) } f(x) \quad \text{where } x^T = [x_1, x_2, \ldots x_N] \qquad \text{(P11.2a)}$$

Subject to:

$$g(x) \leq 0 \quad \text{where } g^T = [g_1, g_2, \ldots g_m]$$

$$h(x) = 0 \quad \text{where } h^T = [h_1, h_2, \ldots h_E]$$

The vector x is referred to as the vector of optimization variables. It is also possible to describe the constraints only as inequality constraints, i.e.,

$$\min \text{ (or max) } f(x) \quad \text{where } x^T = [x_1, x_2, \ldots x_N] \qquad \text{(P11.2b)}$$

Subject to:

$$g(x) \leq 0$$

Program (P11.2b) is general enough to include the following cases:

- Equality constraints: since an equality constraint can be represented by two inequality constraints:

$$g(x) \leq 0 \quad \text{and} \quad -g(x) \leq 0 \qquad (11.1)$$

Both constraints are feasible only when they both become equality constraints.
- Inequality constraints that are greater than or equal to zero: the greater than sign can always be transformed to less than sign by multiplying the constraint by a negative sign, i.e.,

$$g(x) \geq 0 \quad \text{is equivalent to } -g(x) \leq 0 \qquad (11.2)$$

- A constant term on the right-hand side can be moved to the left-hand side leaving a zero, i.e.,

$$g(x) \leq \text{Constant is equivalent to } g(x) - \text{Constant} \leq 0 \qquad (11.3)$$

Formulating process integration and design tasks as optimization problems or mathematical programming models has several advantages:

- It provides an effective framework for solving large-scale problems and highly interactive tasks that cannot be readily addressed by graphical and algebraic techniques.

- An optimization formulation may be parametrically solved to establish optimality conditions that can be used in developing visualization and algebraic solutions.
- In properly formulating an optimization problem, it is necessary to establish a model that accurately describes the task using mathematical relationships that capture the essence of the problem. In so doing, many important interactions are explicitly described with the result of providing better understanding of the system.
- When the solution of a mathematical program is implemented using computer-aided tools, it is possible to effectively examine what-if scenarios and conduct sensitivity analysis.

On the other hand, some of the possible disadvantages of purely mathematical approaches to process integration (compared to graphical and algebraic tools) include:

- Local versus global solution: For some classes of optimization problems, obtaining best possible (global) solutions may be quite tedious and illusive. Figure 11-1 is an illustration of an unconstrained non-linear minimization problem with an objective function that exhibits multiple optimum points (two local minima and one global minimum). The relevant solution is the global optimum point. In fact, a feasible point (such as point a) may outperform the two local minima. Clearly, the task becomes far more complex for problems involving multiple optimization variables and constraints. The area of

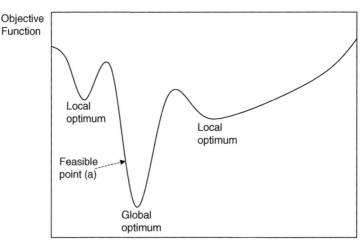

FIGURE 11-1 LOCAL VERSUS GLOBAL OPTIMAL SOLUTIONS FOR A MINIMIZATION PROBLEM

global optimization is an active research area and much progress is anticipated over the next decade.

- The solution of an optimization problem may not shed light on the key characteristics of the system and the overall insights.
- Without certain level of expertise, it is difficult to embed the designer's insights and preferences into the mathematical programming process. Until mathematical optimization becomes a common tool in engineering practice, engineers will tend to favor algebraic (e.g., spreadsheet) and visualization tools. This issue should be resolved with reforms in engineering curricula and the development of more user-friendly and effective optimization platforms.

11.2 CLASSIFICATION OF OPTIMIZATION PROGRAMS

An optimization problem in which the objective function as well as all the constraints are linear is called a linear program (LP); otherwise it is termed as a non-linear program (NLP). A linear program can be mathematically described as:

$$f(x_1, x_2, \ldots, x_n) = c_1 x_1 + c_2 x_2 + \cdots + c_n x_n \qquad \text{(P11.3)}$$

Subject to:

$$g_i(x_1, x_2, \ldots, x_n) \leq a_{i1} x_1 + a_{i2} x_2 + \cdots a_{in} x_n$$

where c_i and a_{ij} ($i = 1,2, \ldots, m; j = 1,2, \ldots, n$) are known constants.

In a vector form, an LP can be expressed as

$$\text{minimize} \quad z = C^{\mathrm{T}} X \qquad \text{(P11.4)}$$

Subject to:

$$AX \leq B$$

where B and C are vectors of constants and A is a matrix of constants.

The nature of optimization variables also affects the classification of optimization programs. If all the variables in the mathematical program are integers (e.g., 0,1,2,3), the optimization program is referred to as an Integer Program (IP). The most commonly used integer variables are the zero/one binary integer variables. An optimization formulation that contains continuous (real) variables (e.g., pressure, temperature, or flowrate) as well as integer variables is called a mixed-integer program (MIP). Depending on the linearity or non-linearity of MIPs, they are designated as mixed-integer linear programs (MILPs) and mixed-integer non-linear programs (MINLPs).

11.3 FORMULATION OF OPTIMIZATION MODELS

The mathematical formulation of an optimization problem entails the following steps:

1. Determine the Objective Function
   - Determine the quantity to be optimized (the objective function).
   - Identify the decision (optimization) variables that are needed to describe the objective function.
   - Express the objective function as a mathematical function in terms of the optimization variables.
2. Develop Your Game Plan to Tackle the Problem
   - Determine an approach to address the problem.
   - What is your rationale?
   - What are the key concepts necessary to transform your thoughts and approach into a working formulation?
3. Develop the Constraints
   - Transform your approach to a mathematical framework.
   - Develop a search space that is rich enough to embed the solution alternatives.
   - Identify all explicit relations, restrictions, and limitations that are needed to describe the approach. Express them mathematically as equality and inequality constraints.
   - Incorporate subtle constraints (e.g., non-negativity or integer requirements on the input variables).
4. Improve Formulation
   - Avoid highly non-linear constraints or terms that lead to difficulties in convergence or solution.
   - Use insights to simplify formulation or to provide useful bounds on variables.
   - Enhance the clarity of the model to become easy to understand, to debug, and to reveal important information.

■■■■■   EXAMPLE 11-1 OPTIMIZATION OF SOLVENT PURCHASE AND MIXING

Owing to the recent environmental regulations, the scrubbing of hydrogen sulfide from gaseous emissions is becoming a common industrial practice. One of the efficient solvents is monoethanolamine "MEA". In the chemical plant, there are two variable grades of MEA. The first grade consists of 80 wt.% MEA and 20 wt.% weight water. Its cost is 85 cent/kg. The second grade consists of 68 wt.% MEA and 32 wt.% water. Its cost is 60 cent/kg. It is desired to mix the two grades so as to obtain an MEA solution that contains no more than 25 wt.% water. What is the optimal mixing ratio of the two grades which will minimize the cost of MEA solution (per kg)?

SOLUTION

Consider a basis of 1.0 kg of MEA solution. What is our objective function? Here, we are trying to minimize the cost of the MEA solution. What is it a function of? It is a function of how much should be purchased from each of the two grades.

Let:

$x1$ be amount of grade 1 (kg)
$x2$ be amount of grade 2 (kg)

These are the two decision (optimization) variables. Additionally, let us define our objective function as:

$z =$ cost of 1 kg of MEA solution (cents)

The objective function can be mathematically described as follows:

$$\text{minimize } z = 85\ x1 + 60\ x2 \tag{11.4}$$

Now, what is our game plan to tackle the problem. Cost has been addressed by the objective function. Selecting the cheaper solvent (grade 2) violates the quality requirements described by water content. Therefore, in order to develop the constraints, let us first determine our approach. Each grade contributes to a certain amount of water. Therefore, our approach should entail tracking water and insuring that the quality constraint is satisfied. This can be accomplished by imagining a mixing operation where the two grades are mixed and a component material balance on water can be written to describe the mixing operation (Figure 11-2). The resulting water content should satisfy the 25% water-content constraint. Additionally, we should include an overall material balance which relates the basis (1.0 kg of MEA solution) to the quantities of the two grades.

Therefore, the following constraints should be used:

Constraint on water content

$$0.20\ x1 + 0.32\ x2 \leq 0.25 \tag{11.5}$$

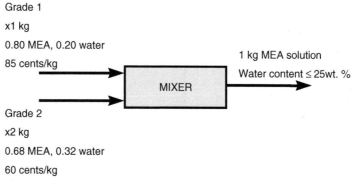

Grade 1

x1 kg

0.80 MEA, 0.20 water

85 cents/kg

1 kg MEA solution

Water content ≤ 25wt. %

MIXER

Grade 2

x2 kg

0.68 MEA, 0.32 water

60 cents/kg

FIGURE 11-2   REPRESENTATION OF THE SOLVENT MIXING PROBLEM

Mass conservation (overall material balance)

$$x1 + x2 = 1.00 \tag{11.6}$$

There are also two subtle constraints: the non-negativity constraints

$$x1 \geq 0 \tag{11.7}$$

$$x2 \geq 0 \tag{11.8}$$

Because of the small size of this problem, it can be solved graphically. Let us first represent the feasibility region by considering all the constraints. Constraint (11.5) is represented by the region lying below the linear expression:

$$0.20\ x1 + 0.32\ x2 = 0.25 \tag{11.9}$$

The region lying below this line is represented by Figure 11-3.

Now, let's add two more constraints (11.7) and (11.8) and represent them on Figure 11-4. These two constraints restrict the search in the first quadrant.

Now, we add the constraint (11.6), which is represented by a line segment as shown by Figure 11-5a. The intersection of this line segment with the rest of the constraints to yield the feasibility region shown by the heavy line segment shown in Figure (11-5 b).

With the feasibility region determined, we can now search for the optimal solution which must lie within the feasibility region. In order to determine z*,

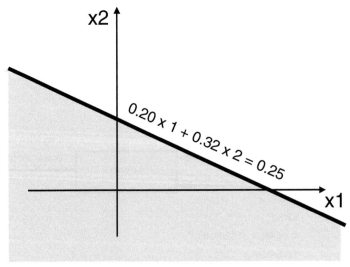

FIGURE 11-3 REPRESENTING THE WATER BALANCE

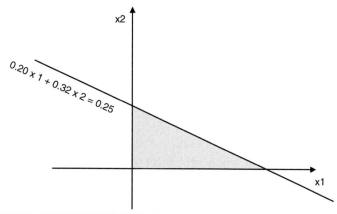

FIGURE 11-4   ADDING THE NON-NEGATIVITY CONSTRAINTS

(a)

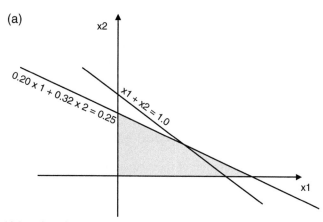

FIGURE 11-5a   ADDING THE OVERALL BALANCE CONSTRAINT

(b)

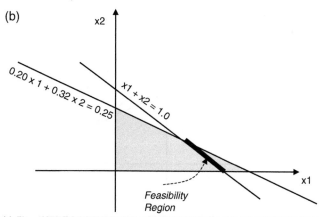

FIGURE 11-5b   INTERSECTING THE CONSTRAINTS TO FORM THE FEASIBILITY
REGION

it is important to notice that the objective function can be represented on an x2 versus x1 diagram as a family of straight lines described by

$$z = 85\,x1 + 60\,x2 \tag{11.10a}$$

or

$$x2 = -\frac{85}{60}x2 + \frac{z}{60} \tag{11.10b}$$

This is a family of straight lines whose slope is $-85/60$ and intercept is $z/60$.

Therefore, to determine the minimum value of z, we select different values of z and plot the corresponding straight lines representing the objective function. For instance, by choosing $z = 85$ and then $z = 80$, we obtain the objectives straight lines $85 = 85x1 + 60\,x2$ and $80 = 85\,x1 + 60\,x2$ represented by the dashed lines on Figure 11-6.

It can be seen from Figure 11-6 that the minimum value of z (referred to as z^*) will be assumed at the upper endpoint of the feasible segment, which is the intersection of the two lines:

$$x1 + x2 = 1.00 \tag{11.6}$$

and

$$0.20\,x1 + 0.32\,x2 = 0.25 \tag{11.11}$$

Simultaneous solution of these equations gives

$$x1^* = 0.58 \tag{11.12a}$$

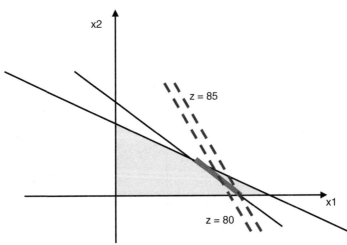

FIGURE 11-6 PLOTTING THE OBJECTIVE FUNCTION BY VARYING THE INTERCEPT

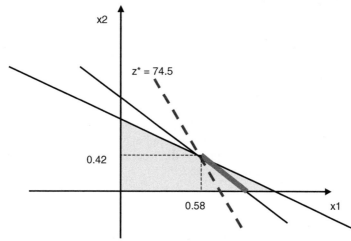

FIGURE 11-7   LOCATING OPTIMAL SOLUTION

and

$$x2 = 0.42 \qquad\qquad (11.12b)$$

Hence,

$$z^* = 85 \times 0.58 + 60 \times 0.42 = 74.5 \text{ cents/kg} \qquad (11.13)$$

The results are shown by Figure 11-7.

This problem can also be solved using computer-aided optimization tools such as the software package LINGO (see Appendix II for details on LINGO). The mathematical formulation is described by:

```
min = 85*x1 + 60*x2;                                          (P11.5)
0.2*x1 + 0.32*x2 < 0.25;
x1 + x2 = 1;
x1 > 0;
x2 > 0;
end
```

Solving with LINGO, we get:

```
Objective value: 74.58333
Variable        Value
X1              0.5833333
X2              0.4166667
```

EXAMPLE 11-2 REDUCTION OF DISCHARGE IN A TIRE-TO-FUEL PLANT

This case study was described in detail in Example 2-2 and is adapted from El-Halwagi (1997) and Noureldin and El-Halwagi (2000). It involves a processing facility that converts scrap tires into fuel via pyrolysis. Figure 11-8 is a simplified flowsheet of the process. The amount of water generated by chemical reaction is a function of the reaction temperature, T_{rxn}, through the following correlation:

$$W_{rxn} = 0.152 + (5.37 - 7.84 \times 10^{-3} T_{rxn})e^{(27.4 - 0.04 T_{rxn})} \qquad (11.14)$$

where W_{rxn} is in kg/s and T_{rxn} is in K. At present, the reactor is operated at 690 K which leads to the generation of 0.12 kg water/s. In order to maintain acceptable product quality, the reaction temperature should be maintained within the following range:

$$690 \leq T_{rxn}(K) \leq 740 \qquad (11.15)$$

Wastewater leaves the plant from three outlets: W_1 from the decanter, W_2 from the seal pot, and W_3 with the wet cake. Fresh water is used in two locations: shredding and the seal pot. The flowrate of water-jet makeup depends on the applied pressure coming out of the compression stage "P_{comp}" via the following expression:

$$G_1 = 0.47 e^{-0.009 P_{comp}} \qquad (11.16)$$

where G_1 is in kg/s and P_{comp} is in atm. In order to achieve acceptable shredding, the jet pressure may be varied within the following range:

$$70 \leq P_{comp}(atm) \leq 95 \qquad (11.17)$$

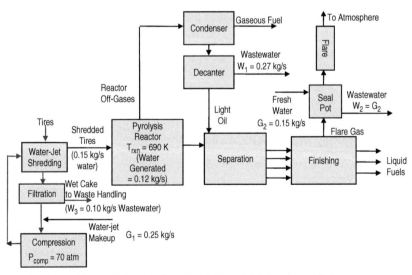

FIGURE 11-8 SIMPLIFIED FLOWSHEET OF TIRE-TO-FUEL PLANT

At present, P_{comp} is 70 atm which requires a water-jet make-up flowrate of 0.25 kg/s.

The flowrate of the water stream passing through the seal pot is referred to as G_2 and it has a flowrate of 0.15 kg/s. An equivalent flowrate of wastewater stream is withdrawn from the seal pot, i.e.,

$$W_2 = G_2 \tag{11.18}$$

The water lost in the cake is related to the mass flowrate of the water-jet makeup through:

$$W_3 = 0.4\, G_1 \tag{11.19}$$

At present, the values of W_1, W_2, and W_3 are 0.27, 0.15, and 0.10 kg/s, respectively. The cost of wastewater transportation and treatment is \$0.10/kg leading to a wastewater treatment cost of approximately \$1.33 million/year. If minor process modifications (changes in reaction temperature and pressure of the jet shredder) are allowed (without recycle or rerouting), what is the target for reduction in flowrate of terminal wastewater discharges?

SOLUTION

The objective of this problem is to minimize wastewater discharge, i.e.,

$$\text{Minimize } W_1 + W_2 + W_3 \tag{11.20}$$

In order to minimize these three wastewater streams, it is necessary to relate them to the flowrates of fresh water and to the possible variations in the allowed changes: reaction temperature and jet pressure. Although these changes are not directly related to the flowrate of the waste streams, they can be related to the flowrates of fresh water used. Consequently, water balances can be employed to relate the flowrates of the fresh, wastewater, and generation by chemical reaction.

Let us now proceed and transform this approach to mathematical constraints. An overall water balance is given by:

$$G_1 + G_2 + W_{rxn} = W_1 + W_2 + W_3 \tag{11.21}$$

where

$$W_{rxn} = 0.152 + (5.37 - 7.84 \times 10^{-3} T_{rxn}) e^{(27.4 - 0.04 T_{rxn})} \tag{11.14}$$

$$G_1 = 0.47 e^{-0.009 P_{comp}} \tag{11.16}$$

and

$$G_2 = 0.15$$

Next, we relate W_2 and W_3 to G_2 and G_3:

$$W_2 = G_2 \tag{11.18}$$

$$W_3 = 0.4\, G_1 \tag{11.19}$$

Additionally, there are bounds on reaction temperature and jet pressure given by:

$$690 \le T_{\mathrm{rxn}}(\mathrm{K}) \le 740 \tag{11.15}$$

$$70 \le P_{\mathrm{comp}}(\mathrm{atm}) \le 95 \tag{11.17}$$

Therefore, we can formulate the optimization program as the following LINGO input file:

```
min = W1 + W2 + W3;                                           (P11.6)
G1 + G2 + Wrxn = W1 + W2 + W3;
Wrxn = 0.152 + (5.37 - 0.00784*Trxn)*@exp(27.4 - 0.04*
                      Trxn);
G1 = 0.47*@exp(-0.009*Pcomp);
G2 = 0.15;
W2 = G2;
W3 = 0.4*G1;
Trxn > 690;
Trxn < 740;
Pcomp > 70;
Pcomp < 95;
End
```

Using LINGO, we get the following solution:

```
Objective value: 0.4296314

Variable              Value

W1                    0.1996782
W2                    0.1500000
W3                    0.079953
G2                    0.1500000
G1                    0.1998831
PCOMP                 95.0000
WRXN                  0.7974833E-01
TRXN                  709.9490
```

These results are consistent with the ones obtained in Chapter Two (Figure 2-14). The value of G_1 is reduced to 0.20 kg/s and the minimum value of W_3 is 0.08 kg/s. Furthermore, the optimum reaction temperature is 710 K.

EXAMPLE 11-3 TRANSPORTATION PROBLEM

A chemical supplier has received a contract to supply methanol for three plants owned by a manufacturing company. The supplier has production plants in cities 1, 2, 3, and 4 while the consumer (manufacturing company) has plants in towns A, B, and C. Process engineers have estimated that amounts of methanol needed at the consumer plants as sown by Table 11-1.

Daily production capacities of supplier's plants are given by Table 11-2.

The cost of shipping one ton of methanol from each supplying plant to each consuming plant is given in Table 11-3. The number listed at the intersection of a supplier and consumer is the transportation cost per ton.

The objective of this problem is to schedule the shipment from each supplier to consumer in such a manner so as to minimize the total transportation cost within the constraints imposed by plant capacities and requirements.

SOLUTION

The optimization variables are the transportation loads from suppliers to consumers. Table 11-4 illustrates the terms used to denote these

TABLE 11-1 DAILY REQUIREMENTS FOR THE CONSUMER IN EXAMPLE 11-3

Receiving plants	Daily requirements (tons)
A	6
B	1
C	10

TABLE 11-2 DAILY PRODUCTION CAPACITIES OF THE SUPPLIER IN EXAMPLE 11-3

Production facilities	Daily production (tons)
1	7
2	5
3	3
4	6

TABLE 11-3 TRANSPORTATION COSTS, $

Supplier	Consumer A	Consumer B	Consumer C
1	20	10	50
2	30	0	80
3	110	60	150
4	70	10	90

TABLE 11-4 TERMS USED FOR TRANSPORTATION LOADS

Supplier	Consumer A	Consumer B	Consumer C
1	X1A	X1B	X1C
2	X2A	X2B	X2C
3	X3A	X3B	X3C
4	X4A	X4B	X4C

transportation loads (tons/day).

Objective Function   (P11.7)

$$\text{Minimize transportation cost} = 20 \times X1A + 10 \times X1B + 50 \times X1C$$
$$+ 30 \times X2A + 0 \times X2B2B + 80 \times X2C$$
$$+ 110 \times X3A + 60 \times X3B + 150 \times X3C$$
$$+ 70 \times X4A + 10 \times X4B + 90 \times X4C$$

Subject to the following constraints:

Requirements for demand

$$X1A + X2A + X3A + X4A = 6$$
$$X1B + X2B + X3B + X4B = 1$$
$$X1C + X2C + X3C + X4C = 10$$

Availability of supply

$$X1A + X1B + X1C \le 7$$
$$X2A + X2B + X2C \le 5$$
$$X3A + X3B + X3C \le 3$$
$$X4A + X4B + X4C \le 6$$

Nonnegativity constraints

$$X1A \ge 0, X1B \ge 0, \ldots, X4C \ge 0$$

Solving using LINGO, we get the following optimum solution:

```
Objective value: 840.0000

Variable            Value

X1A                 1.000000
X1B                 0.000000
```

X1C	6.000000
X2A	5.000000
X2B	0.000000
X2C	0.000000
X3A	0.000000
X3B	0.000000
X3C	0.000000
X4A	0.000000
X4B	1.000000
■ ■ ■ X4C	4.000000

11.4 THE USE OF 0-1 BINARY-INTEGER VARIABLES

An important and very common use of 0–1 binary integer variables is to represent discrete decisions and what-if scenarios. Consider an event that may or may not occur, and suppose that it is part of the problem to decide between these two alternatives. To model such a dichotomy, we use a binary variable x and let:

$$x = 1 \text{ if the event occurs} \qquad (11.22a)$$

and

$$x = 0 \text{ if the event does not occur} \qquad (11.22b)$$

Additionally, binary integers can be instrumental in modeling logical statements, what-if scenarios, and Boolean logic. Boolean algebra (named after nineteenth century mathematician Boole) is a form of algebra, for which the values are reduced to one of two statements: *True* or *False*. Binary integer variables are naturally suited to represent Boolean algebra because of its discrete nature where 1 may be assigned to True and 0 assigned to False (or the other way around).

There are numerous applications for the use of binary integers. The following are some examples.

EXAMPLE 11-4 THE ASSIGNMENT PROBLEM
A classical optimization problem is the assignment problem. It involves the assignment of n people (or machines or streams, etc.) to m jobs (or spaces or units, etc.). Each job must be done exactly by one person; also, each person can do at most one job. The cost of person i doing job j is $C_{i,j}$. The problem is to assign the people to the jobs so as to minimize the total cost of completing all the jobs.

To formulate this problem, which is known as the assignment problem, we introduce 0–1 variables $X_{i,j}$, $i = 1,2,\ldots,n$; $j = 1,2,\ldots m$ corresponding to the event of assigning person j to do job i. Therefore, the assignment problem can be posed through the following MILP formulation:

$$\text{Minimize assignment cost} - \sum_{i=1}^{n}\sum_{j=1}^{m} C_{i,j} X_{i,j} \qquad \text{(P11.8)}$$

Since exactly one person must do job j, we have the following constraint

$$\sum_{j=1}^{m} X_{i,j} = 1 \quad \text{for} \quad i = 1, 2, \ldots, n$$

Since each person can do no more than one job, we have the following constraint:

$$\sum_{i=1}^{n} X_{i,j} \le 1$$

As an illustration, consider the following assignment problem on plant layout. Four new reactors R1, R2, R3 and R4 are to be installed in a chemical plant. There are four vacant spaces 1, 2, 3, and 4 available. The cost of assigning reactor i to space j (in 10,000 of dollars) is given in Table 11-5. Assign the reactors to the spaces so as to minimize the total cost.

SOLUTION

Let $X_{i,j}$ be a binary integer variable to denote existence (or absence) of reactor i in space j (when $X_{i,j} = 1$, reactor i exists in space j; when $X_{i,j} = 0$, reactor i does not exist in space j).

Objective Function   (P11.9)

Minimize assignment cost $= 150000 \times \text{X11} + 110000 \times \text{X12} + 130000 \times \text{X13}$
$+ 150000 \times \text{X14} + 130000 \times \text{X21} + 90000 \times \text{X22} + 120000 \times \text{X23} + 170000 \times \text{X24}$
$+ 140000 \times \text{X31} + 150000 \times \text{X32} + 100000 \times \text{X33} + 140000 \times \text{X34} + 170000 \times \text{X41}$
$+ 130000 \times \text{X42} + 110000 \times \text{X43} + 160000 \times \text{X44}$

TABLE 11-5 ASSIGNMENT COSTS (EXPRESSED IN 10,000 OF DOLLARS)

Reactor	Space 1	Space 2	Space 3	Space 4
R1	15	11	13	15
R2	13	9	12	17
R3	14	15	10	14
R4	17	13	11	16

Subject to the following constraints:

On reactors: each space must be assigned to one and only one reactor

$$X11 + X12 + X13 + X14 = 1 \text{ for reactor R1}$$

$$X21 + X22 + X23 + X24 = 1 \text{ for reactor R2}$$

$$X31 + X32 + X33 + X34 = 1 \text{ for reactor R3}$$

$$X41 + X42 + X43 + X44 = 1 \text{ for reactor R4}$$

On spaces: each reactor must be assigned to one and only one space (since there are four spaces and four reactors, there is no need to use less than or equal constraint. Instead, we use an equality constraint).

$$X11 + X21 + X31 + X41 = 1 \text{ for space 1}$$

$$X12 + X22 + X32 + X42 = 1 \text{ for space 2}$$

$$X13 + X23 + X33 + X43 = 1 \text{ for space 3}$$

$$X14 + X24 + X34 + X44 = 1 \text{ for space 4}$$

The input file for LINGO can be expressed as (P11.10)

```
min = 150000*X11 + 110000*X12 + 130000*X13 + 150000*X14
+ 130000*X21 + 90000*X22 + 120000*X23 + 170000*X24 + 140000
× X31 + 150000*X32 + 100000*X33 + 140000*X34 + 170000*X41
+ 130000*X42 + 110000*X43 + 160000*X44;
X11 + X12 + X13 + X14 = 1;
X21 + X22 + X23 + X24 = 1;
X31 + X32 + X33 + X34 = 1;
X41 + X42 + X43 + X44 = 1;
X11 + X21 + X31 + X41 = 1;
X12 + X22 + X32 + X42 = 1;
X13 + X23 + X33 + X43 = 1;
X14 + X24 + X34 + X44 = 1;
@bin(X11);
@bin(X12);
@bin(X13);
@bin(X14);
@bin(X21);
@bin(X22);
@bin(X23);
@bin(X24);
@bin(X31);
@bin(X32);
```

```
@bin(X33);
@bin(X34);
@bin(X41);
@bin(X42);
@bin(X43);
@bin(X44);
End
```

This program can be solved using LINGO. The following are the results of the optimal solution.

```
Objective value: 490000.0

Variable              Value

X11                   0.000000
X12                   1.000000
X13                   0.000000
X14                   0.000000
X21                   1.000000
X22                   0.000000
X23                   0.000000
X24                   0.000000
X31                   0.000000
X32                   0.000000
X33                   0.000000
X34                   1.000000
X41                   0.000000
X42                   0.000000
X43                   1.000000
X44                   0.000000
```

Therefore, the minimum total assignment cost associated with this assignment policy is \$490,000. An optimal assignment policy is:

$$\text{Reactor R1 to space 2} \qquad (11.23a)$$

$$\text{Reactor R2 to space 1} \qquad (11.23b)$$

$$\text{Reactor R3 to space 4} \qquad (11.23c)$$

■ ■ ■ $\text{Reactor R4 to space 3} \qquad (11.23d)$

11.5 ENUMERATING MULTIPLE SOLUTIONS USING INTEGER CUTS

It is worth mentioning that there may be more than one solution attaining the optimal value of the objective function. In such cases, it is necessary to enumerate all of these solutions. This can be achieved by using *integer cut* constraints. The basic idea is to add a constraint which forbids the formation of an optimal solution that has already been identified. The optimization program with the additional constraint is run and if the identified solution is equal to the solution without the integer cut, then both solutions are equivalent in terms of the objective function but provide different implementations. The procedure is repeated by adding integer cuts until an inferior solution is obtained or no solution is found.

As an illustration, let us re-consider the reactor assignment example, the optimal solution involves the following matches:

$$X12 = 1 \tag{11.24a}$$

$$X21 = 1 \tag{11.24b}$$

$$X34 = 1 \tag{11.24c}$$

$$X43 = 1 \tag{11.24d}$$

Therefore, the following integer-cut constraint may be added to the optimization formulation:

$$X12 + X21 + X34 + X43 \leq 3 \tag{11.25}$$

Consequently, in order for this constraint to be satisfied, the combination of these four assignments must be excluded from any subsequent optimization. By adding this constraint and re-running LINGO, we get the following optimal solution:

```
Objective value: 490000.0

Variable          Value

X11               0.000000
X12               0.000000
X13               0.000000
X14               1.000000
X21               0.000000
X22               1.000000
X23               0.000000
X24               0.000000
X31               1.000000
X32               0.000000
X33               0.000000
X34               0.000000
X41               0.000000
X42               0.000000
X43               1.000000

X44               0.000000
```

This solution provides the same value of the objective function but yields another integer combination:

$$X14 = 1 \qquad (11.26a)$$

$$X22 = 1 \qquad (11.26b)$$

$$X31 = 1 \qquad (11.26c)$$

$$X43 = 1 \qquad (11.26d)$$

Hence, we add the following integer cut:

$$X14 + X22 + X31 + X43 \leq 3 \qquad (11.27)$$

By re-running LINGO with the two integer cuts, we get:

```
Objective value: 490000.0

Variable          Value

X11               1.000000
X12               0.000000
X13               0.000000
X14               0.000000
X21               0.000000
X22               1.000000
X23               0.000000
X24               0.000000
X31               0.000000
X32               0.000000
X33               0.000000
X34               1.000000
X41               0.000000
X42               0.000000
X43               1.000000

X44               0.000000
```

This solution provides the same value of the objective function but yields another integer combination:

$$X11 = 1 \qquad (11.28a)$$

$$X22 = 1 \qquad (11.28b)$$

$$X34 = 1 \qquad (11.28c)$$

$$X43 = 1 \qquad (11.28d)$$

Hence, we add the following integer cut:

$$X11 + X22 + X34 + X43 \leq 3 \qquad (11.29)$$

By re-running LINGO, we get

```
Objective value: 500000.0

Variable              Value

X11                 1.000000
X12                 0.000000
X13                 0.000000
X14                 0.000000
X21                 0.000000
X22                 1.000000
X23                 0.000000
X24                 0.000000
X31                 0.000000
X32                 0.000000
X33                 1.000000
X34                 0.000000
X41                 0.000000
X42                 0.000000
X43                 0.000000

X44                 1.000000
```

Since the identified optimal solution of the last program is $500,000 which is higher than the already found solution of $490,000, then we stop and there are no additional alternatives yielding the same value of the global optimum solution.

11.6 MODELING DISCONTINUOUS FUNCTIONS AND WHAT-IF SCENARIOS USING INTEGER VARIABLES

Discontinuous functions can be used to describe abrupt changes over a certain decision variable. For instance, consider a cost model which favors the selection of technology A up to a certain production capacity (F_{Switch}), then moves to technology B for production capacities immediately above F_{Switch}, i.e.,

$$\text{Cost} = \text{Cost\_A} \quad \text{for } F \leq F_{\text{Switch}} \qquad (11.30a)$$

$$\text{Cost} = \text{Cost\_B} \quad \text{for } F > F_{\text{Switch}} \qquad (11.30b)$$

These cost functions may be constant numbers (e.g., Figure 11-9a) or variable depending on the production capacity (e.g., Figure 11-9b).

Binary integer variables can be used to model these functions. For instance, the cost function may be transformed to the following mixed-integer formulation using a binary integer variable (I):

$$\text{Cost} = \text{Cost\_A} \times I + \text{Cost\_B} \times (1 - I) \tag{11.31a}$$

$$\text{where } I = 1 \quad \text{for } F \leq F_{\text{Switch}} \tag{11.31b}$$

and

$$I = 0 \quad \text{for } F > F_{\text{Switch}} \tag{11.31c}$$

Let us examine this formulation:

When $I = 1$,

$$\text{Cost} = \text{Cost\_A} \times 1 + \text{Cost\_B} \times (1 - 1) = \text{Cost\_A} \tag{11.32a}$$

and when $I = 0$,

$$\text{Cost} = \text{Cost\_A} \times 0 + \text{Cost\_B} \times (1 - 0) = \text{Cost\_B} \tag{11.32b}$$

But how do we model the conditions that assign the values of I to be zero or one based on the production capacity? Consider the following constraints:

$$(F_{\text{Switch}} - F) \times (2 \times I - 1) \geq 0 \tag{11.33a}$$

When $F \leq F_{\text{Switch}}$, I is forced to be 1 (otherwise, if it is zero, the value of $2 \times I - 1$ becomes -1 which renders the right-hand side negative, and therefore infeasible). On the other hand, when $F > F_{\text{Switch}}$, the term $(F_{\text{Switch}} - F)$ is negative and I is forced to be zero so that $(2 \times I - 1)$ becomes negative and the product of the two negative terms satisfies the non-negative inequality constraint.

Equation (11.33a) contains the term $(F_{\text{Switch}} - F)^*(2^* I - 1)$ which is non-convex because of the bilinear nature of the product of the two unknowns: F and I. Non-convex terms may prevent the solver from identifying the global solution.

Alternatively, the same effect of equation (11.33a) may be modeled using a linear formulation. Let us designate the lower and upper bounds on the feasible values of F to be L and U, respectively. Both L and U are constants. If these constants are not available for the particular system, one may assume some lower and upper bounds (e.g., zero and a very large number). Consider the following linear constraint:

$$(L - F_{\text{Switch}})^* I < F - F_{\text{Switch}} \leq (U - F_{\text{Switch}})^*(1 - I) \tag{11.33b}$$

When $F \leq F_{\text{Switch}}$, I is forced to be 1, otherwise, if it is zero, the value of $(L - F_{\text{Switch}})^* I$ becomes zero which violates the constraint

(a)   Cost

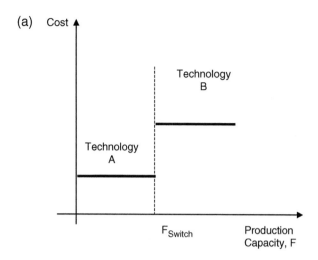

Technology
B

Technology
A

F_{Switch} Production
Capacity, F

(b)   Cost

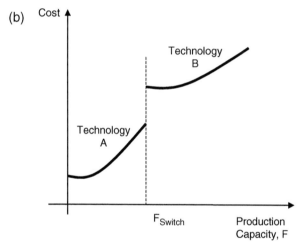

Technology
B

Technology
A

F_{Switch} Production
Capacity, F

FIGURE 11-9   DISCONTINUOUS FUNCTIONS (a. CONSTANT COSTS, b. VARIABLE COSTS)

$(L - F_{\text{Switch}})*I < F - F_{\text{Switch}}$. On the other hand, when $F > F_{\text{Switch}}$, the term $(F - F_{\text{Switch}})$ is positive and I is forced to be zero, otherwise, if I is 1 then the term $(U - F_{\text{Switch}})*(1 - I)$ becomes zero which is a violaiton of the statement that $F - F_{\text{Switch}} \leq (U - F_{\text{Switch}})*(1 - I)$.

EXAMPLE 11-5 MAXIMIZING REVENUE OF GAS PROCESSING FACILITY
A processing facility utilizes an industrial gas in making several products. The value added by the industrial gas is described by the following function:

$$\text{Value added (\$/h)} = 14.0 \times F^{0.8} \qquad (11.34)$$

where F is the flowrate of purchased gas in MM SCF/h (million standard cubic feet per hour).

The processing facility purchases the gas from an adjacent supplier. According to the contractual arrangement between the facility and the supplier, the supplier provides the gas at \$4.0/MM SCF if the purchase is less than 300 MM SCF/h and \$3.5/MM SCF if the purchase exceeds 300 MM SCF/h.

What should be the flowrate of purchased gas necessary in order to maximize the gas-processing revenue (described as value added – purchased gas cost)?

SOLUTION

The objective function is given by:

$$\text{Revenue} = \text{Value added} - \text{Purchased gas cost} \qquad (11.35)$$

The purchased gas cost can be described using a binary integer variable (I) defined as:

$$I = 1 \text{ if } F \geq 300 \qquad (11.36a)$$

and

$$I = 0 \text{ if } F < 300 \qquad (11.36b)$$

Hence,

$$\text{Purchased gas cost} = 4.0 \times F \times I + 3.5 \times F \times (1 - I) \qquad (11.37)$$

We can now formulate the problem as an MINLP. The LINGO input file is given below.

```
max = value_added - Gas_cost;                          (P11.11)
Value_added = 14*F^0.8;
Gas_Cost = (4.0*I + 3.5*(1-I))*F;
(300 - F)*(2*I - 1) >= 0;
@bin(I);
```

The solution to this program gives the following results.

```
Objective value: 293.6013
    Variable              Value
VALUE_ADDED            1468.006
   GAS_COST            1174.405
          F             335.5443
          I             0.000000
```

Alternatively, the program may be formulated using the linear constraint (11.33b) as follows:

```
max = value_added - Gas_cost;
```

```
Value_added = 14*F^0.8;
Gas_Cost = (4.0*I + 3.5*(1-I))*F;
F - 300 < = (500 - 300)*(1 - I);
(0 - 300)*I < = F - 300;
@bin(I);
```

The solution to this program gives the following results which are identical to ones obtained earlier using the bilinear constraint.

```
Objective value: 293.6013
      Variable                Value
   VALUE_ADDED             1468.006
      GAS_COST             1174.405
             F              335.5443
             I              0.000000
```

■ ■ ■

11.7 PROBLEMS

11.1 A polymer plant consists of two production lines which yield two polymer products: A and B. Both lines use the same monomer (M) as a feedstock. Due the formation of byproducts and process losses, each ton produced of either polymer requires four tons of M. The maximum supply of monomer M to the plant is 75 tons/h. The process requires refrigeration and heating. The consumption of these utilities required for both products are listed in Table 11-6.

Available for service are 80 MM Btu/h of refrigeration and 60 MM Btu/h of heating. The net profits for producing A and B are 1000 and 1500 $/ton produced, respectively. What is the optimal production rate of each polymer that will maximize the net profit of the plant?

11.2 A polymer plant consists of two production lines which yield two polymer products: A and B. Both lines use the same monomer (M) as a feedstock. Due to the formation of byproducts and process losses, each ton produced of either polymer requires four tons of M. The maximum supply of monomer M to the plant is 45 tons/h. The process requires refrigeration and heating. The consumption of these utilities required for both products are listed in Table 11-7.

Available for service are 60 MM Btu/h of refrigeration and 75 MM Btu/h of heating. The net profits for producing A and B are 1000 and 1500 $/ton produced, respectively. What is the optimal production rate of each polymer that will maximize the net profit of the plant?

11.3 The two plants described in problems (11.1) and (11.2) are now owned by the same parent company which is trying to maximize the combined net profit of both plants. The parent company has relaxed the restrictions on monomer supply to each plant (75 tons/h for the first plant and 45 tons/h for the second plant). Instead, the parent company can provide a combined supply of the monomer to both plants of 120 tons/week at most. Therefore, it is necessary to optimally distribute the available monomer between the

TABLE 11-6 UTILITY REQUIREMENTS FOR THE TWO POLYMERS (MM Btu/h)

Utility	Polymer A	Polymer B
Refrigeration	4	2
Heating	2	5

TABLE 11-7 UTILITY REQUIREMENTS FOR THE TWO POLYMERS (MM Btu/h)

Utility	Polymer A	Polymer B
Refrigeration	5	3
Heating	5	6

two plants. What are the optimal production capacities of polymers A and B in each plant which will maximize the combined net profit of both plants? Comment on the results of problems 11.1–11.3.

11.4 A major oil company wants to build a refinery that will be supplied from three port cities. Port B is located 300 km east and 400 km north of Port A, while Port C is 400 km east and 100 km south of Port B. Determine the location of the refinery so that the total amount of pipe required to connect the refinery to the ports is minimized.

11.5 Three ethanol-supplying plants (S_1, S_2, and S_3) are used to provide four customers (T_1, T_2, T_3, and T_4) with their ethanol requirements. The yearly capacities of suppliers S_1, S_2, and S_3 are 135, 56, and 93 tons/year, respectively. The yearly requirements of the customers T_1, T_2, T_3, and T_4 are 62, 83, 39, and 91 tons/year, respectively. The unit costs \$/ton for supplying each customer from each supplier are given in the Table 11-8 (a dash in the table indicates the impossibility of certain suppliers for certain customers).

What are the optimal transportation quantities from suppliers to customers which will minimize the total transportation cost?

11.6 Three chemical engineering students are working in the same group for a design project. The design project involves three tasks; literature search, simulation, and report writing. Since the three students did not get along well, they decided that each person will do one of three assigned tasks without

TABLE 11-8 UNIT TRANSPORTATION COSTS (\$/TON) FOR PROBLEM 11.5

Supplier/Customer	T_1	T_2	T_3	T_4
S_1	132	–	97	103
S_2	85	91	–	–
S_3	106	89	100	98

any cooperation with the other two. In an attempt to identify the best way to distribute tasks among the three students, each one of them indicated the number of hours that he/she estimates will take him/her to finish each task. These estimates are given in Table 11-9.

In order to minimize the total person-hours involved in the project, who should do what? Can there be more than one solution? If yes, what are they?

11.7 A chemical plant produces polyamide synthetic membranes which can be used in pervaporation modules. In order to insure the quality of the product, at least $1500\,m^2$ of the polymeric membrane must be checked in an 8-h shift. The plant has two grades inspectors (I and II) to undertake quality control inspection. Grade I inspector can check $20\,m^2$ in an hour with an accuracy of 96%. Grade II inspector checks $14\,m^2$ an hour with an accuracy of 93%. The maximum available number of grade I and II inspectors are 10 and 15 per shift, respectively. The hourly wages of grade I inspectors are $20 per hour while those of grade II inspectors are $16 per hour. Any error made by an inspector cost the plant $12 per m^2. Find the optimal assignment of inspectors that minimizes the daily inspection cost.

11.8 A chemical plant has budgeted $250,000 for the development of new waste-treatment systems. Seven waste-treatment systems are being considered. Each system can handle a certain capacity (indicated in Table 11-10) and accomplish the required waste-treatment task for that particular capacity. The projected capital costs for these systems are given below. The operating costs for all the systems are almost identical. Which systems should the plant select?

11.9 A plant produces a pharmaceutical ingredient in two shifts (a regular working shift and an overtime shift) to meet specific demands of other plant units for the present and the future. Over the next four months, the

TABLE 11-9 HOURS NEEDED FOR EACH STUDENT FOR EACH TASK

Task/Student	A	B	C
i	20	25	30
ii	10	10	10
iii	15	20	20

TABLE 11-10 CAPACITY AND COST OF EACH WASTE TREATMENT ALTERNATIVE

System Description	A Biological treatment	B Membrane separation	C Adsor- ption	D Oxida- tion	E Precipi- tation	F Strip- ping	G Solvent extraction
Capacity (tons/hr)	20	17	15	15	10	8	5
Cost $1000	145	92	70	70	84	14	47

TABLE 11-11 PRODUCTION CAPACITIES AND DEMAND (TONS PER MONTH) FOR PROBLEM 11.8

	Month 1	Month 2	Month 3	Month 4
Production during regular shift	110	150	190	160
Production during overtime shift	60	75	105	80
Demand	80	240	290	210

production capacities during regular working and overtime as well as monthly demands (tons product per month) are given in Table 11-11.

The cost of production is $2.0 per kg of the ingredient if done in regular working or $2.6 per kg of hydrochloric acid if done in overtime. The produced ingredient can be stored before delivery at a cost of $0.40 per month per kg of the ingredient. What is the optimal production schedule of the product so as to meet present and future demands at minimum cost?

11.8 REFERENCES

Biegler, L.T., and Grossmann, I.E. 2004, 'Retrospective on Optimization', *Comp. Chem. Eng.*, vol. 28, 1169-1192.

Diwekar, U.M. 2003, *Introduction to Applied Optimization*, Kluwer Academic Publishers, Dordrecht, The Netherlands.

Edgar, T.F., and Himmelblau, D.M. 2001, *Optimization of chemical processes*, McGraw Hill, New York, 2nd Edition.

El-Halwagi's M.M. 1997, *Pollution Prevention Through Process Integration*, Academic Press, San Diego.

Floudas, C.A. 1999, *Deterministic Global Optimization*, Kluwer Academic Publishers, Dordrecht, The Netherlands.

Floudas C.A., Akrotirianakis, I.G., Caratzoulas, S., Meyer, C.A., and Kallrath, J. 2004, 'Global Optimization in the Twenty First Century', *European Symposium on Computer-Aided Process Engineering (ESCAPE)-14*, eds. A. Barbosa-Povoa, and H. Matos, Elsevier, pp. 23-51.

Floudas, C.A., and Pardalos, P.M. 2001, *Encyclopedia of Global Optimization*, Kluwer Academic Publishers, Dordrecht, The Netherlands.

Grossmann, I.E. 1996, Editor, *Global Optimization in Engineering Design*, Kluwer Academic Publisher, Dordrecht, The Netherlands.

Grossmann, I.E., and Biegler, L.T. 2004, 'Part II. Future Perspective on Optimization', *Comp. Chem. Eng.*, vol. 28, 1193-1218

Noureldin, M.B., and El-Halwagi, M.M. 2000, 'Pollution-Prevention Targets through Integrated Design and Operation', *Comp. Chem. Eng.*, vol. 24, 1445-1453.

Tawarmalani, M., and Sahinidis, N. V. 2003, *Convexification and Global Optimization in Continuous and Mixed Integer Nonlinear Programming*, Kluwer Academic Publishers, Dordrecht, The Netherlands.

12

MATHEMATICAL APPROACH TO DIRECT RECYCLE

Chapters Three and Six presented the material recycle pinch diagram and the cascade analysis for the graphical and the algebraic targeting of direct recycle problems. As mentioned in Chapter Six, graphical tools may become cumbersome for problems with numerous sources or sinks and for problems with a wide range of flowrate or load scale. Additionally, algebraic techniques cannot easily handle problems with multiple fresh resources. Therefore, it is beneficial to develop a mathematical programming approach to the targeting of material recycle. Such approach can also be integrated with other optimization techniques. This chapter provides an optimization-based formulation for the mathematical solution of the direct recycle problem. First, the problem addressed in Chapters Three and Six will be re-stated. Then, a structural representation of the solution alternatives will be presented. Next, the mathematical programming formulation will be discussed along with its solution applied to a case study.

12.1 PROBLEM STATEMENT

The problem can be expressed as follows:

Given a process with:

- A set of process sinks (units): SINKS = $\{j=1,2,\ldots,N_{\text{Sinks}}\}$. Each sink requires a feed with a given flowrate, G_j, and a composition, z_j^{in}, that satisfies the following constraint:

$$z_j^{\min} \leq z_j^{\text{in}} \leq z_j^{\max} \quad \text{where } j = 1, 2, \ldots, N_{\text{Sinks}} \tag{12.1}$$

where $z_j^{\min}$ and $z_j^{\max}$ are given lower and upper bounds on admissible compositions to unit j.

- A set of process sources: SOURCES $= \{i=1,2,\ldots, N_{\text{Sources}}\}$ which can be recycled/reused in process sinks. Each source has a given flowrate, W_i, and a given composition, y_i.
- Available for service is a set of fresh (external) resources: FRESH $= \{r=1,2,\ldots, N_{\text{Fresh}}\}$ which can be purchased to supplement the use of process sources in sinks. The cost of the *rth* fresh resource is C_r (\$/kg) and its composition is x_r. The flowrate of each fresh resource (F_r) is to be determined so as to minimize the total cost of the fresh resources.

12.2 PROBLEM REPRESENTATION

The first step in the analysis is to represent the problem through a source–sink representation shown in Figure 12-1. (e.g., El-Halwagi 1997). Each source is split into fractions (of unknown flowrate) that are allocated to the various sinks. An additional sink is placed to account for unrecycled/unreused material. This sink is referred to as the "waste" sink. The fresh sources are also allowed to split and are allocated to the sinks. Clearly, there is no need to allocate a portion of a fresh resource to the waste sink.

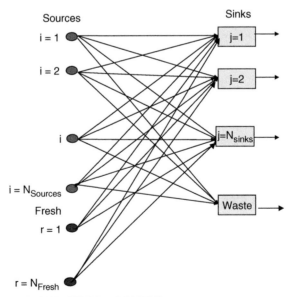

FIGURE 12-1   SOURCE–SINK ALLOCATION

12.3 OPTIMIZATION FORMULATION

The objective is to minimize the cost of the fresh resources, i.e.,

$$\text{Minimize} \sum_{r=1}^{N_{\text{Fresh}}} C_r \times F_r \qquad (12.2a)$$

If the intent is to minimize the flowrate of the fresh, then the objective function can be expressed as:

$$\text{Minimize} \sum_{r=1}^{N_{\text{Fresh}}} F_r \qquad (12.2b)$$

Each source, i, is split into N_{Sinks} fractions that can be assigned to the various sinks (Figure 12-2). The flowrate of each split is denoted by $w_{i,j}$. Additionally, one split is forwarded to the waste sink. The stream accounts for the unrecycled flowrate and is denoted by $w_{i,\text{waste}}$. Similarly, each fresh source is split into N_{Sinks} fractions that can be assigned to the various sinks. The flowrate of each fresh split is denoted by $f_{r,j}$.

Therefore, the splitting constraint can be written as:

$$W_i = \sum_{j=1}^{N_{\text{Sinks}}} w_{i,j} + w_{i,\text{waste}} \quad i = 1, 2, \ldots, N_{\text{Sources}} \qquad (12.3)$$

A similar constraint can be written for the splitting of the rth fresh resource:

$$F_r = \sum_{j=1}^{N_{\text{Sinks}}} f_{r,j} \quad r = 1, 2, \ldots, N_{\text{Fresh}} \qquad (12.4)$$

$$\text{Waste} = \sum_{i=1}^{N_{\text{sources}}} w_{i,\text{waste}} \qquad (12.5)$$

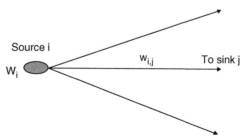

FIGURE 12-2 SPLITTING OF SOURCES

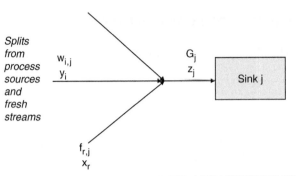

Splits from process sources and fresh streams

$w_{i,j}$
y_i

G_j
z_j

Sink j

$f_{r,j}$
x_r

FIGURE 12-3   MIXING OF SPLIT FRACTIONS AND ASSIGNMENT TO SINKS

Next, we examine the opportunities for mixing these splits and assigning them to sinks. Figure 12-3 shows the mixing of the split fractions into a feed to the jth sink. The split fractions come from the process sources and the fresh streams.

Mixing for the jth sink:

$$G_j = \sum_{i=1}^{N_{\text{Sources}}} w_{i,j} + \sum_{r=1}^{N_{\text{Fresh}}} f_{r,j} \quad \text{where } j = 1, 2, \ldots, N_{\text{Sinks}} \tag{12.6}$$

$$G_j \times z_j^{\text{in}} = \sum_{i=1}^{N_{\text{Sources}}} w_{i,j} \times y_i + \sum_{r=1}^{N_{\text{Fresh}}} f_{r,j} \times x_r \quad \text{where } j = 1, 2, \ldots, N_{\text{Sinks}} \tag{12.7}$$

where

$$z_j^{\text{min}} \leq z_j^{\text{in}} \leq z_j^{\text{max}} \quad \text{where } j = 1, 2, \ldots, N_{\text{Sinks}} \tag{12.1}$$

It is worth pointing out that in cases of multiple components, the same formulation can be used by writing constraints 12.1 and 12.7 for all components.

To ensure the non-negativity of the fresh flows and the fractions of source allocated to a sink and for flow of fresh resources, the following constraints are written:

$$f_{r,j} \geq 0 \quad \text{where } r = 1, 2, \ldots, N_{\text{Fresh}} \text{ and } j = 1, 2, \ldots, N_{\text{sinks}} \tag{12.8}$$

$$w_{i,j} \geq 0 \quad \text{where } i = 1, 2, \ldots, N_{\text{Fresh}} \text{ and } j = 1, 2, \ldots, N_{\text{sinks}} \tag{12.9}$$

The foregoing formulation is a linear program which can be solved globally to identify the minimum cost (or flowrate) of the fresh streams, an optimum assignment of process sources to sinks, and the discharged waste.

▬ TABLE 12-1 SOURCE DATA FOR THE VINYL ACETATE EXAMPLE

Source	Flowrate (kg/h)	Inlet mass fraction
Bottoms of absorber II	1400	0.14
Bottoms of primary Tower	9100	0.25

▬ TABLE 12-2 SINK DATA FOR THE VINYL ACETATE EXAMPLE

Sink	Flowrate (kg/h)	Minimum inlet mass fraction	Maximum inlet mass fraction
Absorber I	5100	0.00	0.05
Acid tower	10,200	0.00	0.10

▬ EXAMPLE 12-1 TARGETING FOR ACETIC ACID USAGE IN A VINYL
ACETATE PLANT

To illustrate the mathematical formulation for recycle strategies, let us revisit Example 3-4 on the recovery of AA from a VAM facility. The data for the problem are shown in Tables 12-1 and 12-2.

What is the target for minimum usage (kg/h) of fresh (pure) acetic acid in the process if segregation, mixing, direct recycle are used? Can you find more than one strategy to reach the target?

SOLUTION

The objective here is to minimize the usage of fresh acetic acid. Let us designate the flowrate of acetic acid as FreshAA. Therefore the objective function can be expressed as:

$$\text{Objective: Minimize FreshAA} \tag{12.10}$$

When considering direct recycle, the problem has two process sources (bottoms of absorber II and bottoms of primary tower) and two process sinks (absorber I and acid tower). Additionally, unrecycle process sources are fed to the waste treatment system. Referring to the flowrate of absorber II as $W\text{Abs2}$ and to the flowrate of the primary tower as $W\text{Primary}$, and to the flowrate from source I to sink j as $w_{i,j}$, the source splitting constraints can be written as:

$$W\text{Abs2} = w11 + w12 + \text{Waste1} \tag{12.11}$$

$$W\text{Primary} = w21 + w22 + \text{Waste2} \tag{12.12}$$

where Waste1 and Waste2 represent the unrecycled flowrates of sources 1 and 2. Similarly, the fresh flowrate is split between sinks 1 and 2. Hence:

$$\text{FreshAA} = \text{FreshAA1} + \text{FreshAA2} \tag{12.13}$$

where FreshAA1 and FreshAA2 are the flowrates of the fresh acetic acid used in sinks 1 and 2, respectively.

Next, we represent the mixing of the split sources and assign them to sinks. Let the flowrate entering absorber I and the acid tower be referred to as $GAbs1$ and $GAcid$ and the inlet composition of acetic acid be referred to as $zAbs1$ and $zAcid$. Therefore, the overall and component material balances at the inlets of the two process sinks are written as follows:

$$GAbs1 = w11 + w21 + FreshAA1 \tag{12.14}$$

$$GAcid = w12 + w22 + FreshAA2 \tag{12.15}$$

$$GAbs1 \times zAbs1 = w11 \times 0.14 + w21 \times 0.25 \tag{12.16}$$

$$GAcid \times zAcid = w12 \times 0.14 + w22 \times 0.25 \tag{12.17}$$

Similarly, the total flowrate going to waste treatment is given by:

$$Waste = Waste\,1 + Waste\,2 \tag{12.18}$$

Additional constraints are also needed to insure the non-negativity of the variables and the proper bounds on the feed to the sinks. The following is a LINGO program of the above formulation:

```
min = FreshAA;
! Splitting of process sources and fresh;
WAbs2 = w11 + w12 + Waste1;
WPrimary = w21 + w22 + Waste2;
FreshAA = FreshAA1 + FreshAA2;
! Mixing at the inlet of the sinks;
GAbs1 = w11+ w21+ FreshAA1;
GAcid = w12 + w22 + FreshAA2;
GAbs1*zAbs1 = w11*0.14 + w21*0.25;
GAcid*zAcid = w12*0.14 + w22*0.25;
Waste = Waste1 + Waste2;
! Problem data and nonnegativity constraints;
WAbs2 = 1400;
WPrimary = 9100;
GAbs1 = 5100;
GAcid = 10200;
zAbs1 >= 0.00;
zAbs1 <= 0.05;
zAcid >= 0.00;
zAcid <= 0.10;
w11>= 0;
W21 >= 0;
W12 >= 0;
```

```
W22 >= 0;
FreshAA1 >=0;
FreshAA2 >= 0;
End
```

Solving using LINGO, we get the following solution:

```
Objective value: 9584.000

          Variable              Value
          FRESHAA            9584.000
              W11            0.000000
              W12            1400.000
           WASTE1            0.000000
            WABS2            1400.000
              W21            1020.000
              W22            3296.000
           WASTE2            4784.000
         WPRIMARY            9100.000
         FRESHAA1            4080.000
         FRESHAA2            5504.000
            GABS1            5100.000
            GACID            10200.00
            ZABS1            0.0500000
            ZACID            0.1000000
            WASTE            4784.000
```

Therefore, the target for minimum fresh AA is 9584 kg/h and the target of minimum waste discharge is 4784 kg/h. These results are consistent with the ones determined graphically in Chapter Three and algebraically in Chapter Six. The implementation of this solution is shown in Figure 12-4a.

As mentioned in Chapter Three, it is possible in some cases to have multiple (sometimes infinite) direct recycle strategies that can be implemented to attain the same target. Mathematical programming can be used to generate such solutions. For instance, let us examine the case when all of the second-absorber-tower source is assigned to the first absorber. This can be accomplished by adding the following constraint to the previous optimization formulation:

```
W11 = 1400;
```

Solving the revised program, we get:

```
Objective value: 9584.000

          Variable              Value
          FRESHAA            9584.000
              W11            1400.000
```

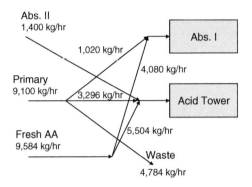

FIGURE12-4a DIRECT RECYCLE CONFIGURATION WHEN ALL OF THE SECOND-ABSORBER-TOWER SOURCE IS FED TO ACID TOWER

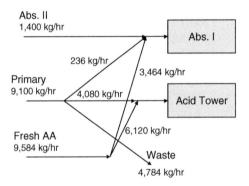

FIGURE 12-4b DIRECT RECYCLE CONFIGURATION WHEN ALL OF THE SECOND-ABSORBER-TOWER SOURCE IS FED TO THE FIRST ABSORBER

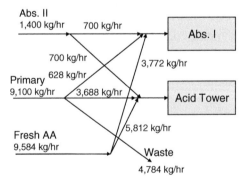

FIGURE 12-4c DIRECT RECYCLE CONFIGURATION WHEN HALF OF THE SECOND-ABSORBER-TOWER SOURCE IS FED TO THE FIRST ABSORBER

```
         W12           0.000000
      WASTE1           0.000000
       WABS2           1400.000
         W21           236.0000
         W22           4080.000
      WASTE2           4784.000
    WPRIMARY           9100.000
    FRESHAA1           3464.000
    FRESHAA2           6120.000
       GABS1           5100.000
       GACID           10200.00
       ZABS1           0.0500000
       ZACID           0.1000000
       WASTE           4784.000
```

This solution is shown by Figure 12-4b.

An additional solution can be determined by assigning some percentage (e.g., 50%) of the first source to the first sink. This can be realized by adding the following constraint to the original optimization formulation:

```
W11 = 0.5*1400;
```

Solving the revised program, we get:

```
Objective value:9584.000
```

```
    Variable              Value
     FRESHAA           9584.000
         W11           700.0000
         W12           700.0000
      WASTE1           0.000000
       WABS2           1400.000
         W21           628.0000
         W22           3688.000
      WASTE2           4784.000
    WPRIMARY           9100.000
    FRESHAA1           3772.000
    FRESHAA2           5812.000
       GABS1           5100.000
       GACID           10200.00
       ZABS1           0.0500000E-01
       ZACID           0.1000000
       WASTE           4784.000
```

This solution is shown by Figure 12-4c.

Other solutions can be generated by assigning various values of a split assigned to a sink and resolving the revised optimization program.

12.4 INTERACTION BETWEEN DIRECT RECYCLE AND THE PROCESS

What happens when the recycle of a source to a sink impacts the flowrate and/or composition of other sources? In such cases, it is necessary to incorporate a process model that tracks the consequences of recycle. This interaction between process synthesis and analysis provides an effective synergism whereby synthesis generates insights, targets, and solution strategies while analysis responds to synthesis by giving input/output relations and by predicting process performance as a result of the synthesized changes. This synergism is shown in Figure 12-5.

To illustrate this interaction, let us consider the following case study where the recycle of sources impacts the compositions of other sources.

EXAMPLE 12-2 DIRECT RECYCLE FOR ETHYL CHLORIDE PROCESS (EL-HALWAGI ET AL., 1996; EL-HALWAGI 1997)

Ethyl chloride (C_2H_5Cl) can be manufactured by catalytically reacting ethanol and hydrochloric acid. Figure 12-6 is a schematic representation of the facility. There are two integrated processes in the facility: the ethanol plant and the ethyl chloride plant. Catalytic hydration of ethylene is used for ethylene production. Compressed ethylene is heated with water and reacted to form ethanol. Two separation units are used for purification of ethanol: distillation followed by pervaporation. The resulting wastewater from distillation is treated in a biotreatment facility. Ethyl chloride is produced by reacting ethanol with hydrochloric acid in a multiphase reactor. The reaction takes place primarily in the liquid phase. A byproduct of the reaction is chloroethanol (CE given by the chemical formula: C_2H_5OCl). This side reaction will be referred to as the *oxychlorination reaction*. The rate of chloroethanol generation via oxychlorination (approximated by a pseudo zero-order reaction) is given

$$r_{oxychlorination} = 6.03 \times 10^{-6} \, kg \, chloroethanol/s \tag{12.19}$$

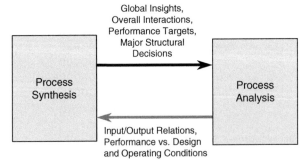

FIGURE 12-5 SYNERGISM BETWEEN PROCESS SYNTHESIS AND PROCESS ANALYSIS

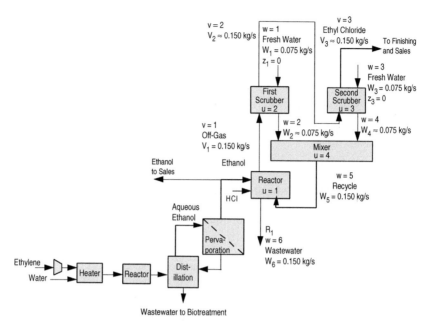

FIGURE 12-6 FLOWSHEET OF THE CHLOROETHANOL CASE STUDY (EL-HALWAGI ET AL., 1996)

FIGURE 12-6 FLOWSHEET OF THE CHLOROETHANOL CASE STUDY (EL-HALWAGI ET AL., 1996)

whereas ethyl chloride is one of the least toxic of all chlorinated hydrocarbons, CE is a toxic pollutant. The off-gas from the reactor is scrubbed with water in two absorption columns. Each scrubber contains two sieve plates and has an overall column efficiency of 65% (i.e., NTP = 1.3). The fresh water used in both scrubbers is pure. Following the scrubber, ethyl chloride is finished and sold. The aqueous streams leaving the scrubbers are mixed and recycled to the reactor. A fraction of the CE recycled to the reactor is reduced to ethyl chloride. This side reaction will be called the *reduction reaction*. The rate of CE depletion in the reactor due to this reaction can be approximated by the following pseudo first-order expression:

$$r_{\text{reduction}} = 0.090\, z_5 \text{ kg chloroethanol/s} \qquad (12.20)$$

where z_5 is in units of mass fraction.

The compositions of CE in the gaseous and liquid effluents of the ethyl chloride reactor are related through an equilibrium distribution coefficient as follows:

$$\frac{y_1}{z_6} = 5 \qquad (12.21)$$

In order to reduce the discharge of CE in wastewater, it is desired to consider segregation, mixing, and direct recycle within the ethyl chloride process. Three liquid sources are to be considered for recycle: the reactor effluent and the bottom liquids from the two scrubbers (W_6, W_2, and W_4). The recycled streams may be fed to three process sinks: the reactor and the two scrubbers. The following process constraints should be considered for recycle:

$$\text{Composition of aqueous feed to reactor} \quad \leq 65 \text{ ppmw CE} \qquad (12.22)$$

$$\text{Composition of aqueous feed to first scrubber} \quad \leq 8 \text{ ppmw CE} \qquad (12.23)$$

$$\text{Composition of aqueous feed to second scrubber} \quad = 0 \text{ ppmw CE} \quad (12.24)$$

$$0.090 \leq \text{Flowrate of aqueous feed to reactor (kg/s)} \quad \leq 0.150 \qquad (12.25)$$

$$0.075 \leq \text{Flowrate of aqueous feed to first scrubber (kg/s)} \quad \leq 0.090 \qquad (12.26)$$

$$0.075 \leq \text{Flowrate of aqueous feed to second scrubber (kg/s)} \leq 0.085 \qquad (12.27)$$

The objective of this case study is to determine the target for minimizing the total load (flowrate × composition) of CE discharged in terminal waste-water of the plant using segregation, mixing, and recycle strategies.

SOLUTION

In this case study, there are four process sinks. There are also nine sources: three gaseous and six liquid. The flowrates of all gaseous and liquid sources are shown in Figure 12-6 where the flowrate of a liquid is referred to as W while the flowrate of the gas is designated by V. Overall material balances around the four sinks provide the following flowrates of the carrier gas and liquid:

$$V_1 = V_2 = V_3 = 0.150 \text{ kg/s} \qquad (12.28)$$

$$W_1 = W_2 = W_3 = W_4 = 0.075 \text{ kg/s} \qquad (12.29)$$

$$W_5 = W_6 = 0.150 \text{ kg/s} \qquad (12.30)$$

Next, it is important to derive a process model which provides the appropriate level of details needed for tracking the effects of direct recycle. The following are the details of the model development.

Component Material Balance for Chloroethanol around the Reactor

$$\begin{array}{l}
\text{Chloroethanol in recycled reactants} \\
+ \text{Chloroethanol generated due to oxychlorination reaction} \\
= \text{Chloroethanol in off-gas} + \text{Chloroethanol in wastewater } W_i \\
+ \text{Chloroethanol depleted by reduction reaction.}
\end{array} \qquad (12.31)$$

Hence,

$$W_5 z_5 + 6.03 \times 10^{-6} = V_1 y_1 + W_6 z_6 + 0.09 z_5 \qquad (12.32)$$

But, as discussed in the problem statement, the compositions of CE in the gaseous and liquid effluents of the ethyl chloride reactor are related through an equilibrium distribution coefficient as follows:

$$\frac{y_1}{z_6} = 5 \qquad (12.27)$$

Combining equations (12.27) and (12.32) and using the values of V_1, W_5, and W_6 to get

$$0.180 \, y_1 - 0.060 \, z_5 = 6.03 \times 10^{-6} \qquad (12.33a)$$

with y_1 and z_5 in mass fraction units. Therefore, the previous equation can be rewritten as

$$0.180 \, y_1 - 0.060 \, z_5 = 6.030 \qquad (12.33b)$$

with y_1 and z_5 in units of ppmw. *Component Material Balance for Chloroethanol around the First Scrubber*

$$V_1 \, y_1 + W_1 \, z_1 = V_2 \, y_2 + W_2 \, z_2 \qquad (12.34)$$

Substituting for the values of V_1, V_2, W_1, and W_2 and noting that z_1 is zero (fresh water), we get

$$2(y_1 - y_2) - z_2 = 0 \qquad (12.35)$$

The Kremser equation can be used to model the scrubber:

$$\text{NTP} = \frac{\ln[(1 - (HV_1/W_1))(y_1 - Hz_1/y_2 - Hz_1) + (HV_1/W_1)]}{\ln(W_1/HV_1)} \qquad (12.36)$$

where Henry's coefficient $H = 0.1$ and NTP $= 1.3$. Therefore,

$$1.3 = \frac{\ln\left(0.8(y_1/y_2) + 0.2\right)}{\ln(5)}$$

i.e.,

$$y_2 = 0.10 \, y_1 \qquad (12.37)$$

Similar to equations (12.36) and (12.37), one can derive the following two equations for the second scrubber:

$$2(y_2 - y_3) - z_4 = 0 \qquad (12.38)$$

and

$$y_3 = 0.10 \, y_2 \qquad (12.39)$$

Component Material Balance for Chloroethanol around the Mixer

$$W_2z_2 + W_4z_4 = W_5z_5 \qquad (12.40)$$

By plugging the values of W_1, W_2, and W_3 into equation (12.40), we get

$$z_2 + z_4 - 2z_5 = 0 \qquad (12.41)$$

A summary of the derived process model is given below (with all compositions in ppmw):

For the reactor,

$$0.180y_1 - 6.030 = 0.060 \, z5$$

$$y_1 - 5z_6 = 0$$

For the first scrubber

$$2y_2 + z_2 = 2y_1$$

$$y_2 = 0.10 \, y_1$$

For the second scrubber,

$$2y_3 + z_4 = 2y_2$$

$$y_3 = 0.10 \, y_2$$

For the mixer,

$$2z_5 = z_2 + z_4$$

This model can be formulated as the following LINGO program:

model:

```
0.180*y1 - 0.060*z5 = 6.030;
y1 - 5*z6 = 0.0;
2*y2 + z2 - 2*y1 = 0.0;
y2 - 0.10*y1 = 0.0;
```

```
2*y3 + z4 - 2*y2 = 0.0;
y3 - 0.10*y2 = 0.0;
2*z5 - z2 - z4 = 0.0;
end
```

The is an **LP** which can be globally solved to give the following compositions (in ppmw CE) before any process changes:

$$y_1 = 50.0$$
$$y_2 = 5.0$$
$$y_3 = 0.5$$
$$z_2 = 90.0$$
$$z_4 = 9.0$$
$$z_5 = 49.5$$
$$z_6 = 10.0$$

Next, a source–sink representation is developed to allow the consideration of the segregation, mixing, and direct recycle candidate strategies for the problem. This is shown in Figure 12-7. Each source is split into several fractions that can be fed to a sink. The flowrate of the streams passed from source w to sink j is referred to as f_{wj}. The terms F_j^{in}, Z_j^{in}, and F_j^{out} represent the inlet flowrate, inlet composition, and outlet flowrate of the streams associated with unit j. Unrecycled streams are allocated to biotreatment for

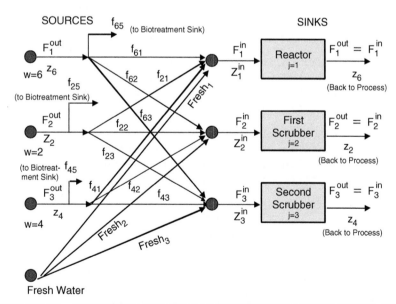

FIGURE 12-7   STRUCTURAL REPRESENTATION OF SEGREGATION, MIXING, AND DIRECT RECYCLE OPTIONS FOR CE CASE STUDY

waste treatment. Their flowrates are referred to as f_{w5} ($j = 5$ is the biotreatment sink). Finally, the flowrate of the fresh water used in the jth sink is referred to as Fresh$_j$.

Since the flowrates of the streams fed to the sinks are to be determined as part of the optimization problem, the process model developed earlier should be revised as follows to allow for variable flowrate of the streams:

$$(0.15 + 0.2 \times F_1^{in})y_1 - (F_1^{in} - 0.09)Z_1^{in} = 6.030 \tag{12.42}$$

$$0.15(y_1 - y_2) - F_2^{in}(z_2 - Z_2^{in}) = 0.0 \tag{12.43}$$

$$\left(\frac{F_2^{in}}{0.015}\right)^{1.3} - \frac{[(1 - (0.015/F_2^{in})](y_1 - 0.1Z_2^{in})}{y_2 - 0.1\,Z_2^{in}} - \frac{0.015}{F_2^{in}} = 0.0 \tag{12.44}$$

$$0.15(y_2 - y_3) - F_3^{in}(z_4 - Z_3^{in}) = 0.0 \tag{12.45}$$

$$\left(\frac{F_3^{in}}{0.015}\right)^{1.3} - \frac{[(1 - (0.015/F_3^{in})](y_2 - 0.1\,Z_3^{in})}{y_3 - 0.1\,Z_3^{in}} - \frac{0.015}{F_3^{in}} = 0.0 \tag{12.46}$$

The revised process model can be formulated as the following LINGO program:

model:

```
! Objective is to minimize load of CE in terminal wastewater
streams
min = F65*z6 + F25*z2 + F45*z4;

! Water balances around inlets of sinks
F1In - f21 - f41 - f61 - Fresh1 = 0.0000;
F2In - f22 - f42 - f62 - Fresh2 = 0.0000;
F3In - f23 - f43 - f63 - Fresh3 = 0.0000;

! CE balances around inlets of sinks
F1In*Z1In - f21*z2 - f41*z4 - f61*z6 = 0.0000;
F2In*Z2In - f22*z2 - f42*z4 - f62*z6 = 0.0000;
F3In*Z3In - f23*z2 - f43*z4 - f63*z6 = 0.0000;

! Water balances around sinks
F1In - F1Out = 0.0000;
F2In - F2Out = 0.0000;
F3In - F3Out = 0.0000;

! Water balances for sources to be split
F1Out - f65 - f61 - f62 - f63 = 0.0000;
F2Out - f25 - f21 - f22 - f23 = 0.0000;
F3Out - f45 - f41 - f42 - f43 = 0.0000;
```

```
! Path-diagram equations
(0.15+0.2*F1In)*y1 - (F1In - 0.09)*Z1In = 6.030;
0.15*(y1 - y2) - F2In*(z2 - Z2In) = 0.0000;
(F2In/0.015)^1.3 - (1 - 0.015/F2In)*(y1 - 0.1*Z2In)/
(y2 - 0.1*Z2In) - 0.015/F2In = 0.0000;
0.15*(y2 - y3) - F3In*(z4 - Z3In) = 0.0000;
(F3In/0.015)^1.3 - (1 - 0.015/F3In)*(y2 - 0.1*Z3In)/
(y3 - 0.1*Z3In) - 0.015/F3In = 0.0;
y1 - 5*z6 = 0.0000;

! Restrictions on what can be recycled
Z1In <= 65.0000;
Z2In < 8.0000;
Z3In = 0.0000;
F1In <= 0.1500;
F1In >= 0.0900;
F2In <= 0.0900;
F2In >= 0.07500;
F3In <= 0.0850;
F3In >= 0.0750;

! If solution gives negative flows or compositions, the non-
negativity constraints can then ! be added
end
```

This is a non-linear program which can be solved using LINGO to get the following results:

```
Objective value: 0.4881247

    Variable        Value
    F65             0.0000000E+00
    Z6              7.178572
    F25             0.0000000E+00
    Z2              60.96842
    F45             0.7500000E-01
    Z4              6.508329
    F1IN            0.9000000E-01
    F21             0.9000000E-01
    F41             0.0000000E+00
    F61             0.0000000E+00
    FRESH1          0.0000000E+00
    F2IN            0.9000000E-01
    F22             0.0000000E+00
    F42             0.0000000E+00
    F62             0.9000000E-01
    FRESH2          0.0000000E+00
    F3IN            0.7500000E-01
```

F23	0.0000000E+00
F43	0.0000000E+00
F63	0.0000000E+00
FRESH3	0.7500000E-01
Z1IN	60.96842
Z2IN	7.178572
Z3IN	0.0000000E+00
F1OUT	0.9000000E-01
F2OUT	0.9000000E-01
F3OUT	0.7500000E-01
Y1	35.89286
Y2	3.618950
Y3	0.3663237

These results can be used to construct the solution as shown in Figure 12-8. The solution involves segregating the two liquid bottoms of the scrubbers, feeding the effluent of the first scrubber to the reactor, recycle the aqueous effluent of the reactor to the first scrubber, and disposing of the second scrubber effluent as the terminal wastewater stream. When this solution involving segregation, mixing, and direct recycle is used, the target for minimum CE discharge is determined to be 0.488 kg/s.

The solution shown in Figure 12-8 is based on integrating process resources within the ethyl chloride plant. If the scope of the case study is expanded to incorporate integration between the ethanol plant and the

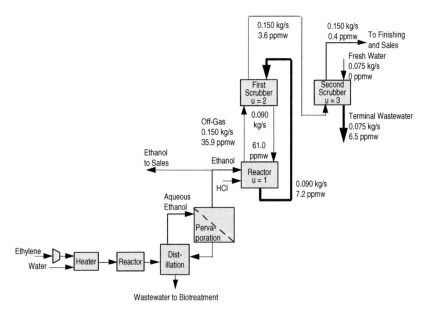

FIGURE 12-8 SOLUTION TO CE CASE STUDY USING SEGREGATION, MIXING, AND DIRECT RECYCLE (EL-HALWAGI 1997)

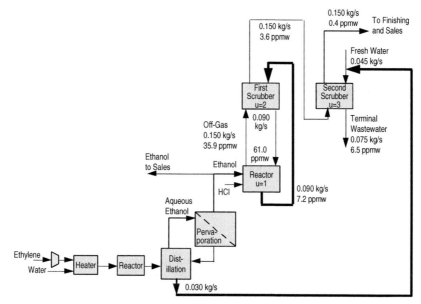

FIGURE 12-9   SOLUTION TO CE CASE STUDY USING SEGREGATION, MIXING, AND INTER-PROCESS DIRECT RECYCLE (EL-HALWAGI ET AL., 1996, EL-HALWAGI 1997)

ethyl chloride plant, then the wastewater leaving the distillation column (0.030 kg/s) can be used to replace some of the fresh water. This solution is shown in Figure 12-9.

12.5 PROBLEMS

12.1 Solve problem 3.1 using the mathematical programming method.
12.2 Solve problem 3.2 using the mathematical programming method.
12.3 Solve problem 3.3 using the mathematical programming method.
12.4 Solve problem 3.4 using the mathematical programming method.
12.5 Solve the food processing case study (Example 3-6) using the mathematical programming method.

12.6 SYMBOLS

F	flowrate of the fresh source
$f_{r,j}$	flowrate of the split of fresh r assigned to sink j
G	sink (unit) flowrate, mass/time
$N_{sources}$	number of process streams (or sources)
N_{sinks}	number of process units (sinks)
W	sink (unit) flow, mass or volume/time
$w_{i,j}$	flowrate of the split of source i assigned to sink j
x	contaminant composition of fresh source

y	contaminant composition of process streams (or sources)
z	contaminant composition of process streams (or sources)

Subscripts

i	index for sources
j	index for sinks
r	index for fresh
waste	index for waste

12.7 REFERENCES

El-Halwagi, M.M. 1997, *Pollution Prevention through Process Integration: Systematic Design Tools*, Academic Press, San Diego.

El-Halwagi, M.M., Hamad, A.A., and Garrison, G.W. 1996, 'Synthesis of waste interception and allocation networks', AIChE J., vol. 42, no. 11, pp. 3087-3101.

13

■ **MATHEMATICAL TECHNIQUES FOR THE SYNTHESIS OF MASS AND HEAT EXCHANGE NETWORKS**

Chapters Four and Seven provided graphical and algebraic procedures for the synthesis of mass exchange networks (MENs) while Chapter Nine presented the graphical and algebraic techniques for the synthesis of heat exchange networks (HENs). As we discussed in Chapter Nine, there is a strong analogy between the two problems. This chapter presents mathematical programming approaches to the synthesis of HENs and MENs. First, the HEN formulation will be presented and solved to attain minimum heating and cooling utility cost and selecting the optimum utilities. Next, an analogous formulation is presented for MENs to screen the mass separating agents (MSAs), determine the minimum operating cost of the network, and synthesize a network of mass exchange units.

13.1 SYNTHESIS OF HENs

This section is based on the transshipment formulation of Papoulias and Grossmann (1983). According to this formulation, heat is to be transshipped from sources (hot streams) to destinations (cold streams). In order to insure thermodynamic feasibility of heat exchange, heat must go through "warehouses". These are represented by the temperature intervals through which heat transfer is feasible. The construction of these temperature intervals was discussed in Chapter Nine as part of developing the temperature interval diagram (TID). Once the TID is constructed, an optimization-based cascade analysis can be developed as a generalization to the algebraic procedure described in Chapter Nine.

Consider a process with N_H process hot streams, N_C process cold streams, N_{HU} heating utilities, and N_{CU} cooling utilities. The index u is used for process hot streams and heating utilities while the index v is used for process cold streams and cooling utilities. The index z is used for temperature intervals. As mentioned in Chapter Nine, on the TID two corresponding temperature scales are generated: hot and cold separated by a minimum temperature driving force. Streams are represented by vertical arrows extending between supply and target temperatures. Next, horizontal lines are drawn at the heads and tails of the arrows thereby defining a series of temperature intervals $z = 1, 2, \ldots, n_{\text{int}}$. Heat can be transferred within the same interval or passed downwards. Next, the table of exchangeable heat loads (TEHL) is constructed (as described in Chapter Nine) to determine the heat exchange loads of the process streams in each temperature interval. The exchangeable load of the uth hot stream (losing sensible heat) which passes through the zth interval is defined as

$$HH_{u,z} = F_u\, C_{P,u}\, (T_{z-1} - T_z), \tag{13.1}$$

where T_{z-1} and T_z are the hot-scale temperatures at the top and the bottom lines defining the zth interval. On the other hand, the exchangeable capacity of the vth cold stream (gaining sensible heat) which passes through the zth interval is computed through

$$HC_{v,z} = f_v\, C_{P,v}(t_{z-1} - t_z), \tag{13.2}$$

where t_{z-1} and t_z are the cold-scale temperatures at the top and the bottom lines defining the zth interval. By summing up the heating loads and cooling capacities, we get:

$$HH_z^{\text{Total}} = \sum_{\substack{u \text{ passes through interval } z \\ \text{where } u=1,\,2,...,\,N_H}} HH_{u,z} \tag{13.3}$$

and

$$HC_z^{\text{Total}} = \sum_{\substack{v \text{ passes through interval } z \\ \text{and } v=1,2,...,\,N_C}} HC_{v,z} \tag{13.4}$$

Next, we move to incorporating heating and cooling utilities. For temperature interval z, the heat load of the uth heating utility is given by:

$$HHU_{u,z} = FU_u C_{P,u}(T_{z-1} - T_z) \quad \text{where } u = N_H + 1, \ N_H + 2, \ldots, \ N_H + N_{HU}$$

$$(13.5)$$

where FU_u is the flowrate of the uth heating utility. The sum of all heating loads of the heating utilities in interval is expressed as:

$$HHU_z^{Total} = \sum_{\substack{u \text{ passes through interval } z \\ \text{where } u = N_H+1, N_H+2, \ldots, N_H+N_{HU}}} HHU_{u,z} \qquad (13.6)$$

The total heating load of the uth utility in the HEN may be evaluated by summing up the individual heat loads over intervals:

$$QH_u = \sum_z HHU_{u,z} \qquad (13.7)$$

Similarly, the cooling capacities of the vth cooling utility in the zth interval is calculated as follows:

$$HCU_{v,z} = fU_v c_{P,v} (t_{z-1} - t_z) \quad \text{where } v = N_C + 1, \ N_C + 2, \ldots, \ N_C + N_{CU}$$

$$(13.8)$$

where fU_v is the flowrate of the vth cooling utility. The sum of all cooling capacities of the cooling utilities is expressed as:

$$HCU_z^{Total} = \sum_{\substack{v \text{ passes through interval } z \\ \text{where } v = N_C+1, N_C+2, \ldots, N_C+N_{CU}}} HCU_{v,z} \qquad (13.9)$$

The total cooling capacity of the uth utility in the HEN may be evaluated by summing up the individual cooling loads over intervals:

$$QC_v = \sum_z HCU_{v,z} \qquad (13.10)$$

For the zth temperature interval, one can write the following heat balance equation (see Figure 13-1):

$$HH_z^{Total} - HC_z^{Total} = HHU_z^{Total} - HHC_z^{Total} + r_{z-1} - r_z, z = 1, 2, \ldots, n_{int}$$

$$(13.11)$$

where

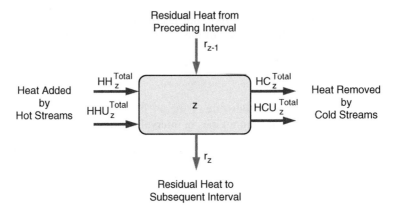

FIGURE 13-1   HEAT BALANCE AROUND TEMPERATURE INTERVAL INCLUDING UTILITIES

$$r_0 = r_{n_{\text{int}}} = 0 \tag{13.12}$$

$$r_z \geq 0, \quad z = 1, 2, \ldots, n_{\text{int}} - 1 \tag{13.13}$$

$$FU_u \geq 0, \quad u = N_H + 1, \ N_H + 2, \ldots, N_H + N_{HU} \tag{13.14}$$

$$fU_v \geq 0 \quad v = N_C + 1, \ N_C + 2, \ldots, N_C + N_{CU.} \tag{13.15}$$

The objective here is to minimize the cost of heating and cooling utilities. Let CH_u designate the cost of the uth heating utility and CC_v designate the cost of the vth cooling utility. If this cost is given in terms of \$/unit flow of utility, then the objective function can be expressed as:

$$\text{Minimize} \sum_{u=N_H+1}^{N_H+N_{HU}} CH_u \times FU_u + \sum_{v=N_C+1}^{N_C+N_{CU}} CC_v \times fU_v \tag{13.16a}$$

On the other hand, if this cost is given in terms of \$/unit heat added or removed by the utility, then the objective function can be expressed as:

$$\text{Minimize} \sum_{u=N_H+1}^{N_H+N_{HU}} CH_u \times QH_u + \sum_{v=N_C+1}^{N_C+N_{CU}} CC_v \times QC_v \tag{13.16b}$$

This objective function can be minimized subject to the set of aforementioned constraints. This formulation is a linear program that can be solved using commercially available software (e.g., LINGO).

EXAMPLE 13-1 USING LINEAR PROGRAMMING FOR UTILITY MINIMIZATION IN A CHEMICAL PLANT

Let us revisit Example 9-1 which involves a chemical processing facility. The process has two process hot streams and two process cold streams. Additionally, two utilities are available for heating and cooling referred to as HU and CU, respectively. The cost of HU is $4/10^6$ kJ and the cost of CU is $5/10^6$ kJ. The stream data are given by Table 13-1. A value of $\Delta T^{min} = 10$ K is used.

The objective of this case study is to use heat integration via the transshipment linear programming model to identify the target for minimum cost of heating and cooling utilities.

SOLUTION

First, the TID is constructed for the process hot and cold streams as well as the heating and cooling utilities. The TID is shown by Figure 13-2. Next, the tables of exchangeable loads for the process hot streams and process cold streams are developed as illustrated by Tables 13-2 and 13-3.

TABLE 13-1 STREAM DATA FOR THE CHEMICAL PROCESS

Stream	Flowrate × specific heat (kW/K)	Supply temperature (K)	Target temperature (K)	Enthalpy change (kW)
H_1	300	460	350	33,000
H_2	500	400	300	50,000
C_1	600	420	490	42,000
C_2	200	320	390	14,000
HU	?	500	460	?
CU	?	290	320	?

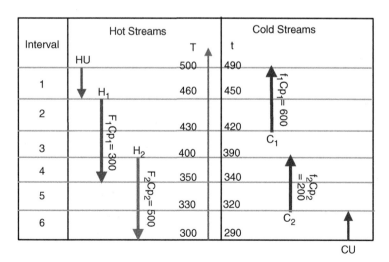

FIGURE 13-2   TID FOR EXAMPLE 13-1

TABLE 13-2 TABLE OF EXCHANGEABLE LOADS FOR THE HOT STREAMS

Interval	Load of H_1 (kW)	Load of H_2 (kW)	Total Load (kW)
1	–	–	–
2	9000	–	9000
3	9000	–	9000
4	15,000	25,000	40,000
5	–	10,000	10,000
6	–	15,000	15,000

TABLE 13-3 TABLE OF EXCHANGEABLE LOADS FOR THE PROCESS COLD STREAMS

Interval	Capacity of C_1 (kW)	Capacity of C_2 (kW)	Total capacity (kW)
1	24,000	–	24,000
2	18,000	–	18,000
3	–	–	–
4	–	10,000	10,000
5	–	4000	4000
6	–	–	–

Based on the aforementioned transshipment model, the following formulation can be developed for the example.

$$\text{Minimize}\,(0.03 \times Q^{\text{min}}_{\text{Heating}} + 0.05 \times Q^{\text{min}}_{\text{Cooling}})3600 \times 8760$$
$$= 94.608 Q^{\text{min}}_{\text{Heating}} + 157.68 Q^{\text{min}}_{\text{Cooling}}$$

Subject to

$$r_1 - Q^{\text{min}}_{\text{Heating}} = -24,000$$

$$r_2 - r_1 = -9000$$

$$r_3 - r_2 = 9000$$

$$r_4 - r_3 = 30,000$$

$$r_5 - r_4 = 6000$$

$$-r5 + Q^{\text{min}}_{\text{Cooling}} = 15,000$$

This formulation can be expressed as the following LINGO program:

```
min = 94.608*QHmin + 157.68*QCmin;
r1 - QHmin = -24000;
r2 - r1 = -9000;
r3 - r2 = 9000;
r4 - r3 = 30000;
r5 - r4 = 6000;
-r5 + QCmin = 15000;
QHmin >= 0;
QCmin >= 0;
r1 >= 0;
r2 >= 0;
r3 >= 0;
r4 >= 0;
r5 >= 0;
```

The solution obtained using LINGO is

```
Objective value: 12.58286E+06

Variable                Value
   QHMIN             33000.00
   QCMIN             60000.00
      r1             9000.000
      r2             0.000000
      r3             9000.000
      r4             39000.00
      r5             45000.00
```

As can be seen from the solution output, there is no heat flow between the second and third intervals (at 430 K on the hot scale and 420 K on the cold scale). This designates the location of the heat exchange pinch point. Also, the solution identifies the minimum heating and cooling utilities to be 33,000 and 60,000 kW, respectively. The minimum utility cost is $12.58 MM/year.

13.2 SYNTHESIS OF MENs

The mathematical formulation for the synthesis of MENs is analogous to the synthesis of HENs. The initial step in the mathematical programming approach to synthesizing MENs is to develop an extended representation of the composition-interval diagram (CID) previously described in Chapter Seven. The main extension is the addition of the external MSAs to the CID. Therefore, $N_S + 1$ composition scales are created. A single composition scale, y, is established for the rich streams and N_s corresponding to composition scales are established for the process and the external MSAs. Each MSA is represented versus its composition axis as a vertical arrow extending between

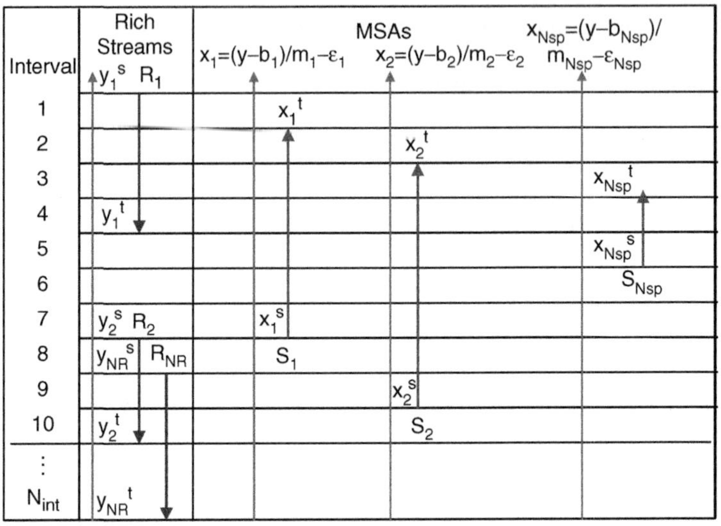

FIGURE 13-3   CONSTRUCTING THE CID

supply and target compositions. Next, horizontal lines are drawn at the heads and tails of the arrows. A schematic representation of the CID is shown in Figure 13-3.

Once the CID is constructed, one can create a table of exchangeable loads (TEL) for the rich streams in which the exchangeable load of the ith rich stream within the kth interval through which it passes is defined as

$$W_{i,k}^R = G_i(y_{k-1} - y_k) \qquad (13.17)$$

By summing up the loads of all rich streams in the kth interval, the collective load of the rich streams within interval k, W_k^R, is evaluated as follows:

$$W_k^R = \sum_{i \, \text{passes through interval} \, k} W_{i,k}^R \qquad (13.18)$$

In evaluating the loads of the MSAs, it is important to recall that the flowrate of each external MSA is unknown. Therefore, for the MSAs the table of exchangeable loads (TEL) is determined per unit mass of the MSAs. In this table, the exchangeable load *per unit mass of the MSA* passing through interval k is determined as follows:

$$W_{j,k}^S = x_{j,k-1} - x_{j,k} \qquad (13.19)$$

Within any composition interval, it is possible to transfer mass from the rich streams to the lean-streams. Additionally, it is possible to transfer residual mass downwards from a composition interval to the

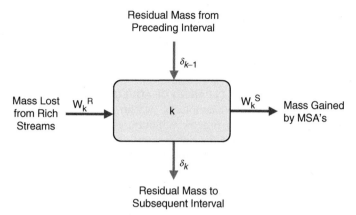

Residual Mass from
Preceding Interval

δ_{k-1}

Mass Lost W_k^R
from Rich
Streams

k

W_k^S Mass Gained
by MSA's

δ_k

Residual Mass to
Subsequent Interval

FIGURE 13-4 FEASIBLE MATERIAL BALANCE AROUND A COMPOSITION INTERVAL

subsequent interval. The material balance around a generic interval is shown by Figure 13-4.

Denoting the cost (\$/kg MSA) and the flowrate of the jth MSA by C_j and L_j, respectively, the problem of minimizing the cost of the MSAs can be formulated through the following optimization program (El-Halwagi and Manousiouthakis, 1990):

$$\min \sum_{j=1}^{N_S} C_j L_j \tag{13.20}$$

subject to

Component material balance around each composition interval

$$\delta_k - \delta_{k-1} + \sum_{j \text{ passes through interval } k} L_j W_{j,k}^S = W_k^R \quad k = 1, 2, \ldots, N_{\text{int}} \tag{13.21}$$

Availability of MSA

$$L_j \leq L_j^c \quad j = 1, 2, \ldots, N_S \tag{13.22}$$

No residual mass entering or leaving the cascade

$$\delta_0 = 0 \tag{13.23}$$

$$\delta_{N_{\text{int}}} = 0 \tag{13.24}$$

Non-negativity constraints

$$\delta_k \geq 0 \quad k = 1, 2, \ldots, N_{\text{int}} - 1 \tag{13.25}$$

$$L_j \geq 0 \quad j = 1, 2, \ldots, N_S \tag{13.26}$$

This transshipment model if a linear program whose solution determines the minimum cost of the MSAs, the optimal flowrate of each MSA, and the residual mass exchange loads and the pinch location (which corresponds to zero residual flow).

EXAMPLE 13-2 CLEANING OF AQUEOUS WASTES

Consider the example of wastewater cleaning previously described by Section 7.4. An organic pollutant is to be removed from two aqueous wastes. The data for the rich streams are given in Table 13-4.

Two process MSAs and two external MSAs are considered for separation. The stream data for the MSAs are given in Table 13-5. The equilibrium data for the pollutant in the four MSAs are given by:

$$y = 0.25x_1 \tag{13.27}$$

$$y = 0.50x_2 \tag{13.28}$$

$$y = 1.00x_3 \tag{13.29}$$

$$y = 0.10x_4 \tag{13.30}$$

The minimum allowable composition difference for any MSA is taken as 0.001 (mass fraction of pollutant).

Determine the minimum load to be removed using an external MSA, the pinch location, and excess capacity of the process MSAs.

TABLE 13-4 DATA OF RICH STREAMS FOR THE WASTEWATER CLEANING EXAMPLE

Stream	Flowrate G_1 (kg/s)	Supply composition of pollutant (mass fraction) y_i^s	Target composition of pollutant (mass fraction) y_i^t
R_1	2.0	0.030	0.005
R_2	3.0	0.010	0.001

TABLE 13-5 DATA OF PROCESS LEAN-STREAMS FOR THE WASTEWATER CLEANING EXAMPLE

Stream	Upper bound on flowrate L_j^C (kg/s)	Supply composition of pollutant (mass fraction) x_j^s	Target composition of pollutant (mass fraction) x_j^t	Cost C_j ($/kg MSA)
S_1	17.0	0.007	0.009	0.000
S_2	1.0	0.005	0.015	0.000
S_3	∞	0.019	0.029	0.010
S_4	∞	0.009	0.029	0.020

SOLUTION

The CID for the problem is constructed as shown in Figure 13-5. Then, the TELs are developed for the rich and the lean-streams as shown in Tables 13-6 and 13-7.

It is now possible to screen the MSAs and determine their optimal flowrates through the following optimization formulation:

```
Min = 0.01*L3 + 0.02*L4;
D1 + 0.010*L3 = 0.020;
D2 - D1 + 0.002*L1 = 0.008;
D3 - D2 = 0.012;
D4 - D3 = 0.010;
D5 - D4 + 0.006*L2 = 0.015;
```

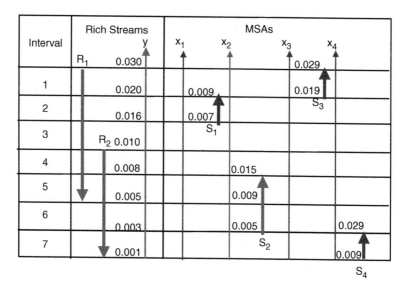

FIGURE 13-5   THE CID FOR THE WASTEWATER CLEANING EXAMPLE

TABLE 13-6 TEL FOR THE RICH STREAMS OF THE WASTEWATER CLEANING EXAMPLE

Interval	Load of R_1 (kg/s)	Load of R_2 (kg/s)	Load of $R_1 + R_2$ (kg/s)
1	0.020	–	0.020
2	0.008	–	0.008
3	0.012	–	0.012
4	0.004	0.006	0.010
5	0.006	0.009	0.015
6	–	0.006	0.006
7	–	0.006	0.006

TABLE 13-7 TEL (kg POLLUTANT/S PER kg OF MSA) FOR THE LEAN-STREAMS
OF THE WASTEWATER CLEANING EXAMPLE

Interval	Load of S_1	Load of S_2	Load of S_3	Load of S_4
1	–	–	0.010	–
2	0.002	–	–	–
3	–	–	–	–
4	–	–	–	–
5	–	0.006	–	–
6	–	0.004	–	–
7	–	–	–	0.020

```
D6 - D5 + 0.004*L2 = 0.006;
- D6 + 0.020*L4 = 0.006;
D1 >= 0.0;
D2 >= 0.0;
D3 >= 0.0;
D4 >= 0.0;
D5 >= 0.0;
D6 >= 0.0;
L1>=0;
L2 >= 0.0;
L3>=0;
L4>=0;
L1 < 17.00;
L2 < 1.00;
End
```

The solution can be obtained using the software LINGO as follows:

```
Objective value: 0.0390000
Variable               Value
      L3           0.000000
      L4           1.950000
      D1           0.2000000E-01
      D2           0.000000
      L1           14.00000
      D3           0.1200000E-01
      D4           0.2200000E-01
      D5           0.3100000E-01
      L2           1.000000
      D6           0.3300000E-01
```

As can be seen from the results, the optimal flowrates of S_1, S_2, S_3, and S_4 are 14.00, 1.00, 0.00, and 1.95 kg/s, respectively. The minimum operating cost of the system is 0.039 kg/s. The residual mass leaving the second interval is zero. Therefore, the mass exchange pinch is located on the line separating the second and the third intervals. As can be seen in Figure 13-5, this location corresponds to a y composition of 0.016. These results are consistent with the solution obtained using the algebraic procedure in Chapter Seven.

■ ■ ■

13.3 PROBLEMS

13.1 Resolve Problem 9.1 using linear programming.
13.2 Resolve Problem 9.2 using linear programming.
13.3 Resolve Problem 9.3 using linear programming.
13.4 Using linear programming, resolve Example 13-2 for the case when the two wastewater streams are allowed to mix.
13.5 Resolve Problem 4.1 using linear programming.
13.6 Resolve Problem 4.3 using linear programming.
13.7 Resolve Problem 4.4 using linear programming.
13.8 Resolve Problem 4.5 using linear programming.
13.9 Resolve Problem 4.8. using linear programming.

13.4 SYMBOLS

C_j	unit cost of the jth MSA including regeneration and makeup (\$/kg of recirculating MSA)
$C_{P,u}$	specific heat of hot stream u [kJ/(kg K)]
$C_{P,v}$	specific heat of cold stream v [kJ/(kg K)]
f	flowrate of cold stream (kg/s)
F	flowrate of hot stream (kg/s)
G_i	flowrate of the ith waste stream
$HC_{v,z}$	cold load in interval z
$HH_{u,z}$	hot load in interval z
i	index of waste streams
j	index of MSAs
k	index of composition intervals
L_j	flowrate of the jth MSA (kg/s)
L_j^C	upper bound on available flowrate of the jth MSA (kg/s)
m_j	slope of equilibrium line for the jth MSA
N_C	number of process cold streams
N_{CU}	number of cooling utilities
N_H	number of process hot streams
N_{HU}	number of process cold streams
N_{int}	number of composition intervals
N_R	number of waste streams
N_S	number of MSAs
R_i	the ith waste stream
r_z	residual heat leaving interval z
S_j	the jth MSA
t	temperature of cold stream (K)
t_v^s	supply temperature of cold stream v (K)

t_v^t target temperature of cold stream v (K)

T_u^s supply temperature of hot stream u (K)

T_u^t target temperature of hot stream u (K)

T temperature of hot stream (K)

U index for hot streams

V index for cold streams

$W_{i,k}^R$ exchangeable load of the ith waste stream which passes through the kth interval

$W_{j,k}^S$ exchangeable load of the jth MSA which passes through the kth interval

W_k^R the collective exchangeable load of the waste streams in interval k

$x_{j,k-1}$ composition of key component in the jth MSA at the upper horizontal line defining the kth interval

$x_{j,k}$ composition of key component in the jth MSA at the lower horizontal line defining the kth interval

x_j^S supply composition of the jth MSA

x_j^t target composition of the jth MSA

y_{k-1} composition of key component in the ith waste stream at the upper horizontal line defining the kth interval

y_k composition of key component in the ith waste stream at the lower horizontal line defining the kth interval

z temperature interval

Greek

ΔT^{min} minimum approach temperature (K)

ε minimum allowable composition difference

13.5 REFERENCES

El-Halwagi, M.M., and Manousiouthakis, V. 1990, 'Automatic synthesis of mass exchange networks with single-component targets', *Chem. Eng. Sci.*, vol. 45, no. 9, pp. 2813-2831.

Papoulias, S.A., and Grossmann, I.E. 1983, 'A structural optimization approach in process synthesis II', Heat recovery networks. *Comput. Chem. Eng.*, vol. 7, no. 6, 707-721.

14

MATHEMATICAL TECHNIQUES FOR MASS INTEGRATION

Chapters Twelve and Thriteen presented mathematical techniques for direct recycle and for mass exchange networks. As mentioned in Chapter Five, mass integration is a much broader concept than direct recycle and mass exchange. Mass integration is a holistic and systematic methodology that provides a fundamental understanding of the global flow of mass within the process and employs this understanding in identifying performance targets and optimizing the allocation, separation, and generation of streams and species. In order to develop detailed mass-integration strategies, Chapter Five presented visualization tools that provide insights and guide the development of solution strategies. The purpose of this chapter is to provide mathematical-programming techniques that can determine cost effective mass integration strategies.

14.1 SOURCE-INTERCEPTION–SINK REPRESENTATION

As mentioned in Chapter Five, the process flowsheet can be described through a source-interception–sink representation (e.g. El-Halwagi and Spriggs 1998; El-Halwagi et al., 1996) as shown in Figure 14-1. For each targeted species, there are sources (streams that carry the species) and process sinks (units that can accept the species). Streams leaving the sinks become, in turn, sources. Therefore, sinks are also generators of the targeted species.

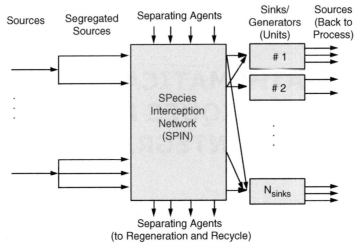

FIGURE 14-1   PROCESS FROM A SPECIES PERSPECTIVE (EL-HALWAGI ET AL., 1996)

Each sink/generator may be manipulated via design and/or operating changes to affect the flowrate and composition of what each sink/generator accepts and discharges. In order to adjust the flows, compositions, and properties of the sources, a species interception network (SPIN) is used. It may be composed of interception devices (e.g., separators) that employ mass and energy separating agents.

The taks of optimizing a source-interception–sink problem may be stated as follows (El-Halwagi et al., 1996; Gabriel and El-Halwagi 2005):

Given a process with:

- A set of process sinks (units): $\text{SINKS} = \{j \mid j = 1, 2, \ldots, N_{\text{sinks}}\}$. Each sink requires a given flowrate, G_j, and a given composition, z_j^{in}, that satisfies the following constraint:

$$z_j^{\min} \leq z_j^{\text{in}} \leq z_j^{\max} \qquad \forall j \in \{1 \ldots N_{\text{sinks}}\} \tag{14.1}$$

  where $z_j^{\min}$ and $z_j^{\max}$ are given lower and upper bounds on acceptable compositions to unit j.
- A set of process sources: $\text{SOURCES} = \{i \mid i = 1, 2, \ldots, N_{\text{sources}}\}$ which can be recycled/reused in process sinks. Each source has a given flowrate, F_i, and a given composition, y_i^{in}.
- A set of interception units: $\text{INTERCEPTORS} = \{k \mid k = 1, 2, \ldots, N_{\text{Int}}\}$ that can be used to remove the targeted species from the sources.

Available for service is a fresh (external) resource (or multiple fresh resources) that can be purchased to supplement the use of process sources.

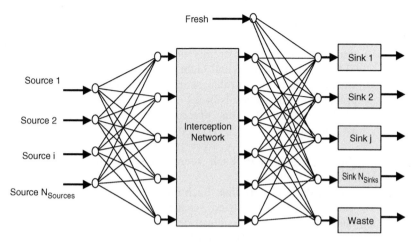

Fresh

Source 1

Source 2

Source i

Source N$_{Sources}$

Interception
Network

Sink 1

Sink 2

Sink j

Sink N$_{Sinks}$

Waste

FIGURE 14-2 STRUCTURAL REPRESENTATION OF THE PROBLEM (GABRIEL AND EL-HALWAGI 2005)

It is desired to develop a mathematical-programming, a formulation that optimizes a certain objective such as minimizing the total annualized cost of the fresh resources and the interception devices. The solution should also determine the optimum allocation of sources to sinks, the interception devices to be selected, and the separation task assigned to each interceptor.

First, a more detailed representation of Figure 14-1 is shown by Figure 14-2. Each source is segregated and split into fractions of unknown flowrate (to be optimized). Those fractions are allocated to the interception network. The intercepted streams are allocated to the sinks. The flowrate of each allocated stream is to be optimized. The unallocated streams are fed to a waste sink.

The objective is to minimize the cost of the fresh resource, interception devices, and waste treatment. Hence, the objective functions can be expressed as:

$$\text{Minimize total annualized cost} =$$
$$C_{\text{Fresh}} \sum_{j=1}^{N_{\text{sinks}}} \text{Fresh}_j + \sum_{k=1}^{N_{\text{int}}} \text{Interception\_Cost}_k \times E_k + C_{\text{waste}} \cdot \text{waste} \quad (14.2)$$

where C_{Fresh} is the cost of the fresh resources (\$/amount of resource), Fresh_j is the amount of fresh resource fed to the jth sink (mass per year). $\text{Interception\_Cost}_k$ is the total annualized fixed equipment cost associated with interception device k. This function may be expressed in terms of flowrate of the streams entering the interception device, its inlet and outlet compositions, and design and operating variables. In general, it is a non-linear non-convex function which is not desirable from a global-optimization perspective. The variable E_k is a binary integer that has the value of 1 or 0 depending on whether or not unit k is used or not, respectively. C_{waste} is the annual waste treatment cost and waste is the total flowrate going to waste. For a description of the mathematical-programming formulation including the constraints of this problem, the reader is referred to El-Halwagi et al., 1996 and Gabriel and El-Halwagi 2005. However, in this section a globally solvable simplified version is presented. The solution provided in this section is based on the procedure described by Gabriel and El-Halwagi (2005). This procedure employs the following simplifying assumptions:

Source Segregation: Each source goes to interception without mixing with other sources.

Interceptor Discretization: Each interceptor is discretized into a number of interceptors with a fixed separation extent (removal efficiencies). The discretization scheme is illustrated by Figure 14-3. Each source is split into several substreams each of which is assigned to a discretized interceptor. The index u is used for the discretized interceptors.

Since the removal efficiency is fixed for each discretized interceptor, it is possible to determine the interception cost (for instance, using the techniques described in Chapter Four). These calculations can be done prior

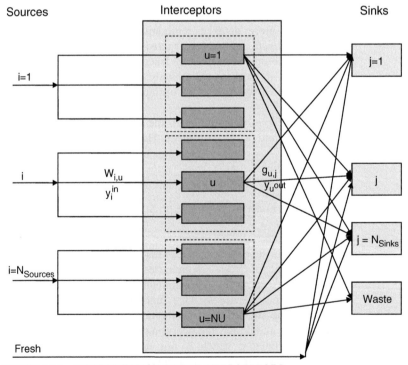

FIGURE 14-3   DISCRETIZATION OF THE INTERCEPTORS

to the optimization formulation. As a result of these pre-synthesis calculations, each interceptor u has a known removal efficiency α_u and cost C_u (\$/Load removed). Consider the case when the interception cost for the uth unit is given by:

$$\text{Interception\_Cost}_u = C_u \times w_u \times y_u^{\text{in}} \times \alpha_u \qquad (14.3)$$

Therefore, the mathematical formulation is given by the following program:

Objective Function:

$$\text{Minimize total annualized cost} =$$

$$C_{\text{Fresh}} \times \sum_{j=1}^{N_{\text{Sinks}}} \text{Fresh}_j + \sum_{u=1}^{\text{NU}} C_u \times \alpha_u \times w_u \times y_u^{\text{in}} + C_{\text{waste}} \times \text{waste} \qquad (14.4)$$

where all the flowrates are given on an annual basis.

Subject to the following constraints:

Splitting of the sources to all the interception devices:

$$F_i = \sum_{u \in U_i} w_u \quad i = 1, 2, \ldots N_{\text{sources}} \qquad (14.5)$$

Pollutant removal in the uth interceptor:

$$y_u^{\text{out}} = (1 - \alpha_u) \times y_u^{\text{in}} \quad u = 1, 2, \ldots \text{NU} \qquad (14.6)$$

Splitting of the sources after the interception devices:

$$w_u = \sum_{j=1}^{N_{\text{sink}}} g_{u,j} + g_{u,\text{waste}} \quad u = 1, 2, \ldots \text{NU} \qquad (14.7)$$

Overall material balance around the mixing point of the feed to the sink:

$$G_j = \text{Fresh}_j + \sum_{u=1}^{\text{NU}} g_{u,j} \quad j = 1 \ldots N_{\text{sinks}} \qquad (14.8)$$

Component material balance around the mixing point of the feed to the sink:

$$G_j \times z_j^{\text{in}} = \text{Fresh}_j \times y_{\text{fresh}} + \sum_{u=1}^{\text{NU}} g_{u,j} \times y_u^{\text{out}} \quad j = 1, 2, \ldots N_{\text{sinks}} \qquad (14.9)$$

Sink constraints:

$$z_j^{\text{min}} \leq z_j^{\text{in}} \leq z_j^{\text{max}} \quad j = 1, 2, \ldots N_{\text{sinks}} \qquad (14.10)$$

The unused flows of sources that have not been used in process sinks are fed to the waste treatment facility before discharge:

$$\text{waste} = \sum_{u=1}^{NU} g_{u,j,\text{waste}} \qquad (14.11)$$

The above formulation is a linear program which can be globally solved to determine the minimum cost of the system, the optimum usage of the fresh resource, and the optimum interception and allocation of sources.

EXAMPLE 14-1 WATER INTERCEPTION AND ALLOCATION

This example is based on the case study given by Gabriel and El-Halwagi (2005). Three sources of wastewater are to be recycled (with or without interception) to two sinks and the rest goes to waste treatment. Table 14-1 summarizes the data for the sources and the sinks. In order to remove the contaminants from the wastewater, three interception technologies are considered: steam stripping, ion exchange using a polymeric resin, and adsorption using activated carbon. Cost data for each technology operating at various pollutant-removal efficiencies on each source are assigned in Tables 14-2–14-4. To reduce the size of the optimization problem, the

TABLE 14-1 PROCESS INFORMATION FOR CASE STUDY (GABRIEL AND EL-HALWAGI 2005)

Sinks	Flow (ton/h)	Maximum inlet concentration (ppm)	Load (kg/h)
1	200	20	4.0
2	80	75	6.0
Sources	**Flow (ton/h)**	**Concentration (ppm)**	**Load (kg/h)**
1	150	10	1.5
2	60	50	3.0
3	100	85	8.5

TABLE 14-2 COST FOR EACH TECHNOLOGY OPERATING AT DIFFERENT CONTAMINANT-REMOVAL EFFICIENCIES FOR SOURCE 1 (GABRIEL AND EL-HALWAGI 2005)

Technology	Index u	Removal efficiency (%)	Cost ($/kg removed)
Stripping	1	10	0.68
	2	50	1.46
	3	90	2.96
Ion Exchange	4	10	0.81
	5	50	1.75
	6	90	3.55
Adsorption	7	10	0.88
	8	50	1.89
	9	90	3.84

TABLE 14-3 COST FOR EACH TECHNOLOGY OPERATING AT DIFFERENT CONTAMINANT-REMOVAL EFFICIENCIES FOR SOURCE 2 (GABRIEL AND EL-HALWAGI 2005)

Technology	Index u	Removal efficiency (%)	Cost ($/kg removed)
Stripping	10	10	0.54
	11	50	1.16
	12	90	2.36
Ion Exchange	13	10	0.65
	14	50	1.40
	15	90	2.84
Adsorption	16	10	0.70
	17	50	1.51
	18	90	3.07

TABLE 14-4 COST FOR EACH TECHNOLOGY OPERATING AT DIFFERENT CONTAMINANT-REMOVAL EFFICIENCIES FOR SOURCE 3 (GABRIEL AND EL-HALWAGI 2005)

Technology	Index u	Removal efficiency (%)	Cost ($/kg removed)
Stripping	19	10	0.45
	20	50	0.97
	21	90	1.97
Ion Exchange	22	10	0.54
	23	50	1.16
	24	90	2.36
Adsorption	25	10	0.59
	26	50	1.26
	27	90	2.56

discretization in these tables is coarser than that in the original case study by Gabriel and El-Halwagi (2005). The cost of fresh water is assigned to be $0.13/ton and the waste treatment cost is $0.22/ton of effluent. A basis of 8000 operating hours per year is selected.

The objective of the case study is to minimize the total cost of the system (including fresh usage, interception, and waste treatment) while satisfying all the process demands.

SOLUTION

There are three discretizations for each technology and three technologies for each source. Therefore, there are 27 discretizations for interception. The aforementioned LP can be formulated and solved using the software LINGO as follows:

```
Min=0.13*(Fresh1+Fresh2+Fresh3)+
0.68*0.1*w1*10*0.001+
1.46*0.5*w2*0.001*10+2.96*0.9*w3*10*0.001+
```

0.81*0.1*w4*10*0.001+1.75*0.5*w5*10*0.001+
3.55*0.9*w6*10*0.001+0.88*0.1*w7*10*0.001+
1.89*0.5*w8*10*0.001+3.84*0.9*w9*10*0.001+
0.54*0.1*w10*50*0.001+
1.16*0.5*w11*50*0.001+2.36*0.9*w12^50*0.001+
0.65*0.1*w13*50*0.001+1.40*0.5*w14*50*0.001+
2.84*0.9*w15*50*0.001+0.70*0.1*w16*50*0.001+
1.51*0.5*w17*50*0.001+3.07*0.9*w18*50*0.001+
0.45*0.1*w19*85*0.001+
0.97*0.5*w20*85*0.001+1.97*0.9*w21*85*0.001+
0.54*0.1*w22*85*0.001+1.16*0.5*w23*85*0.001+
2.36*0.9*w24*85*0.001+0.59*0.1*w25*85*0.001+
1.26*0.5*w26*85*0.001+2.56*0.9*w27*85*0. 001+0.22*Gwaste;
! Splitting of Source 1;
150=wunint1+w1+w 2+w3+w4+w5+w6+w7+w8+w9;
60=wunint2+w10+w11+w12+w13+w14+w15+w16+
w17+w18;
100=wunint3+w19+w20+w21+w22+w23+w24+w25+
w26+w27;
!Mixing of Intercepted Substreams;
Yout1=(1-0.1)*10;
Yout2=(1-0.5)*10;
Yout3=(1-0.9)*10;
Yout4=(1-0.1)*10;
Yout5=(1-0.5)*10;
Yout6=(1-0.9)*10;
Yout7=(1-0.1)*10;
Yout8=(1-0.5)*10;
Yout9=(1-0.9)*10;
Yout10=(1-0.1)*50;
Yout11=(1-0.5)*50;
Yout12=(1-0.9)*50;
Yout13=(1-0.1)*50;
Yout14=(1-0.5)*50;
Yout15=(1-0.9)*50;
Yout16=(1-0.1)*50;
Yout17=(1-0.5)*50;
Yout18=(1-0.9)*50;
Yout19=(1-0.1)*85;
Yout20=(1-0.5)*85;
Yout21=(1-0.9)*85;
Yout22=(1-0.1)*85;
Yout23=(1-0.5)*85;
Yout24=(1-0.9)*85;
Yout25=(1-0.1)*85;
Yout26=(1-0.5)*85;
Yout27=(1-0.9)*85;

```
! Splitting the unintercepted streams from the three sources;
Wunint1 = gunint1_1 + gunint1_2 + gunint1_waste;
Wunint2 = gunint2_1 + gunint2_2 + gunint2_waste;
Wunint3 = gunint3_1 + gunint3_2 + gunint3_waste;
! Splitting of the 27 intercepted substreams;
w1 = g1_1 + g1_2 + g1_waste;
w2 = g2_1 + g2_2 + g2_waste;
w3 = g3_1 + g3_2 + g3_waste;
w4 = g4_1 + g4_2 + g4_waste;
w5 = g5_1 + g5_2 + g5_waste;
w6 = g6_1 + g6_2 + g6_waste;
w7 = g7_1 + g7_2 + g7_waste;
w8 = g8_1 + g8_2 + g8_waste;
w9 = g9_1 + g9_2 + g9_waste;
w10 = g10_1 + g10_2 + g10_waste;
w11 = g11_1 + g11_2 + g11_waste;
w12 = g12_1 + g12_2 + g12_waste;
w13 = g13_1 + g13_2 + g13_waste;
w14 = g14_1 + g14_2 + g14_waste;
w15 = g15_1 + g15_2 + g15_waste;
w16 = g16_1 + g16_2 + g16_waste;
w17 = g17_1 + g17_2 + g17_waste;
w18 = g18_1 + g18_2 + g18_waste;
w19 = g19_1 + g19_2 + g19_waste;
w20 = g20_1 + g20_2 + g20_waste;
w21 = g21_1 + g21_2 + g21_waste;
w22 = g22_1 + g22_2 + g22_waste;
w23 = g23_1 + g23_2 + g23_waste;
w24 = g24_1 + g24_2 + g24_waste;
w25 = g25_1 + g25_2 + g25_waste;
w26 = g26_1 + g26_2 + g26_waste;
w27 = g27_1 + g27_2 + g27_waste;
! Material balances at mixing before entrance of the sinks;
G1 = Fresh1 + gunint1_1 + gunint2_1 + gunint3_1 + g1_1 +
g2_1 + g3_1 + g4_1 + g5_1 + g6_1 + g7_1 + g8_1 + g9_1 +
g10_1 + g11_1 + g12_1 + g13_1 + g14_1 + g15_1 + g16_1 +
g17_1 + g18_1 + g19_1 + g20_1 + g21_1 + g22_1 + g23_1 +
g24_1 + g25_1 + g26_1 + g27_1;
G1*Zin1 = Fresh1*yFresh + gunint1_1*10 + gunint2_1*50 +
gunint3_1*85 + g1_1*Yout1 + g2_1*Yout2 + g3_1*Yout3 +
g4_1*Yout4 + g5_1*Yout5 + g6_1*Yout6 + g7_1*Yout7 +
g8_1*Yout8 + g 9_1*Yout9 + g10_1*Yout10 + g11_1*Yout11 +
g12_1*Yout12 + g13_1*Yout13 + g14_1*Yout14 + g15_1*Yout15 +
g16_1*Yout16 + g17_1*Yout17 + g18_1*Yout18 + g19_1*Yout19
+ g20_1*Yout20 + g21_1*Yout21 + g22_1*Yout22 + g23_1*Yout23
+ g24_1*Yout24 + g25_1*Yout25 + g26_1*Yout26 +
g27_1*Yout27;
```

```
G2 = Fresh2 + gunint1_2 + gunint2_2 + gunint3_2 + g1_2 +
g2_2 + g3_2 + g4_2 + g5_2 + g6_2 + g7_2 + g8_2 + g9_2 +
g10_2 + g11_2 + g12_2 + g13_2 + g14_2 + g15_2 + g16_2 +
g17_2 + g18_2 + g19_2 + g20_2 + g21_2 + g22_2 + g23_2 +
g24_2 + g25_2 + g26_2 + g27_2;
G2*Zin2 = Fresh2*yFresh + gunint1_2*10 + gunint2_2*50 +
gunint3_2*85 + g1_2*Yout1 + g2_2*Yout2 + g3_2*Yout3 +
g4_2*Yout4 + g5_2*Yout5 + g6_2*Yout6 + g7_2*Yout7 +
g8_2*Yout8 + g9_2*Yout9 + g10_2*Yout10 + g11_2*Yout11 +
g12_2*Yout12 + g13_2*Yout13 + g14_2*Yout14 + g15_2*Yout15 +
g16_2*Yout16 + g17_2*Yout17 + g18_2*Yout18 + g19_2*Yout19
+ g20_2*Yout20 + g21_2*Yout21 + g22_2*Yout22 + g23_2*Yout23
+ g24_2*Yout24 + g25_2*Yout25 + g26_2*Yout26 +
g27_2*Yout27;
Gwaste = gunint1_waste + gunint2_waste + gunint3_waste +
g1_waste + g2_waste + g3_waste + g4_waste + g5_waste +
g6_waste + g7_waste + g8_waste + g9_waste + g10_waste +
g11_waste + g12_waste + g13_waste + g14_waste + g15_waste +
g16_waste + g17_waste + g18_waste + g19_waste + g20_waste +
g21_waste + g22_waste + g23_waste + g24_waste + g25_waste +
g26_waste + g27_waste;
! Sink data;
G1 = 200;
G2 = 80;
Zin1 ≤ 20;
Zin2 ≤ 75;
```

This LP can be solved globally. The solution of the L INGO program as given below:

```
Objective value: 6.802500

        Variable              Value
         FRESH1            0.000000
         FRESH2            0.000000
         FRESH3            0.000000
             W1            0.000000
             W2            0.000000
             W3            0.000000
             W4            0.000000
             W5            0.000000
             W6            0.000000
             W7            0.000000
             W8            0.000000
             W9            0.000000
            W10            0.000000
            W11            0.000000
            W12            0.000000
```

W13	0.000000
W14	0.000000
W15	0.000000
W16	0.000000
W17	0.000000
W18	0.000000
W19	52.94118
W20	0.000000
W21	0.000000
W22	0.000000
W23	0.000000
W24	0.000000
W25	0.000000
W26	0.000000
W27	0.000000
GWASTE	30.00000
WUNINT1	150.0000
WUNINT2	60.00000
WUNINT3	47.05882
YOUT1	9.000000
YOUT2	5.000000
YOUT3	1.000000
YOUT4	9.000000
YOUT5	5.000000
YOUT6	1.000000
YOUT7	9.000000
YOUT8	5.000000
YOUT9	1.000000
YOUT10	45.00000
YOUT11	25.00000
YOUT12	5.000000
YOUT13	45.00000
YOUT14	25.00000
YOUT15	5.000000
YOUT16	45.00000
YOUT17	25.00000
YOUT18	5.000000
YOUT19	76.50000
YOUT20	42.50000
YOUT21	8.500000
YOUT22	76.50000
YOUT23	42.50000
YOUT24	8.500000
YOUT25	76.50000
YOUT26	42.50000
YOUT27	8.500000
GUNINT1_1	150.0000

GUNINT1_2	0.000000
GUNINT1_WASTE	0.000000
GUNINT2_1	50.00000
GUNINT2_2	10.00000
GUNINT2_WASTE	0.000000
GUNINT3_1	0.000000
GUNINT3_2	17.05882
GUNINT3_WASTE	30.00000
G1_1	0.000000
G1_2	0.000000
G1_WASTE	0.000000
G2_1	0.000000
G2_2	0.000000
G2_WASTE	0.000000
G3_1	0.000000
G3_2	0.000000
G3_WASTE	0.000000
G4_1	0.000000
G4_2	0.000000
G4_WASTE	0.000000
G5_1	0.000000
G5_2	0.000000
G5_WASTE	0.000000
G6_1	0.000000
G6_2	0.000000
G6_WASTE	0.000000
G7_1	0.000000
G7_2	0.000000
G7_WASTE	0.000000
G8_1	0.000000
G8_2	0.000000
G8_WASTE	0.000000
G9_1	0.000000
G9_2	0.000000
G9_WASTE	0.000000
G10_1	0.000000
G10_2	0.000000
G10_WASTE	0.000000
G11_1	0.000000
G11_2	0.000000
G11_WASTE	0.000000
G12_1	0.000000
G12_2	0.000000
G12_WASTE	0.000000
G13_1	0.000000
G13_2	0.000000
G13_WASTE	0.000000

```
        G14_1           0.000000
        G14_2           0.000000
     G14_WASTE          0.000000
        G15_1           0.000000
        G15_2           0.000000
     G15_WASTE          0.000000
        G16_1           0.000000
        G16_2           0.000000
     G16_WASTE          0.000000
        G17_1           0.000000
        G17_2           0.000000
     G17_WASTE          0.000000
        G18_1           0.000000
        G18_2           0.000000
     G18_WASTE          0.000000
        G19_1           0.000000
        G19_2          52.94118
     G19_WASTE          0.000000
        G20_1           0.000000
        G20_2           0.000000
     G20_WASTE          0.000000
        G21_1           0.000000
        G21_2           0.000000
     G21_WASTE          0.000000
        G22_1           0.000000
        G22_2           0.000000
     G22_WASTE          0.000000
        G23_1           0.000000
        G23_2           0.000000
     G23_WASTE          0.000000
        G24_1           0.000000
        G24_2           0.000000
     G24_WASTE          0.000000
        G25_1           0.000000
        G25_2           0.000000
     G25_WASTE          0.000000
        G26_1           0.000000
        G26_2           0.000000
     G26_WASTE          0.000000
        G27_1           0.000000
        G27_2           0.000000
     G27_WASTE          0.000000
          G1          200.0000
        ZIN1           20.00000
      YFRESH            0.000000
          G2           80.00000
        ZIN2           75.00000
```

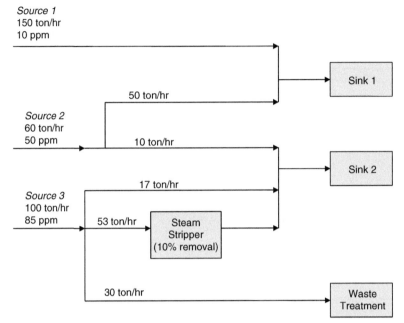

FIGURE 14-4 OPTIMAL SOLUTION TO EXAMPLE 14-1 (GABRIEL AND EL-HALWAGI 2005)

The minimum total annualized cost was found to be \$6.8025/h (or \$54,420/year). No fresh is needed while 30 ton/h of effluent are fed to wastewater treatment. A portion of source 3 (53 ton/h) is intercepted using a stream stripper with 10% removal efficiency. The system configuration is illustrated in Figure 14-4.

14.2 INCORPORATION OF PROCESS MODEL IN MASS INTEGRATION

As mentioned in Chapter Twelve, there are cases when mass-integration strategies impact the original data of the process. Examples of these cases include:

- When changes in one of the targeted streams impact other targeted streams
- When recycle loops are created leading to buildup of certain components to new concentration levels
- When design and operating variables throughout the process are altered
- When new devices (e.g., interceptors) are added within the process

In such cases, it is necessary to incorporate a process model with the appropriate level of details to keep track of the effect of process changes and embed them during the generation of mass-integration strategies.

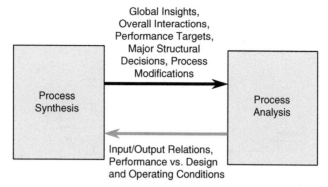

Global Insights,
Overall Interactions,
Performance Targets,
Major Structural
Decisions, Process
Modifications

Process
Synthesis

Process
Analysis

Input/Output Relations,
Performance vs. Design
and Operating Conditions

FIGURE 14-5 INTERACTION BETWEEN PROCESS SYNTHESIS AND PROCESS ANALYSIS

This integration of process synthesis and analysis provides beneficial interaction insights, targets, process changes, technology selection, and solution strategies while analysis responds to synthesis by giving input/output relations and by predicting process performance as a result of the synthesized changes. This synergism is shown in Figure 14-5.

In the following, two examples are solved to demonstrate the interaction between process synthesis and process analysis. In the first example, the effect of changing design and operating variables is discussed. The second example discusses the interaction between the process model and the synthesis of a mass exchange network.

EXAMPLE 14-2 YIELD TARGETING IN ACETALDEHYDE PRODUCTION THROUGH ETHANOL OXIDATION (AL-OTAIBI AND EL-HALWAGI 2004, 2005)

Consider the process of producing acetaldehyde through ethanol oxidation as shown in Figure 14-6. Ethanol is flashed and mixed with air before being fed to the reactor where the following reaction takes place:

$$CH_3CH_2OH + \tfrac{1}{2}O2 \rightarrow CH_3CHO + H_2O$$

One way of tracking the desired product from a certain raw material is to evaluate the reactor yield, $Y_{reactor}$, which is defined as the ratio of mass of acetaldehyde formed in the reactor to mass of ethanol fed to the reactor. The reactor yield is given by:

$$Y_{reactor} = 0.33 - 4.2 \times 10^{-6} \times (T_{rxn} - 580)^2 \qquad (14.12)$$

where T_{rxn} is the reactor temperature (K). Currently, the reaction temperature is 442 K leading to a reactor yield of 0.25 kg acetaldehyde formed in the reactor per kilogram ethanol fed to the reactor.

The rest of the process involves a separation train to purify and recover the product. Two scribbers are used; in the first scrubber, a mixture of fresh and recycled ethanol is used; in the second scrubber, water is

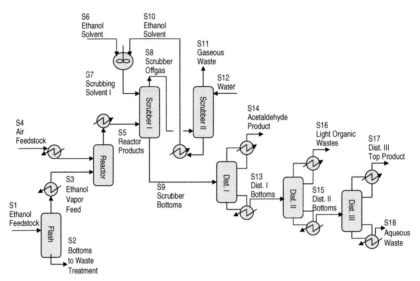

FIGURE 14-6 SCHEMATIC REPRESENTATION OF ACETALDEHYDE PROCESS (AL-OTAIBI AND EL-HALWAGI 2004)

used. Both ethanol and acetaldehyde are recovered along with impurities. The off-gas leaving the second scrubber, mostly nitrogen and trace amounts of oxygen, acetaldehyde, ethanol, and water is discharged as a gaseous waste. The bottom stream of the first scrubber is fed to a distillation sequence. In the first column, acetaldehyde is recovered as the top product. The second column separates light organic wastes as the overhead product. The third column is an ethanol recovery unit where ethanol (with some water) is separated as the overhead product and is subsequently fed to a boiler to utilize its heating value. The bottom stream is an aqueous waste which is sent to a biotreatment facility.

The objective of this case study is to maximize the overall process yield without adding new process equipment. Process modifications and direct recycle may be used. Direct recycle is allowed only from the top of the third distillation column to the flash column.

The overall process yield (Process yield) is defined as:

$$\text{Process yield} = \frac{\text{Acetaldehyde in final product stream}}{\text{Fresh ethanol fed to process as feedstock}} \qquad (14.13a)$$

In the definition of the process yield, the use of the phrase "ethanol fed to the process as a feedstock" excludes the use of ethanol for non-reactive purposes such as solvents (e.g., stream S6). Therefore, the process yield is given by:

$$\text{Process yield} = \frac{A14}{E1} \qquad (14.13b)$$

where A14 is the mass flowrate of acetaldehyde in the final acetaldehyde product (in stream S14) and E1 is the mass flowrate of ethanol in the fresh feedstock to the process (in stream S1)[1]. The plant is to produce 100,000 ton/year of acetaldehyde (i.e., A14 = 100,000 ton/year). The present (base case) value of the overall process yield is 0.237. The following flows may be assumed to hold throughout the case study (even after process changes):

- No ethanol in S4, S12, S14, or S16
- E6 = 400 ton ethanol/year
- No acetaldehyde in S1, S2, S4, S6, S12, or S15.

The process model relating the optimization variables to the tracking of ethanol and acetaldehyde as well as the process constraints are given below.

Reactor

The following is the admissible range for reaction temperature:

$$300 = T_{rxn}(K) = 860 \tag{14.14}$$

The reactor yield should be distinguished from the process yield. The reactor yield is given by:

$$Y_{reactor} = \frac{AR}{E_{feed}} \tag{14.15}$$

where E_{feed} is the ethanol fed to the reactor and AR is the generated acetaldehyde in the reactor. The ethanol consumed in the reactor (ER) is related to AR through stoichiometry and molecular weights. Therefore,

$$ER = \left(\frac{46}{44}\right) \times AR \tag{14.16}$$

Flash Column

As a result of flashing ethanol before the reactor, some ethanol is lost in the bottoms of the flash. Ethanol recovery relating ethanol in the flash top to the flash feed is given by:

$$E2 = a \times E1 \tag{14.17}$$

where

$$a = 10.5122 - 0.0274 \times T_{flash} \tag{14.18}$$

---

[1]In this case study, a stream is referred to by S followed by the stream number and the loads (ton/year) of acetaldehyde and ethanol are designated respectively by A and E followed by the stream number.

where T_{flash} is the temperature of the flash drum in kelvin and is bounded by the following constraint:

The range for the flash temperature is:

$$380 = T_{flash}(K) = 384 \qquad (14.19)$$

First Distillation Column

It is possible to relate the amount of acetaldehyde recovered as a top product to the feed to the distillation column and to the reboiler duty through the following expressions:

$$A14 = \beta \times A9 \qquad (14.20)$$

where

$$\beta = 0.14 \times Q_R + 0.89 \qquad (14.21)$$

where Q_R is the reboiler heat duty in megawatt. The range of the reboiler duty is:

$$0.55 = Q_R(MW) = 0.76 \qquad (14.22)$$

For the base case, the reboiler duty is 0.55 MW.

Third Distillation Column

The recovered ethanol can be related to the feed to the column and to reflux ratio through:

$$E17 = \gamma \times E15 \qquad (14.23)$$

where

$$\gamma = 0.653 \times e^{(0.085 \times RR)} \qquad (14.24)$$

where RR is the reflux ratio in the third distillation column. Currently, the reflux ratio for the column is 2.5 and the working range for the reflux ratio is:

$$2.5 = RR = 5.0 \qquad (14.25)$$

SOLUTION

In this case study, yield can be enhanced by manipulating process variables (e.g., flash temperature, reaction temperature, reboiler duty, reflux ratio, etc.) and by directly recycling ethanol from the top of the third distillation column to the flash column (and subsequently to the reactor). The following is the LINGO formulation of the problem:

```
Max = YP;
YP = A14/E1;
! Problem data;
```

```
A14 = 100000;
E4 = 0;
E6 = 400;
E14 = 0;
E16=0;
E12 = 0;
A1 = 0;
A2 = 0;
A4 = 0;
A6 = 0;
A12 = 0;
A15 = 0;
! Reactor Yield;
YR*E3 = AR;
YR = 0.33 - 0.0000042*(TR - 580)*(TR - 580);
TR ≥ 300;
TR ≤ 860;
ER = (46/44)*AR;
! Balances for Ethanol and acetaldehyde;
E1 + E4 + E6 + E12 = ER + E11 + E18 + E14 + E2+E16 + EW;
A1 + A4 + A6 + A12 + AR = A11 + A18 + A14 + A2+A16 + AW;
E15 = E17 + E18;
A15 = A17 + A18;
E5 = E3 - ER + E4;
ER ≤ E3;
! EG is the gross feed to the flash and EC is the recycled
ethanol to the flash from the third distillation column;
EG = EC + E1;
EG = E2 + E3;
E9 = E14 + E16 + E15;
! EW is the unrecycled (waste) ethanol from the top of third
distillation column;
E17 = EC + EW;
A17 = AC + AW;
AG + A4 + AR = A5 + A2;
A5 + A6 + A12 = A11 + A18+A16 + A14 + A17;
! Flash recovery;
E2 = Alfa*EG;
Alfa = -0.0274*TF + 10.5122;
TF ≥ 380;
TF ≤ 384;
! Distillation recovery;
Gama = 0.653*@exp(0.085*RR);
E17 = Gama*E15;
RR ≥ 2.5;
RR=5;
RR ≤ 5;
```

```
Beta = 0.14*QR + 0.89;
A14 = Beta*A9;
A9 = A5 + A6 + A12 − A11;
A9=A14+A15 | A16;
QR ≤ 0.76;
QR ≥ 0.55;
```

The program can be solved using LINGO. The solution shows that the overall process yield can be maximized to 95.5%. The optimal values of some of the optimization values are:

Reaction temperature: 580 K
Flash temperature: 383.7 K
Reboiler heat duty for the first distillation column: 0.76 MW
Reflux ratio for third distillation column: 5.0
Ethanol recycled from the top of third distillation column to the flash column: 199,367 ton/year.

The following is the solution to the LINGO program.

```
Objective value: 0.9545792
Variable                Value
      YP            0.9545792
     A14            100000.0
      E1            104758.2
      E4            0.000000
      E6            400.0000
     E14            0.000000
     E16            0.000000
     E12            0.000000
      A1            0.000000
      A2            0.000000
      A4            0.000000
      A6            0.000000
     A12            0.000000
     A15            0.000000
      TR            580.0000
      YR            0.3300000
      E3            304125.2
      AR            100361.3
      ER            104923.2
     E11            0.000000
     E18            235.0225
      E2            0.000000
      EW            0.000000
     A11            0.000000
     A18            0.000000
```

A16	361.3007
AW	0.000000
E15	199602.0
E17	199367.0
A17	0.000000
E5	199202.0
EG	304125.2
EC	199367.0
E9	199602.0
AC	0.000000
AG	0.000000
A5	100361.3
ALFA	0.000000
TF	383.6569
GAMA	0.9988225
RR	5.000000
BETA	0.9964000
QR	0.7600000
A9	100361.3

■ ■ ■

EXAMPLE 14-3 INTERCEPTION OF ETHYL CHLORIDE PROCESS (EL-HALWAGI ET AL., 1996; EL-HALWAGI 1997)

Here, we revisit Example 12-2 on the reduction of chloroethanol discharge from the liquid wastes of an ethyl chloride process, the process is shown in Figure 14-7. In Example 14-2, the focus was on direct recycle. In this example, the emphasis is on interception. Six mass separating agents (MSAs) are available for the removal of CE from the process streams: three from gaseous stream and three from liquid stream. The MSAs are listed in Tables 14-5 and 14-6 along with their data.

The objective of this case study is to minimize the cost of interception while reducing the composition of CE in the aqueous discharge to 7 ppm.

SOLUTION

The nominal case was modeled in Chapter Twelve. The process model and the results are summarized below:

```
model:
0.180*y1 − 0.060*z5 = 6.030;
y1 − 5*z6 = 0.0;
2*y2 + z2 − 2*y1 = 0.0;
y2 − 0.10*y1 = 0.0;
2*y3 + z4 − 2*y2 = 0.0;
y3 − 0.10*y2 = 0.0;
2*z5 − z2 − z4 = 0.0;
end
```

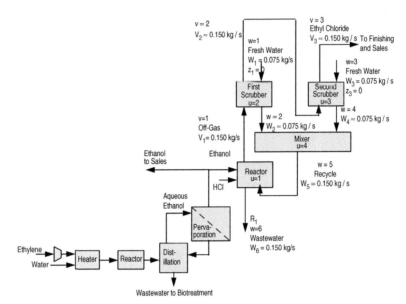

FIGURE 14-7   FLOWSHEET OF CE CASE STUDY (EL-HALWAGI ET AL., 1996)

TABLE 14-5 DATA FOR MSAs THAT CAN REMOVE CE FROM GASEOUS STREAM (EL-HALWAGI ET AL., 1996)

Stream	Description	Supply composition x_j^s (ppmw)	Target composition x_j^t (ppmw)	m_j	ε_j ppmw	c_j \$/kg MSA	c_j^r \$/ kg CE removed
SV_1	Polymeric resin	2	10	0.03	5	0.08	10,000
SV_2	Activated carbon	5	30	0.06	10	0.10	4,000
SV_3	Oil	200	300	0.80	20	0.05	500

TABLE 14-6 DATA FOR MSAs THAT CAN REMOVE CE FROM LIQUID STREAM (EL-HALWAGI ET AL., 1996)

Stream	Description	Supply composition x_j^s (ppmw)	Target composition x_j^t (ppmw)	m_j	ε_j (ppmw)	c_j \$/kg MSA	c_j^r \$/ kg CE removed
SW_1	Zeolite	3	15	0.09	15	0.70	58,333
SW_2	Air	0	10	0.10	100	0.05	5000
SW_3	Steam	0	15	0.80	50	0.12	8000

The solution to this model gives the following compositions (in ppmw CE) prior to interception:

$$y_1 = 50.0$$
$$y_2 = 5.0$$

$y_3 = 0.5$
$z_2 = 90.0$
$z_4 = 9.0$
$z_5 = 49.5$
$z_6 = 10.0$

As can be seen from the results, the terminal wastewater node ($w = 6$) has 10 ppm of CE. The objective is to reduce it to 7 ppm. Instead of focusing on interception the terminal waste, it is important to consider intercepting the various process streams (both gaseous and liquid). The challenge is that when a stream within the process is intercepted and its CE content is reduced, the composition of CE throughout the rest of the process will consequently change. Therefore, it is necessary to integrate the synthesis of a mass exchange network with the process model which tracks the changes in CE throughout the process. This interaction is shown by Figure 14-8 which illustrates the integration of data between the process model and the mass exchange pinch diagram. Consider the process streams (source or rich streams) to be intercepted. Upon interception, these streams change composition from y_v to the intercepted compositions y_v^{int}. Each intercepted stream is processed through the model. According to the process model equations, this interception propagates throughout the whole process model, affecting the composition of the other streams. In turn, this propagation in the

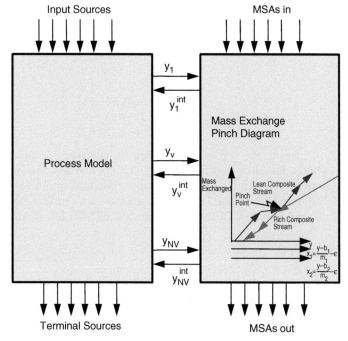

FIGURE 14-8   INTERACTION OF PROCESS MODEL AND MASS EXCHANGE PINCH DIAGRAM (EL-HALWAGI ET AL., 1996)

process model affects the pinch diagram by changing the rich composite curve. Consequently, the lean composite stream is adjusted to respond to the changes in the waste composite stream.

    The implementation of this approach to the case study is detailed by El-Halwagi (1997). For instance, when the off-gas leaving the reactor is intercepted using activated carbon, the flowing LINGO model can be developed:

```
min = 0.10*3600*8760*LCarbon;
0.180*y1 − 0.060*z5 = 6.030;
y1 − 5*z6 = 0.0;
2*y2 + z2 − 2*yINT1 = 0.0;
y2 − 0.10*yINT1 = 0.0;
2*y3 + z4 − 2*y2 = 0.0;
y3 − 0.10*y2 = 0.0;
2*z5 − z2 − z4 = 0.0;
z6 ≤ 7.0;
25*LCarbon − 0.150*(35.0 − yINT1) = 0.0;
end
```

The solution of this program is given below.

```
Objective value: 576248.8
```

Variable	Value
LCARBON	0.1827273
Y1	35.00000
Z5	4.499997
Z6	7.000000
Y2	0.4545451
Z2	8.181812
YINT1	4.545451
Y3	0.4545451E-01
Z4	0.8181812

    Again, the rest of the solution formulation is detailed by El-Halwagi (1997). Based on this formulation, the optimal solution is found to involve a single interception using activated-carbon adsorption to separate CE from the gaseous stream leaving the reactor and reduce its composition to $y_1^{int} = 4.55$ ppmw CE at a minimum cost of \$576,250 /year.

14.3 PROBLEMS

14.1.   Solve Problem 5.1 using mathematical programming.
14.2.   Solve Problem 5.2 using mathematical programming.
14.3.   Pulp and paper plants are characterized by the use of large quantities of water. A common process for pulping is the kraft process. Figure 14-9. is a schematic representation of a kraft process. In this process (Lovelady et al., 2006), wood chips are cooked in a caustic solution referred to as the white

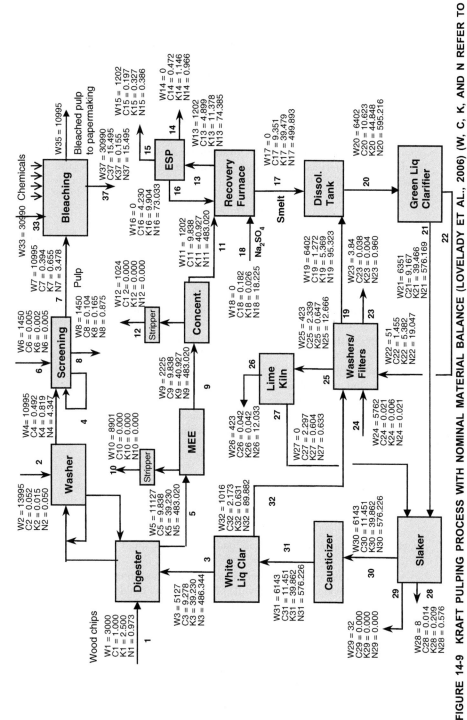

FIGURE 14-9 KRAFT PULPING PROCESS WITH NOMINAL MATERIAL BALANCE (LOVELADY ET AL., 2006) (W, C, K, AND N REFER TO FLOWRATES (TON/DAY) OF WATER, CHLORIDE, POTASSIUM, AND SODIUM, RESPECTIVELY)

liquor (composed primarily of NaOH and Na_2S). The reaction takes place in a digester. The produced pulp is wasted in brown stock washers and then fed to papermaking plant where it is bleached and formed into paper sheets. After brown stock washing, the spent cooking solution is referred to as weak black liquor. This liquor is concentrated by a number of evaporators and a condenser to provide a strong black liquor. The strong black liquor is burned in a recovery boiler to generate heat and chemicals (referred to as smelt). The smelt from the recovery furnace is composed primarily of Na_2S, Na_2CO_3, and some Na_2SO_4. The smelt is dissolved to form green liquor which is clarified to remove any undissolved materials (known collectively as dregs). The clarified green liquor is fed to a slaker where lime and water react to form slaked lime:

$$CaO + H_2O \rightarrow Ca(OH)_2$$

Slaking is followed by a causticizing reaction between the slaked lime and sodium carbonate to yield white liquor:

$$Ca(OH)_2 + Na_2CO_3 \rightarrow 2NaOH + CaCO_3$$

The white liquor is fed back to the digester for cooking the wood chips.

A significant amount of water is used and discharged. Any attempt to recycle the wastewater must address the issues of buildup of ionic species referred to as non-process elements (NPEs). Failure to account for the buildup of NPEs can lead to serious results including scaling, fouling, corrosion, and even collapse of some units.

A critical constraint on the buildup of NPEs is associated with the "sticky temperature" for the recovery furnace. It can be related to the Cl^-, K^+, and Na^+ through the following constraints:

$$\frac{K_{11} + K_{16} + K_{18}}{39.1} \leq 0.1 \frac{N_{11} + N_{16} + N_{18}}{23} \tag{14.26}$$

$$\frac{C_{11} + C_{16} + C_{18}}{35} \leq 0.02 \left(\frac{N_{11} + N_{16} + N_{18}}{23} + \frac{K_{11} + K_{16} + K_{18}}{39} \right) \tag{14.27}$$

where C_i, N_i, and K_i are the ionic loads of Cl^-, K^+, and Na^+ (respectively) in the ith source.

The objective of this problem is to employ mass-integration techniques to reduce wastewater discharge while alleviating any buildup of ionic species. The following tasks are to be undertaken:

a. Determine the target for minimum water usage and discharge
b. Develop a process model with the appropriate level of details to track water and the three targeted NPEs (*hint*: see Lovelady et al., 2005 for details on deriving the model)
c. Using mathematical programming, determine the optimal direct recycle strategies

d.   Using mathematical programming, develop a mass-integration solution to reach the target identified in part a.

14.4 SYMBOLS

F_i	flow rate of source i
Fresh_j	amount of fresh resource fed to the sink j
G_j	flowrate demand for sink j
$g_{u,j}$	amount of flow from interceptor u going to sink j
$g_{u,j=\text{Waste}}$	flow from interceptor u to waste
N_{int}	number of interceptors
N_{sinks}	number of sinks
N_{sources}	number of sources
NU	total amount of discretized interceptors
waste	total amount of flow going to waste
$w_{i,u}$	amount of flow from source i entering discretized interceptor u
y_{fresh}	impurity of fresh resource
y_i^{in}	given composition of source I
y_u^{out}	outlet composition of interceptor u
z_j^{min}	lower bound composition to sink j
z_j^{in}	inlet composition to sink j
z_j^{max}	upper bound composition to sink

14.5 REFERENCES

Gabriel, F.B. and El-Halwagi, M.M. 2005, 'Simultaneous synthesis of waste interception and material reuse networks: problem reformulation for global optimization', *Env. Prog.*, Vol. 24, no. 2, pp. 171-180.

El-Halwagi, M.M., Hamad, A.A., and Garrison, G.W. 1996, 'Synthesis of waste interception and allocation networks', *AIChE J.*, vol. 42, no. 11, pp. 3087-3101.

El-Halwagi, M.M., 1997, pollution prevention through process integration: Systematic design tools, Academic Press, San Diego.

El-Halwagi, M.M. and Spriggs, H.D. 1998, 'Solve design puzzles with mass integration', *Chem. Eng. Prog.*, vol. 94, no. 8, 25-44.

Lovelady, E.M., El-Halwagi, M.M., and Krishnagopalan, G. 'An integrated approach to the optimization of water usage and discharge in pulp and paper plants', *Int. J. of Environ. and Pollution (IJEP)*, (2006, in press)

15

■■■ PUTTING IT TOGETHER: INITIATIVES AND APPLICATIONS

The book has covered fundamentals, tools, insights, and case studies on process integration (PI) for resource conservation and process optimization. Now, how do you create value by applying PI in an industrial setting and how do you start an actual process integration initiative or a project? This chapter sheds the light on key insights, common pitfalls, and necessary requirements for defining, launching, and implementing successful PI applications.

15.1 ARE THERE OPPORTUNITIES?

Competitive industrial facilities should always seek "continuous process improvement". As such, these companies are constantly looking for ways to make their processes faster, more profitable, safer, cleaner, and higher quality. Process integration provides a unique and attractive framework to meet these objectives. However, because PI is a relatively new field, there are many process improvement projects that are implemented worldwide using brainstorming, heuristics, and trial and error. The pitfalls of these conventional approaches have been discussed in Chapter One. Nonetheless, these approach have contributed and continue to contribute value to the process industries. The question is whether or not there are opportunities within processing facilities and why? The answer is: absolutely yes! Regardless of the type and size of the process, there are always significant opportunities for improvement. Why? El-Halwagi and Spriggs (1998) attribute these opportunities to three main factors:

Organization: Because of the complexity of a processing facility, there is typically the tendency to subdivide the process into smaller, tractable portions. Sometimes, these subdivisions are at the level of sections of the plant and sometimes they are even at the level of individual units. The net result is that the focus shifts to specific, localized, and stratified portions of the process so as to manage the complex in a streamlined manner. This subdivision works against a holistic perspective of the process and quite often leads to missing major opportunities that are only identified when the whole process is understood from an integrated viewpoint and when all the necessary interactions are properly understood and utilized. Additionally, these subdivisions normally lead to solutions that address the symptoms of the problems while failing to address the root causes.

Changing Business Requirements: The process industries are constantly evolving and changing. There are several reasons for such changes. These include changing market conditions, supply, and demand, varying availability, qualities and costs of raw materials and products, growing stringency in environmental regulations, needs for expansion, incorporation of new production lines, replacement of existing units and technologies, and need for safer operations. Even for a process that has been recently "optimized", there are numerous opportunities for improvement because of the changing business requirements.

Integrated Tools: A processing facility is a complex system of technologies, chemistries, units, and streams. Without an integrated approach to optimizing these systems, it becomes extremely difficult to understand the global insights and interactions of the process, identify broad opportunities, and develop site-wide implementation. The process integration tools described in this books are just becoming available for widespread use. They should result in a step change in identifying and realizing opportunities.

15.2 STARTING AND SUSTAINING PI INITIATIVES AND PROJECTS

Tools, procedures, and insights are very important in applying PI but are not sufficient in starting and sustaining PI initiatives and projects. In this context, personal initiatives, human resources, work processes, and working cultures are all necessary. The following are key building blocks that are needed to start and sustain a successful PI initiative (Dunn and El-Halwagi 2003):

- Understand the company's short- and long-term goals and align your PI initiative with them.

- Before selecting a specific set of problems and projects, perform a preliminary targeting analysis to determine where opportunities exist, their potential magnitude, and priority areas of work. At this stage, use approximate data that are appropriate for a targeting exercise. Remember that targets are upper bounds on performance. While targets are excellent metrics to identify and prioritize opportunities, actual implementation may not reach the exact targets.
- Review any relevant work undertaken in the company to be aware of previous efforts and to avoid redundancy.
- Define the tasks needed and the required human, technical, and financial resources.
- Describe anticipated constraints, unique aspects of the process, and potential challenges.
- Get enthusiastic support from senior management.
- Get buy-in and enthusiastic support from managers, process engineers, and operators.
- Recruit local champions from among the process experts and the stakeholders.
- Form task-driven teams that involve the right members.
- Encourage an open environment which fosters creativity and out-of-the-box integrated thinking where the dominating culture is "how do we make it happen?" instead of "why it won't work". It is critical in the early stages of a PI project to "grow the tree" of ideas. Later, this tree will be trimmed and the appropriate projects will be defined with all the necessary details.
- Measure, analyze, use process integration tools to develop, improve, synthesize, feedback, refine, and sustain projects and strategies. Do not collect more data than you should. The tools covered in this book are associated with certain data that enable the analysis to proceed forward. Typically, little data are needed in the beginning and as the PI analysis proceeds, specific data are needed as guided by PI.
- Develop aggregate and composite representations of the process and the solutions (as described throughout the book). Present the proposed changes in a way which focuses on gained insights, is easy to follow, and highlights the key characteristics of the findings.
- Focus on the key findings and the broad issues that are emerging.
- Consult with relevant individuals all along to capture process know-how, ensure that appropriate details are included and hurdles are overcome. This is particularly important in detailing the solutions and in getting buy-in and approval of the proposed changes.

Another important aspect is to take initiative and not be discouraged by resisting responses. The following is a table of "top ten" statements and attitudes described by Dunn and El-Halwagi (2003).

■ **TABLE 15-1 EXAMPLES OF DISCOURAGING ATTITUDES ABOUT PROCESS INTEGRATION AND SUGGESTED RESPONSES (DUNN AND EL-HALWAGI 2003)**

Discouraging attitudes	Response
We don't have the resources to support this process integration initiative.	Let us create resources that match the anticipated results or let us do the best we can within the available resources.
We have tried something similar before and it did not work.	Let us study the previous effort and see indeed if no more progress can be made.
These concepts will not work in my plant. We have a very unique operation.	There is now a track record of tens of very successful process integration projects that have applied to a wide variety of industrial processes; each of which is unique in its own right.
Has anyone else applied for it before?	See previous response.
Our process is too big/too small for this approach.	See previous response.
I am the process expert; there is no way that someone else can do better.	Let us incorporate your experience in a process integration framework. Time and again, track record has indicated that when proper process experience is incorporated into a process integration framework, significant and intuitively non-obvious benefits have accrued.
You really don't understand the issues and problems that we face.	See previous two responses.
Sounds great but you need to speak to someone else.	Get suggestions on "the someone else" but also see if there is a legitimate role for the individual.
I don't wish to participate in an initiative where I don't feel comfortable with the tools and techniques.	Provide appropriate training to develop the proper comfort level and understanding.
Not now! We will include it in our long-term strategic planning.	Each day without process integration implies missed opportunities.

15.3 EXAMPLES OF INDUSTRIAL APPLICATIONS

Process integration procedures and tools do indeed work and can lead to remarkable results. Table 15-2 presents examples of successful industrial applications. The applications span a wide range of processing industries.

A common theme in the results summarized in Table 15-2 is that PI provides a unique framework for identifying attractive and significant opportunities.

It is hoped that this book has introduced you to the key elements of PI and has motivated you to take initiative and start your own applications to generate value, enhance productivity, reduce pollution, conserve resources,

TABLE 15-2 EXAMPLES OF INDUSTRIAL APPLICATIONS (TAKEN FROM EL-HALWAGI AND SPRIGGS 1998 AND DUNN AND EL-HALWAGI 2003)

Type of process	Project objectives	Motivation	Approach	Key results
Specialty chemicals process	−Water-usage reduction −Solvent-usage reduction −Yield enhancement	Profitability enhancement and debottlenecking	Mass integration with direct recycle and interception	−Water-usage reduction: 33% −Solvent-usage reduction: 25% −Yield enhancement: 8%
Chemical and polymer complex	−Reduction of losses of volatile organic compounds (VOCs) −Reduction of thermal load of wastewater effluent	Meet environmental regulations, reduce losses, and allow future expansions	Heat and mass integration	−VOC losses reduced by 50% −Thermal load reduced by 90% −Payback period: 1.5 years
Specialty chemicals process	Debottlenecking of the process and hydrogen management	Soldout product with no additional capacity and significant cost for hydrogen consumption	Systematic elimination of two primary bottlenecks and sitewide integration of hydrogen generation, usage, and discharge	12% additional capacity and 25% reduction in hydrogen cost with a payback period of less than one year
Kraft pulping process	Water management and conservation	High usage of water and buildup of non process elements upon recycle	Sitewide tracking of water and non-process elements followed by a mass integration study for water minimization	55% Reduction in water usage with a payback period of less than two years
Resin production facility	Production debottlenecking	Soldout product with more market demands but a capped production capacity (bottleneck)	Mass-integration techniques to determine subtle causes of process bottlenecks and eliminate them at minimum cost	Increase in capacity by process debottlenecking: 4% (> $1million/year additional revenue)

(Continued)

TABLE 15-2 EXAMPLES OF INDUSTRIAL APPLICATIONS (TAKEN FROM EL-HALWAGI AND SPRIGGS 1998 AND DUNN AND EL-HALWAGI 2003)

Type of process	Project objectives	Motivation	Approach	Key results
Organic chemicals production process	Identification of sitewide water stream recycle opportunities to reduce river water discharges	Pressure from local environmentalists and the need to meet more stringent environmental permit requirements	Sitewide tracking of water followed by a mass integration study for water recycle opportunities and potential land treatment and reverse osmosis treatment of select wastewater streams	Nine process designs selected for implementation, including one separation system resulting in 670 gpm wastewater reduction with a payback period of one year
Polymer and monomer production processes	Identification of sitewide energy conservation opportunities to reduce energy costs	Reduction in operating costs for manufacturing processes and the need for additional steam generation for production capacity expansion.	Sitewide tracking of energy usage followed by a heat integration study to identify energy conservation opportunities	A heat exchange network and utility optimization process design implemented resulting in a 10% reduction in site utility costs, a 10% reduction in site wastewater hydraulic load and a 5% production capacity increase. Annual savings are in excess of $2.5 million/year.
Specialty chemicals production process	Identification of sitewide energy conservation opportunities to reduce energy costs	High operating costs for utilities	Sitewide tracking of energy usage followed by a heat integration study to identify energy conservation opportunities	Five process designs implemented leading to a 25% reduction in energy usage with a payback period of less than one year

Metal finishing process	Reduce cost of industrial solvent	Major solvent losses leading to a large operating cost and environmental problems	Synthesis of an energy-efficient heat-induced separation network	Recovery of 80% of lost solvent with a payback period of three years
Papermaking process	Recovery of lost fibers and management of water system	7% losses of purchased fibers during processing and high usage of water	Integrated matching of properties of broke fibers with demands of paper machines (property integration).	Recovery and reuse of 60% of lost fibers and reduction in water usage by 30% with a payback period of less than one year
Polymer production processes	Identification of sitewide wastewater stream recycle opportunities	Future expansion (Wastewater discharge system expected to exceed its maximum capacity during the production process expansion)	Sitewide tracking of water followed by a mass integration study for water recycle opportunities and reverse osmosis treatment of select wastewater streams	24 process designs implemented resulting in a 30% reduction in site wastewater discharge and with a payback period of less than one year
Petrochemical facility	Develop power cogeneration strategies and optimize utility systems	Significant usage of steam for process uses and high cost of power usage	Energy integration with emphasis on combined heat and power optimization	25% reduction in steam cost and cogeneration of 20% of power requirement for the process. Payback period is four years.

and contribute to a sustainable development. There are abundant opportunities within the process industries and it is up to you to identify them
and transform them into positive changes for the future. Start now and
good luck!

15.4 REFERENCES

Dunn, R.F. and El-Halwagi, M.M. 2003, 'Process Integration Technology
 Review: Background and Applications in the Chemical Process Industry',
 J. Chem. Tech. And Biotech, vol. 7, pp. 1011-1021.
El-Halwagi, M.M. and Spriggs, H.D. 1998, 'Solve Design Puzzles with Mass
 Integration', *Chem. Eng. Prog.*, vol. 94, August, pp. 25-44.

APPENDIX I: CONVERSION FACTORS

LENGTH

1 Å (angstrom)	$= 10^{-10}\,\text{m}$
	$= 0.1\,\text{nm}$
1 ft	$= 0.3048\,\text{m}$
1 in.	$= 2.540\,\text{cm}$
1 km	$= 0.6214\,\text{mile}$
1 m	$= 3.2808\,\text{ft}$
	$= 39.37\,\text{in.}$
1 micron	$= 10^{-6}\,\text{m}$
	$= 10^{-4}\,\text{cm}$
	$= 10^{-3}\,\text{mm}$
	$= 1\,\mu\text{m (micrometer)}$
1 mile	$= 1.6093\,\text{km}$
	$= 5280\,\text{ft}$
1 yd	$= 3\,\text{ft}$
	$= 0.9144\,\text{m}$

AREA

1 acre	$= 4047\,\text{m}^2$
	$= 4840\,\text{yd}^2$
	$= 43560\,\text{ft}^2$

$$1 \, ft^2 \qquad\qquad = -0.0929 \, m^2$$
$$1 \, hectare \qquad = 10^4 \, m^2$$
$$1 \, in.^2 \qquad\qquad = 6.4516 \times 10^{-4} \, m^2$$
$$1 \, mile^2 \qquad\quad = 2.58998811 \times 10^6 \, m^2$$
$$1 \, yd^2 \qquad\qquad = 0.8361 \, m^2$$

VOLUME

$$1 \, bbl \, (barrel) \quad = 5.6146 \, ft^3$$
$$= 42 \, US \, gal$$
$$= 0.15899 \, m^3$$
$$1 \, ft^3 \qquad\qquad\;\; = 28.317 \, L$$
$$= 0.028317 \, m^3$$
$$= 7.481 \, US \, gal$$
$$1 \, in.^3 \qquad\qquad = 16.387 \, cm^3$$
$$= 1.6387 \times 10^{-5} \, m^3$$
$$1 \, L \, (Liter) \qquad = 1000 \, cm^3$$
$$1 \, m^3 \qquad\qquad\;\, = 1000 \, L$$
$$= 264.17 \, US \, gal$$
$$1 \, US \, gal \qquad\;\; = 4 \, qt$$
$$= 3.7854 \, L$$
$$1 \, yd^3 \qquad\qquad = 0.764555 \, m^3$$

MASS

$$1 \, kg \qquad\qquad\quad = 1000 \, gm$$
$$= 2.2046 \, lb_m$$
$$1 \, lb_m \qquad\qquad = 453.59 \, gm$$
$$= 0.45359 \, kg$$
$$= 16 \, oz \, (ounces)$$
$$= 7000 \, grains$$
$$1 \, oz \, (ounce) \qquad = 0.02835 \, kg$$
$$1 \, ton \, (metric)/tonne \;\; = 1000 \, kg$$
$$1 \, ton \, (short) \qquad = 2000 \, lb_m$$
$$= 907 \, kg$$
$$1 \, ton \, (long) \qquad = 2240 \, lb_m$$
$$= 1016 \, kg$$

DENSITY

$$1 \, gm/cm^3 \qquad\quad = 62.427961 \, lb_m/ft^3$$
$$1 \, kg/m^3 \qquad\qquad = 0.062428 \, lb_m/ft^3$$
$$1 \, lb_m/ft^3 \qquad\quad\; = 16.018 \, kg/m^3$$

PRESSURE

1 atm	$= 760$ mm Hg at $0°$C (millimeters mercury pressure)
	$= 760$ torr
	$= 29.921$ in. Hg at $0°$C
	$= 14.696$ psia
	$= 1.01325 \times 10^5$ N/m^2
	$= 1.01325 \times 10^5$ Pa (pascal)
	$= 101.325$ kPa
1 bar	$= 1 \times 10^5$ N/m^2
	$= 100$ kPa
1 Pa (pascal)	$= 1$ N/m^2
	$= 1$ kg/(m$^{-1} \cdot$s^{-2})
1 psia	$= 6.89476 \times 10^3$ N/m^2
1 torr	$= 1$ mm Hg
	$= 133.322$ Pa

ENERGY

1 Btu	$= 252.16$ cal (thermochemical)
	$= 778.17$ lb$_f$.ft
	$= 1.05506$ kJ
1 cal (thermochemical)	$= 4.1840$ J
1 J (Joule)	$= 1$ N.m
	$= 1$ kg.m^2/s^2
	$= 10^7$ gm.cm^2/s^2 (erg)
	$= 9.48 \times 10^{-4}$ Btu
	$= 0.73756$ lb$_f$.ft
1 quadrillion Btu	$= 10^{15}$ Btu
	$= 2.93 \times 10^{11}$ kWhr
	$= 172 \times 10^6$ barrels of oil equivalent
	$= 36 \times 10^6$ metric tons of coal equivalent
	$= 0.93 \times 10^{12}$ cubic feet of natural gas equivalent

POWER

1 hp (horsepower)	$= 0.74570$ kW
1 kW	$= 1000$ W
1 quadrillion Btu/year	$= 0.471 \times 10^6$ barrels of oil equivalent/day
1 W (watt)	$= 1$ J/s

VISCOSITY

1 cp (centipoise)	$= 10^{-2}$ gm/cm.s (poise)
	$= 10^{-3}$ kg/m.s
	$= 10^{-3}$ N.s/m^2
	$= 10^{-3}$ Pa.s
	$= 6.7197 \times 10^{-4}$ lb$_m$/ft.s

TEMPERATURE (CONVERSION OF CERTAIN VALUES OF TEMPERATURE)

$$\text{Kelvin (K): } x \text{ K} \quad = (x - 273.15) \text{ °C}$$
$$= 1.8 \times x - 459.67 \text{ °F}$$
$$= 1.8 \times x \text{ °Rankine}$$

$$\text{Celcius (or centigrade) (°C): } x \text{ °C}$$
$$= (x + 273.15) \text{ K}$$
$$= 1.8 \times x + 32 \text{ °F}$$
$$= 1.8 \times (x + 273.15) \text{ °Rankine}$$

$$\text{Fahrenheit (°F): } x \text{ °F} \quad = (5/9)(x + 459.67) \text{ K}$$
$$= (5/9)(x - 32) \text{ °C}$$
$$= (x + 459.67) \text{ °Rankine}$$

$$\text{Rankine (°R): } x \text{ °Rankine}$$
$$= (5/9)x \text{ K}$$
$$= (5/9)(x - 491.67) \text{ °C}$$
$$= (x - 459.67) \text{ °F}$$

SPECIFIC HEAT CAPACITY

$$1 \frac{\text{cal}}{\text{gm} \cdot \text{°C}} \qquad = 1 \frac{\text{Btu}}{\text{lb} \cdot \text{°F}}$$
$$= 4.18 \frac{\text{kJ}}{\text{kg} \cdot \text{K}}$$

APPENDIX II: LINGO SOFTWARE

LINGO is a computer-aided optimization software that solves linear, non-linear, and mixed integer linear and non-linear programs. It is developed by LINDO Systems Inc. and may be ordered from http://www.lindo.com. A trial version for assessing the software may be downloaded from the LINDO Systems web site. The following is a brief description of how to get started on using LINGO. Additional information can be obtained through the Help menu on the LINGO software or from Schrage (1999).

AII.1 CREATING A NEW FILE

From the File menu, use the New command. An empty 'untitled' window will be displayed. On this window, you can type the model.

AII.2 MODEL FORMAT

The following are some basic information on writing a LINGO optimization model. The general form of the model is composed of an objective function and a number of constraints. The objective function is either to be minimized or maximized. These two options are expressed as follows:

Min = Objective function;
or
Max = Objective function;

Next, we write the set of constraints. Each constraint is followed by a semicolon ';'.

Finally, the program is finished by typing End.

Comment lines may be added to the program as follows:

! Comment line;

All the statements between the exclamation mark '!' and the semicolon ';' will be ignored by LINGO during computations.

If the line for minimizing or maximizing an objective function is not included, LINGO will solve the model as a set of equations provided that the degrees of freedom are appropriate. In writing constraints, the equalities and inequalities can be described as follows:

=   The expression to the left must equal the one on the right.

<=   The expression to the left must be less than or equal to the expression on the right.

>=   The expression to the left must be greater than or equal to the expression on the right.

<   The expression to the left must be strictly less than the expression on the right.

>   The expression to the left must be strictly greater than the expression on the right.

In writing constraints, the following symbols are used for mathematical operations:

+   Addition
−   Subtraction
*   Multiplication
/   Division
^   Power

For example, let us consider the following simple minimization program:

Minimize $2*x + y^2$

Subject to the following constraints

$x \geq 2.5$

$y \geq 4.0$

In terms of LINGO modeling language, the program can be written as follows:

```
Min = 2*x + y^2;
x >= 2.5;
y >= 4.0;
End
```

AII.3 SOLVING A PROGRAM

From the LINGO menu, choose solve. LINGO will display a LINGO Solver Status window and a Reports window which shows the following solution to the previous program:

```
Objective value:  21.00000

        Variable        Value

            X         2.500000
            Y         4.000000
```

It indicates that the minimum value of the objective function is 21.0 and that the optimal solution for x and y is 2.5 and 4.0, respectively.

AII.4 MATHEMATICAL FUNCTIONS

The following are some of the mathematical functions used by LINGO:

@ABS(X)	Returns the absolute value of X.
@EXP(X)	Returns the constant e (2.718281…) to the power X.
@LOG(X)	Returns the natural logarithm of X.
@SIGN(X)	Returns -1 if X is less than 0, returns $+1$ if X is greater than or equal to 0.
@BND(L, X, U)	Limits the variable or attribute X to greater or equal to L, and less than or equal to U.
@BIN(X)	Limits the variable or attribute X to a binary integer value (0 or 1).
@GIN(X)	Limits the variable or attribute X to only integer values.

For example, if x in the previous program is to assume only an integer value, the program should be extended to:

```
Min = 2*x + y^2;
x >= 2.5;
y >= 4.0;
@GIN(x);
End
```

By using the Solve option from the LINGO menu, we get the following results displayed on the Reports window:

```
Objective value:  22.00000

        Variable        Value
            X         3.000000
            Y         4.000000
```

AII.5 DUAL PRICES AND SENSITIVITY ANALYSIS

In the results report, LINGO displays the values of 'dual prices' of constraints. The dual prices have an interesting economic interpretation, and their numerical values are often of interest. To indicate this interpretation, consider the problem

$$\min f(x)$$

such that

$$h_i(x) = b_i \quad i = 1, 2, \ldots, m$$

where the numerical value b_i is regarded as the amount of some scarce resources which is limited according to the constraint $h_i(x)$. The dual prices correspond to the Lagrange multipliers commonly used in optimization theory and applications. As the amount of the resource b_i varies, the optimal solution also changes. Based on characteristics of the Lagrange multipliers, one can show that at the optimum solution and for a small perturbation Δb_i in the value of the right-hand side of the constraint:

$$\frac{\Delta f}{\Delta b_i} \approx -\text{Dual price of constraint } i$$

If we think of the objective function as dollars of return, then it may be interpreted as dollars per unit of the ith source. Another interpretation which follows directly from the above equation is that the dual prices represent sensitivity coefficients. This means that the marginal value of the objective function for a change in the value of the ith resource is given by the negative value of the dual price of the ith constraint. This observation is only valid near the optimum solution and is useful for conducting sensitivity analyses.

Example: Consider the following optimization model:
model:
min = (x1 − 2)^2 + (x2 − 2)^2;
x1 + x2 = 6.00;
End

Optimal solution found at step: 4
Objective value: 2.000000

Variable	Value	Reduced Cost
X1	3.000000	0.0000000E+00
X2	3.000000	0.6029251E−07

Row	Slack or Surplus	Dual Price
1	2.000000	1.000000
2	0.0000000E + 00	−2.000000

It is worth noting that the second row in the program is the constraint x1 + x2 = 6.0 (the objective function is the first row). Therefore, the dual price of the constraint is −2.0. This means that at the optimum solution and for small perturbation of the right-hand side of the constraint, the ratio of change in value of the objective function to the change in the value of the right-hand side of the constraint is $-(-2) = 2$.

For instance, if we increase the right-hand side of the constraint by 0.01 (i.e., b = 6.01), we should get an increase in the objective function by 2*0.01 = 0.02 (i.e., f * = 2.02). Similarly, if we decrease the right-hand side of the constraint by 0.01 (i.e., b = 5.99), we should get a decrease in the objective function by 0.02 (i.e., f* = 1.98). Let us test the model using LINGO:

```
model:
min = (x1 −2)^2 + (x2 − 2)^2;
x1 + x2 = 6.01;
End
```

Optimal solution found at step: 4
Objective value: **2.020050**

Variable	Value	Reduced Cost
X1	3.005000	0.0000000E+00
X2	3.005000	0.6155309E−07
Row	Slack or Surplus	Dual Price
1	2.020050	1.000000
2	−0.2288818E−06	−2.010000

```
model:
min = (x1 −2)^2 + (x2 − 2)^2;
x1 + x2 = 5.99;
End
```

Optimal solution found at step: 4
Objective value: **1.980050**

Variable	Value	Reduced Cost
X1	2.995000	0.0000000E+00
X2	2.995000	−0.5999458E−07
Row	Slack or Surplus	Dual Price
1	1.980050	1.000000
2	−0.2288818E−06	−1.990000

These results confirm the meaning of dual prices. It is worth noting that such observations are valid for small perturbations in the value of b_i. For instance, if we increase b by 1,0, we get the following model:

```
model:
min = (x1 − 2)^2 + (x2 − 2)^2;
x1 + x2 = 7;
End
```

Optimal solution found at step: 4
Objective value: 4.500000

Variable	Value	Reduced Cost
X1	3.500000	0.0000000E + 00
X2	3.500000	0.1111099E − 06

Row	Slack or Surplus	Dual Price
1	4.500000	1.000000
2	0.0000000E + 00	−3.000000

Notice that when we varied b by 1.0, f^* changed by 2.5 (not 2.0). Consequently, the sensitivity analysis should be carried out for small perturbations in b_i.

AII.6 REFERENCES

Schrage, L. (1999), *Optimization Modeling with LINGO*, 5th edn, LINDO Systems Inc., Chicago.

Index

Printed and bound by CPI Group (UK) Ltd, Croydon, CR0 4YY

08/05/2025

01864819-0001